T0281455

Undergraduate Lecture Notes in Physics

Undergraduate Lecture Notes in Physics (ULNP) publishes authoritative texts covering topics throughout pure and applied physics. Each title in the series is suitable as a basis for undergraduate instruction, typically containing practice problems, worked examples, chapter summaries, and suggestions for further reading.

ULNP titles must provide at least one of the following:

- An exceptionally clear and concise treatment of a standard undergraduate subject.
- A solid undergraduate-level introduction to a graduate, advanced, or non-standard subject.
- A novel perspective or an unusual approach to teaching a subject.

ULNP especially encourages new, original, and idiosyncratic approaches to physics teaching at the undergraduate level.

The purpose of ULNP is to provide intriguing, absorbing books that will continue to be the reader's preferred reference throughout their academic career.

Series editors

More information about this series at http://www.springer.com/series/8917

Maurice H.P.M. van Putten

Introduction to Methods of Approximation in Physics and Astronomy

Springer

Maurice H.P.M. van Putten
Department of Physics and Astronomy
Sejong University
Seoul
Republic of Korea (South Korea)

ISSN 2192-4791 ISSN 2192-4805 (electronic)
Undergraduate Lecture Notes in Physics
ISBN 978-981-10-9742-3 ISBN 978-981-10-2932-5 (eBook)
DOI 10.1007/978-981-10-2932-5

Softcover reprint of the hardcover 1st edition 2017

Printed on acid-free paper

This Springer imprint is published by Springer Nature
The registered company is Springer Nature Singapore Pte Ltd.
The registered company address is: 152 Beach Road, #21-01/04 Gateway East, Singapore 189721, Singapore

To my parents

Preface

Modern astronomy reveals an evolving Universe rife with transient sources, mostly discovered—few predicted—in multi-wavelength observations. Our window of observations now includes electromagnetic radiation, gravitational waves and neutrinos. For the practicing astronomer, these are highly interdisciplinary developments that pose a novel challenge to be well-versed in astroparticle physics and data analysis. In realizing the full discovery potential of these *multimessenger* approaches, the latter increasingly involves high-performance supercomputing.

These lecture notes developed out of lectures on mathematical-physics in astronomy to advanced undergraduate and beginning graduate students. They are organized to be largely self-contained, starting from basic concepts and techniques in the formulation of problems and methods of approximation commonly used in computation and numerical analysis. This includes root finding, integration, signal detection algorithms involving the Fourier transform and examples of numerical integration of ordinary differential equations and some illustrative aspects of modern computational implementation. In the applications, considerable emphasis is put on fluid dynamical problems associated with accretion flows, as these are responsible for a wealth of high energy emission phenomena in astronomy.

The topics chosen are largely aimed at phenomenological approaches, to capture main features of interest by effective methods of approximation at a desired level of accuracy and resolution. Formulated in terms of a system of algebraic, ordinary or partial differential equations, this may be pursued by perturbation theory through expansions in a small parameter or by direct numerical computation. Successful application of these methods requires a robust understanding of asymptotic behavior, errors and convergence. In some cases, the number of degrees of freedom may be reduced, e.g., for the purpose of (numerical) continuation or to identify secular behavior. For instance, secular evolution of orbital parameters may derive from averaging over essentially periodic behavior on relatively short, orbital periods. When the original number of degrees of freedom is large, averaging over dynamical time scales may lead to a formulation in terms of a system in approximately thermodynamic equilibrium subject to evolution on a secular time scale by a regular or singular perturbation.

In modern astrophysics and cosmology, gravitation is being probed across an increasingly broad range of scales and more accurately so than ever before. These observations probe weak gravitational interactions below what is encountered in our solar system by many orders of magnitude. These observations hereby probe (curved) spacetime at low energy scales that may reveal novel properties hitherto unanticipated in the classical vacuum of Newtonian mechanics and Minkowski spacetime. Dark energy and dark matter encountered on the scales of galaxies and beyond, therefore, may be, in part, revealing our ignorance of the vacuum at the lowest energy scales encountered in cosmology. In this context, our application of Newtonian mechanics to globular clusters, galaxies and cosmology is an approximation assuming a classical vacuum, ignoring the potential for hidden low energy scales emerging on cosmological scales. Given our ignorance of the latter, this poses a challenge in the potential for unknown systematic deviations. If of quantum mechanical origin, such deviations are often referred to as anomalies. While they are small in traditional, macroscopic Newtonian experiments in the laboratory, they same is not a given in the limit of arbitrarily weak gravitational interactions.

We hope this selection of introductory material is useful and kindles the reader's interest to become a creative member of modern astrophysics and cosmology.

This book would not have been possible without numerous in-class interactions that largely shaped the choice of materials and method of presentation. The author gratefully thanks the many students who participated in this development. Some of this work is supported by the National Research Foundation of Korea under Grant Nos. 2015R1D1A1A01059793 and 2016R1A5A1013277.

Seoul, Republic of Korea (South Korea) Maurice H.P.M. van Putten

Contents

Part I
Preliminaries

Chapter 1
Complex Numbers

We begin with a review of complex numbers.[1] Functions of a complex variable are at the root of numerous methods and techniques of formulating and solving problems in physics and engineering. Complex variables are an extension of the real numbers $\mathbb{R}$ as follows.

The imaginary number

$$\sqrt{-1} = i \tag{1.1}$$

is introduced to solve the algebraic equation

$$1 + x^2 = 0 \tag{1.2}$$

by promoting the unknown x to a complex number z,

$$z = x + iy \tag{1.3}$$

expressed in a real part $x = \mathrm{Re}\, z$ and an imaginary part $y = \mathrm{Im}\, z$ (Fig. 1.1).

The rules for handling complex numbers are a natural extension of the familiar rules for multiplication and addition of real numbers in $\mathbb{R}$. For two complex numbers $z_k = x_k + iy_k$ $(k = 1, 2)$, we shall have

$$\begin{aligned} z_1 + z_2 &= x_1 + x_2 + i(y_1 + y_2), \\ z_1 z_2 &= (x_1 + iy_1)(x_2 + iy_2) \equiv x_1x_2 - y_1y_2 + i(x_1y_2 + x_2y_1). \end{aligned} \tag{1.4}$$

[1] An excellent reference is [1], some of our examples are guided by this source.

M.H.P.M. van Putten, *Introduction to Methods of Approximation in Physics and Astronomy*, Undergraduate Lecture Notes in Physics,
DOI 10.1007/978-981-10-2932-5_1

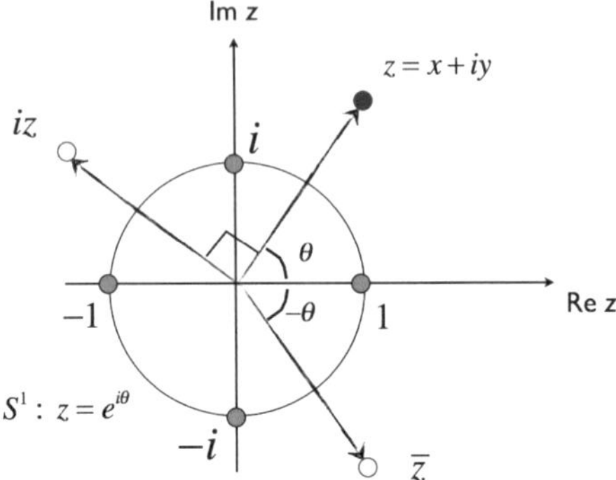

Fig. 1.1 The complex plane $\mathbb{C}$ comprises the numbers $z = x + iy$ with real part $x = \mathrm{Re}\, z$ along the x-axis (the real numbers $\mathbb{R}$) and imaginary part $y = \mathrm{Im}\, z$ along the y-axis (the imaginary numbers $i\mathbb{R}$). In polar coordinates, $z = re^{i\theta} = r(\cos\theta + i\sin\theta)$, where $r = |z|$ denotes the modulus θ denotes the argument. θ is determined up to multiples of 2π. The unit circle S^1 is given by $|z| = \sqrt{x^2 + y^2} = 1$. Multiplication of two complex numbers produces addition of their polar angles, e.g., the argument of iz is $\theta + \frac{\pi}{2}$. The complex conjugate $\bar{z}$ of z is defined by reflection about the real axis, i.e., $\theta \to -\theta$

We hereby preserve the familiar commutative and distributive laws

$$\begin{aligned} z_1 + z_2 &= z_2 + z_1, \\ (z_1 + z_2) + z_3 &= z_1 + (z_2 + z_3), \\ (z_1 z_2) z_3 &= z_1 (z_2 z_3), \\ z_1 (z_2 + z_3) &= z_1 z_2 + z_1 z_3. \end{aligned} \tag{1.5}$$

To illustrate, the product iz appears as a rotation over $\frac{\pi}{2}$ with magnification 1 (Fig. 1.1).

As points in the two-dimensional plane, complex numbers can be written also in polar form

$$z = r\left(\cos\theta + i\sin\theta\right), \tag{1.6}$$

where $r = |z|$ is the *norm* or *modulus* of z, defined by the distance to the origin, and θ is the *argument* of z, given by the angle of the vector pointing to z relative to the x-axis. Here, we adhere to the Euclidean norm of the two-dimensional plane,

$$|z| = \sqrt{x^2 + y^2}. \tag{1.7}$$

On the unit circle, complex numbers reduce to

$$z = \cos\theta + i\sin\theta. \tag{1.8}$$

If z_k are two elements of S^1 with arguments θ_k $(k = 1, 2)$, expanding

$$(\cos\theta_1 + i\sin\theta_1)(\cos\theta_2 + i\sin\theta_2) \tag{1.9}$$

shows the *addition of angles* in multiplication:

$$z_1 z_2 = \cos(\theta_1 + \theta_2) + i\sin(\theta_1 + \theta_2). \tag{1.10}$$

This result is reminiscent of addition of exponents in $e^{x_1}e^{x_2} = e^{x_1+x_2}$ in the calculus of a functions of a real variable. Based on the defining properties of the exponential function, the same extends to imaginary exponents,

$$e^{i\theta_1}e^{i\theta_2} = e^{i(\theta_1+\theta_2)} \tag{1.11}$$

and the identity

$$1 = e^{2\pi ki} \quad (k\epsilon\mathbb{Z}). \tag{1.12}$$

In $\mathbb{C}$, therefore, 1 is covered by $\mathbb{Z}$.

The above is made concise by *Euler's formula*

$$e^{i\theta} = \cos\theta + i\sin\theta, \tag{1.13}$$

where the exponential function on the left hand side is defined by the extension

$$e^x = \sum_{n=0}^{\infty}\frac{1}{n!}x^n \rightarrow e^z = \sum_{n=0}^{\infty}\frac{1}{n!}z^n, \tag{1.14}$$

that is well-defined for all $z\epsilon\mathbb{C}$. By the triangle inequality induced by the Euclidean norm (1.7),

$$|z_1 + z_2| \leq |z_1| + |z_2|, \tag{1.15}$$

we have

$$\left|\sum_{n=0}^{N}\frac{1}{n!}z^n\right| \leq \left|\sum_{n=0}^{N}\frac{1}{n!}r^n\right| = e^r < \infty \tag{1.16}$$

whenever $|z| = r$ is finite. In passing to e^z in (1.14), therefore, the extension (1.14) is well-defined. (The *radius of convergence* of (1.14) remains infinite, Sect. 1.4.) For $z = i\theta$, the series (1.14) expressed in its real and imaginary parts readily recovers the defining series expansions for $\cos\theta$ and $\sin\theta$, here seen to be the projections onto the real and, respectively, imaginary axis of complex numbers on S^1.

Special cases of Euler's formula are

$$e^{0i}=1,\ e^{\frac{\pi}{2}i}=i,\ e^{\pi i}=-1,\ e^{\frac{3}{2}\pi i}=-i,\ e^{2\pi i}=1,\ \cdots \tag{1.17}$$

Here, the dots refer refer to multiplicities in the argument by 2π. The third highlights the intimate connection between the irrational numbers e and π, made famous in Euler's identity

$$e^{i\pi}+1=0. \tag{1.18}$$

This is a deep relation between the irrational constants e and π by the exponential function (1.14), and important for the appearance of e and π in numerous branches of mathematics, physics and finance. Computationally, (1.18) defines π given e, or e given π, as may be seen by numerical root finding methods.

By Euler's formula, we obtain complex numbers in polar form

$$z=re^{i\theta},\quad r=|z|,\quad \theta=\arg z=\operatorname{Arg} z+2\pi k, \tag{1.19}$$

where $k\epsilon\mathbb{Z}$. Here, *arg* refers to the argument of z, the inclination of an arrow pointing to z with respect to the real line, and Arg refers to the *principle value*, i.e.,

$$-\pi\leq \operatorname{Arg} z<\pi. \tag{1.20}$$

In considering the product of two complex numbers, we have

$$\begin{aligned}&|z_1z_2|=|z_1||z_2|,\\ &\arg z_3=\arg z_1z_2=\arg z_1+\arg z_2.\end{aligned} \tag{1.21}$$

In this notation, the product iz appears as a rotation over $\frac{\pi}{2}$ with magnification 1 since $\arg i=\frac{\pi}{2}$ (mod 2π),[2] i.e.:

$$\arg(iz)=\arg z+\arg i=\arg z+\frac{\pi}{2},\quad |iz|=|i||z|=|z|. \tag{1.22}$$

Complex numbers satisfy all the properties you may expect to hold by inspection of the complex plane with the Euclidean norm (1.7). Notably, we have the triangle inequality

$$||z_1|-|z_2||\leq|z_1+z_2|\leq|z_1|+|z_2|, \tag{1.23}$$

To see this, note that $|z_1|=|z_1+z_2-z_2|\leq|z_2|+|z_1-z_2|$, whereby $|z_1|-|z_2|\leq|z_1-z_2|$. Since the right hand side is invariant under switching 1 and 2, (1.23) follows.

[2]The notation mod 2π is often used to denote $+2\pi k$, $k\,\epsilon\,\mathbb{Z}$.

Complex numbers give a remarkably easy derivation of some non-trivial trigonometric identifies, that may express higher harmonics produced by nonlinearities. Consider the *de Moivre's formula*[3]

$$z^n = r^n e^{in\theta} = r^n(\cos n\theta + i\sin n\theta). \tag{1.24}$$

For $n = 2$, we obtain familiar results by equating the real and imaginary parts of $(\cos\theta + i\sin\theta)^2 = \cos 2\theta + i\sin 2\theta$, giving

$$\cos^2\theta - \sin^2\theta = \cos 2\theta,\quad 2\cos\theta\sin\theta = \sin 2\theta, \tag{1.25}$$

where the first can also be written as $\cos 2\theta = 2\cos^2\theta - 1 = 1 - 2\sin^2\theta$ in view of $\cos^2\theta + \sin^2\theta = 1$.[4]

Example 1.1 For $n = 3$, equating real and imaginary parts of $(\cos\theta + i\sin\theta)^3 = \cos^3\theta + 3i\cos^2\theta\sin\theta - 3\cos\theta\sin^2\theta - i\sin^3\theta$ obtains

$$\cos 3\theta = 4\cos^3\theta - 3\cos\theta,\quad \sin 3\theta = 3\sin\theta - 4\sin^3\theta, \tag{1.26}$$

where we used $\cos^2\theta + \sin^2\theta = 1$ once more. Identities such as these can be useful in solving cubic equations.

1.1 Quotients of Complex Numbers

The quotient of two complex numbers

$$z_3 = \frac{z_1}{z_2} \tag{1.27}$$

is well-defined provided that $z_2 \neq 0$. Since $z_1 = z_2 z_3$, $\arg z_1 = \arg z_2 + \arg z_3$, we have

$$\arg\left(\frac{z_1}{z_2}\right) = \arg z_1 - \arg z_2. \tag{1.28}$$

Equivalently, we may write

$$\frac{z_1}{z_2} = \frac{z_1\bar{z}_2}{z_2\bar{z}_2} = K z_1\bar{z}_2, \tag{1.29}$$

[3]Abraham de Moivre (1667–1754).

[4]Generalized to all integers $n \geq 2$ by Franciscus Viete (1540–1603).

where $K = |z_2|^{-2} \in \mathbb{R}$. Since $\arg K = 0$ and $\arg \bar{z}_2 = -\arg z_2$, (1.29) gives

$$\arg\left(\frac{z_1}{z_2}\right) = \arg z_1 - \arg z_2 \tag{1.30}$$

once more. In fact, (1.29) shows explicitly

$$\frac{z_1}{z_2} = \frac{(x_1 + iy_1)(x_2 - iy_2)}{x_2^2 + y_2^2} = \frac{x_1x_2 + y_1y_2}{x_2^2 + y_2^2} + i\frac{y_1x_2 - x_1y_2}{x_2^2 + y_2^2}, \tag{1.31}$$

whereby

$$\tan\left[\arg\left(\frac{z_1}{z_2}\right)\right] = \frac{y_1x_2 - x_1y_2}{x_1x_2 + y_1y_2}. \tag{1.32}$$

1.2 Roots of Complex Numbers

If n is a positive integer, then $\xi^n = 1$ defines n roots of 1. To see this, we recall the multiplicity $2\pi k$ in the argument of 1 in (1.12):

$$\xi^n = 1 = e^{2\pi i k} : \ \xi = e^{\frac{2\pi i k}{n}} \ (k \in \mathbb{Z}). \tag{1.33}$$

The n-th root of unity, therefore, has n solutions on $\mathbb{C}$ (Fig. 1.2). The same can also be seen from de Moivre's formula (1.24),

$$\xi = |\xi|(\cos\theta + i\sin\theta) : \ \ \xi^n = |\xi|^n(\cos n\theta + \sin n\theta). \tag{1.34}$$

Equating real and imaginary parts, we have

$$|\xi|^n \cos n\theta = 1, \ \ |\xi|^n \sin n\theta = 0 : \ \ \sin n\theta = 0, \ \ \cos n\theta = 1, |\xi| = 1. \tag{1.35}$$

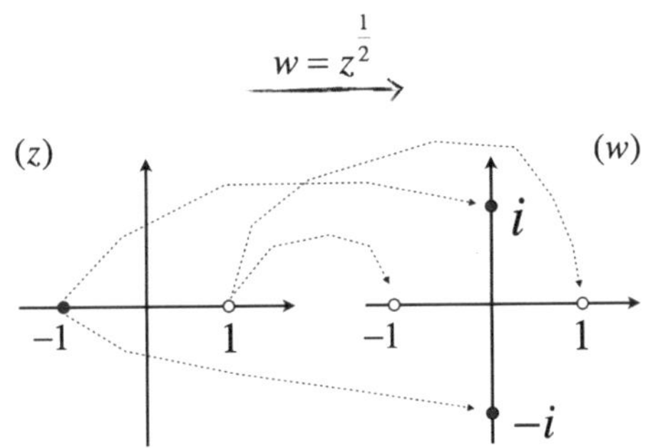

Fig. 1.2 The square root $w = z^{\frac{1}{2}}$ is a two-valued function in the complex plane, taking $z = 1$ into the image $\{1, -1\}$ and $z = -1$ into the image $\{i, -i\}$

The solutions

$$n\theta = 2\pi k \ \ (k \in \mathbb{Z}) \tag{1.36}$$

represent n distinct roots distributed uniformly over S^1,

$$\theta_n = \frac{2\pi}{n} k \ \ (k = 0, 1, 2 \cdots n - 1). \tag{1.37}$$

Example 1.2 [1]. The complex number $z = x + iy$ and its conjugate $\bar{z} = x - iy$ are distinct whenever $y \neq 0$. As variables, they are independent, i.e., $(z, \bar{z})$ have a one-to-one relation to the independent variables (x, y). This can be used to describe the equation of a line in the complex plane in terms of $(z, \bar{z})$, instead of using (x, y). Let $z_0 \neq 0$, arg $z_0 = \theta$. Then

$$w = \frac{z_0}{\bar{z}_0}\bar{z}: \ \ \arg w = \arg\left(\frac{z_0}{\bar{z}_0}\right) + \arg \bar{z} = 2\theta - \phi, \tag{1.38}$$

where we put arg $z = \phi$, We also have $|w| = \left|\frac{z_0}{\bar{z}_0}\right| |\bar{z}| = |z|$. The equation

$$z = w \tag{1.39}$$

hereby defines a condition on the argument of z, given by $\phi = 2\theta - \phi$, i.e., arg $z = \theta$: the straight line through the origin and z_0.

1.3 Sequences and Euler's Constant

Sequences of complex numbers $\{z_n\}_{n=0}^{\infty}$ may converge to a number z_* in the complex plane (Fig. 1.3). If so, we write

$$\lim_{n\to\infty} z_n = z_*, \tag{1.40}$$

meaning that for any $\epsilon > 0$, there exists an integer N such that $|z_n - z_*| < \epsilon$ for all $n > N$. Colloquially, we say that the *tail* $\{z_n\}_{n>N}$ of our sequence approaches L to within arbitrarily small distances.

The mathematical background of convergence is given by Cauchy. We say that $\{a_n\}_{n=0}^{\infty}$ *is a Cauchy sequence* if for every $\epsilon > 0$ there exists an N such that

$$|a_n - a_m| < \epsilon \ \ (n, m > N). \tag{1.41}$$

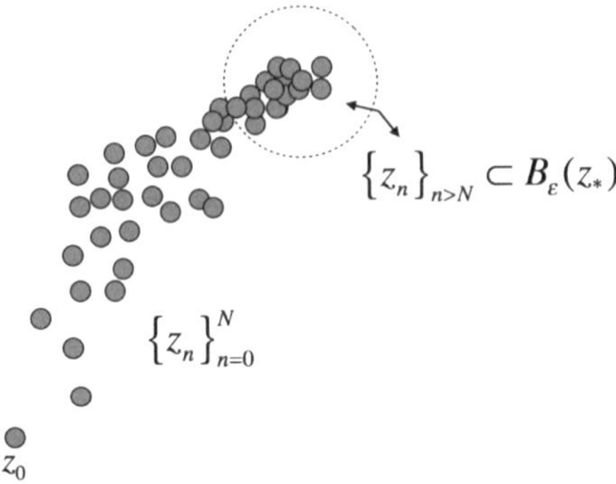

Fig. 1.3 A sequence $\{z_n\}_{k=0}^{\infty}$ is convergent to a complex number z_* if for any $\epsilon > 0$, there exists $N > 0$ such that the tail $\{z_n\}_{n>N}$ lies completely in the ball $B_\epsilon(z_*) : |z - z_*| < \epsilon$ around z_*

We can think of a Cauchy sequence as an (infinite) set of points that increasingly cluster closer and closer, as we move further into the tail with increasing N (Fig. 1.3).

Every Cauchy sequence is bounded. If $\epsilon > 0$, we may choose N such that (1.41) holds. Next, we choose $m > N$, whereby

$$|a_n| \leq |a_n - a_m| + |a_m| < \epsilon + |a_m| \ \ (n > N), \tag{1.42}$$

showing that the tails cluster around a_m within a distance ϵ.

Bounded sequences need not converge, like $a_n = (-1)^n$. Nevertheless, they contain convergent sub-sequences, known as the Boltzano-Weierstrass theorem. Without loss of generality, consider a sequence bounded in the unit interval $I_0 = [0, 1]$, i.e., $0 \leq a_n \leq 1$ for all n. We can think of I_0 as the union of two halves,

$$I_0 = \left[0, \frac{1}{2}\right] \bigcup \left[\frac{1}{2}, 1\right]. \tag{1.43}$$

One of the two halves contains an infinite sub-sequence) $\{a_n\}_{n=0}^{\infty}$, which we shall refer to as I_1:

$$\alpha_1 = 0 \text{if } I_1 = \left[0, \frac{1}{2}\right], \ \ \alpha_1 = 1 \text{ if } I_1 = \left[\frac{1}{2}, 1\right], \tag{1.44}$$

where α_1 labels the choice of left or right half of I_0. Let us also choose n_1 such that $a_{n_1} \epsilon I_1$. Repeating, we obtain a sub-sequence $\{\alpha_n\}_{n=1}^{\infty}$ with convergent sum

$$s = \lim_{k \to \infty} s_k, \ \ s_k = \Sigma_{n=1}^{k} \alpha_n 2^{-n}. \tag{1.45}$$

A subsequence $\{a_{n_k}\}_{k=1}^{\infty}$ with $a_{n_k} \epsilon I_k$ satisfies $a_{n_k} \to s$, since $\left|a_{n_k} - s\right| \leq 2^{-k}$. The same argument can be seen to hold in $\mathbb{C}$.

By the above, our Cauchy sequence $\{a_n\}_{n=0}^{\infty}$ has a convergent subsequence, $a_{n_k} \to a$ as $k \to \infty$ for some $a \,\epsilon\, \mathbb{C}$. Let $\epsilon > 0$ and choose N such that (1.41) holds. Next, choose M such that $\left|a_{n_k} - a\right| < \epsilon$ $(k > M)$. Then

$$|a_k - a| \leq |a_k - a_{n_k}| + |a_{n_k} - a| < \epsilon + \epsilon = 2\epsilon \tag{1.46}$$

for all $k > \max(N, M)$. Since ϵ is arbitrary, $a_k \to a$ as $k \to \infty$, showing a is the limit of $\{a_n\}_{n=0}^{\infty}$, i.e., every Cauchy sequence in $\mathbb{C}$ converges to a limit.[5]

It can be shown that if a sequence converges to some z_*, then so do the algebraic and geometric means

$$A_n = \frac{z_1 + z_2 + \cdots + z_n}{n} \to z_*, \quad B_n = (z_1 z_2 \ldots z_n)^{\frac{1}{n}} \to z_* \tag{1.47}$$

in the limit as n approaches infinity. (In the second, we assume $z_n \neq 0$ for all n.)

Example 1.3. Consider

$$z_n = \frac{n}{n-1}. \tag{1.48}$$

Evidently, $z_n \to 1$ in the limit as n approaches infinity. Their products satisfy

$$n = \frac{2}{1} \times \frac{3}{2} \times \cdots \times \frac{n-1}{n-2} \times \frac{n}{n-1} \equiv z_0 \times z_1 \cdots \times z_n. \tag{1.49}$$

Following (1.47), we have

$$\lim_{n\to\infty} n^{\frac{1}{n}} = \lim_{n\to\infty} \left(\frac{n}{n-1}\right)^{\frac{1}{n}} = \lim_{n\to\infty} \left(1 + \frac{1}{n}\right)^{\frac{1}{n}} = 1. \tag{1.50}$$

The same can be seen following $n^{\frac{1}{n}} = e^{\frac{\log n}{n}}$ and the fact that $n^{-1}\log n \to 0$ as n approaches infinity.

Example 1.4. Consider

$$z_n = \left(\frac{1+n}{n}\right)^n = \left(1 + \frac{1}{n}\right)^n, \tag{1.51}$$

satisfying

$$\lim_{n\to\infty} z_n = \lim_{n\to\infty} \left(1 + \frac{1}{n}\right)^n = e = 2.71\ldots. \tag{1.52}$$

[5] We say that $\mathbb{C}$ is *complete*.

Their products satisfy

$$\left[\frac{2}{1} \times \left(\frac{3}{2}\right)^2 \times \left(\frac{4}{3}\right)^3 \times \cdots \times \left(\frac{n}{n-1}\right)^{n-1} \times \left(\frac{n+1}{n}\right)^n\right]^{\frac{1}{n}} \equiv (n+1)\left(\frac{1}{n!}\right)^{\frac{1}{n}}. \quad (1.53)$$

By (1.47), it follows that

$$\lim_{n\to\infty} (n+1)\left(\frac{1}{n!}\right)^{\frac{1}{n}} = \lim_{n\to\infty} z_n = e. \quad (1.54)$$

Taking the logaritm of both sides of (1.54), we have

$$\log(n+1) - \frac{1}{n}\log n! \simeq 1 : \quad \log n! \simeq n\log n - n \quad (1.55)$$

for large n. Up to next order, the result is known as *Stirling's formula*

$$n! \simeq \sqrt{2\pi n}\, n^n e^{-n}. \quad (1.56)$$

An interesting example of a convergent sequence is

$$s_n = 1 + \frac{1}{2} + \frac{1}{3} + \cdots + \frac{1}{n} - \log n, \quad s_n \to \gamma, \quad (1.57)$$

that converges to *Euler's constant* $\gamma = 0.5772\ldots$. To see this, we note the following.

1. s_n is a decreasing sequence, since

$$\begin{aligned} s_{n+1} &= 1 + \tfrac{1}{2} + \tfrac{1}{3} + \cdots + \tfrac{1}{n+1} - \log(n+1) \\ &= \left[1 + \tfrac{1}{2} + \tfrac{1}{3} + \cdots + \tfrac{1}{n} - \log n\right] + \Delta_n \\ &= s_n + \Delta_n, \end{aligned} \quad (1.58)$$

where $\Delta_n = \frac{1}{n+1} - \left[\log(n+1) - \log n\right]$, satisfying

$$\begin{aligned} \Delta_n &= \frac{\frac{1}{n}}{1+\frac{1}{n}} - \log\left(1 + \tfrac{1}{n}\right) \\ &= \tfrac{1}{n}\left[1 - \tfrac{1}{n} + O(n^{-2})\right] - \left[\tfrac{1}{n} - \tfrac{1}{2n^2} + O(n^{-3})\right] \\ &= -\tfrac{1}{2n^2} + O(n^{-3}). \end{aligned} \quad (1.59)$$

Combining (1.58–1.59), $s_{n+1} < s_n$ when n is sufficiently large.

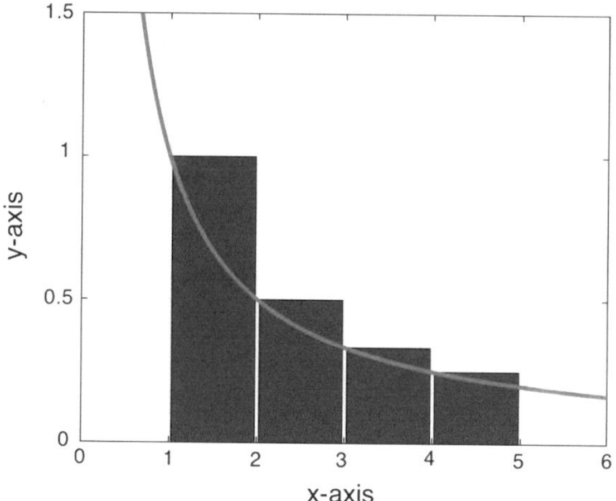

Fig. 1.4 The function $y(x) = 1/x$ and an approximation of its integral over $1 \leq x \leq 5$ by a majoring Riemann sum, expressed by a block function

2. s_n is bounded below by zero, as may be seen as follows. As illustrated in Fig. 1.4, recall that

$$1 + \frac{1}{2} + \frac{1}{3} > \int_1^4 \frac{dx}{x}, \tag{1.60}$$

where the left hand side is a majoring Riemann sum to the integral on the right. Therefore,

$$s_n' = 1 + \frac{1}{2} + \frac{1}{3} + \cdots + \frac{1}{n} - \int_1^{n+1} \frac{dx}{x} > 0, \tag{1.61}$$

and hence

$$s_n > s_n' > 0. \tag{1.62}$$

Since a monotonically decreasing sequence bounded from below is convergent, $s_n \to \gamma$ for some γ. Numerical evaluation shows $\gamma = 0.5772\ldots$.

1.4 Power Series and Radius of Convergence

A power series with coefficients a_n is an object of the form

$$\sum_{n=0}^{\infty} a_n z^n. \tag{1.63}$$

This is meaningful, provided that it converges for some $z \neq 0$ in the sense that the partial sums

$$S_N(z) = \sum_{n=0}^{N} a_n z^n \tag{1.64}$$

converge (cf. Fig. 1.3).

If (1.64) converges for z_0, then $|a_n z_0^n|$ approaches zero in the limit as $n \to \infty$. Hence, the $a_n z_0^n$ are bounded, say by A: $|a_n z_0^n| < A$ for all $n \geq 0$. If z satisfies $|z| < |z_0|$, then $r = \left|\frac{z}{z_0}\right| < 1$ and hence

$$\left|\sum_{n=N}^{M} a_n z^n\right| \leq \sum_{n=N}^{M} |a_n z^n| = \sum_{n=N}^{M} |a_n z_0^n| \left|\frac{z}{z_0}\right|^n \leq \frac{Ar^N}{1-r}. \tag{1.65}$$

The difference

$$S_N(z) - S_M(z) = \sum_{n=N}^{M} a_n z^n \tag{1.66}$$

in the tail of $S_N(z)$ can therefore be made arbitrarily small by choice of sufficiently large N,[6] whereby (1.64) exists. Likewise, if (1.64) is *not* convergent for z_0, then it is not convergent for any z satisfying $|z| > |z_0|$. In (1.65), we may take M to infinity. The result shows that $\sum_{n=N}^{\infty} a_n z^n$ is *absolutely convergent* in $|z| < |z_0|$.

A powerful property of (1.64), therefore, is the existence of a *radius of convergence* R, with convergence in a disk ($|z| < R$) and no convergence outside ($|z| > R$). According to the above, convergence inside is absolute. Convergence on the circle $|z| = R$ depends on the detailed properties of the a_n, e.g., in the form of specific bounds or gaps (many of the a_n being zero), that may lead to convergence on some but not all points on $|z| = R$. This is an interesting topic, which falls outside the scope of this introduction.

When the radius of convergence R is positive,

$$f(z) = \sum_{n=0}^{\infty} a_n z^n \tag{1.67}$$

defines a complex function within a disk of radius R about the origin. It constitutes the *Taylor series* of $f(z)$ about $z = 0$.

[6] By this property, we say that $S_N(z)$ is a *Cauchy sequence*.

Example 1.5. The radius of convergence of the Neumann series

$$\frac{1}{1-z} = \sum_{n=0}^{\infty} z^n \tag{1.68}$$

is $R = 1$. This see this, take any $z = x_1 < 1$ $(0 < x_1 < 1)$ and (1.68) is readily seen to converge by elementary considerations. If $x_2 > 1$, we have a series x_2^n whose elements do not convergence to zero, prohibiting their sum to converge. Accordingly, we have absolute convergence for $|z| \leq x_1 < 1$ and no convergence for $|z| \geq x_2 > 1$. Since $0 < x_1 < 1 < x_2$ are arbitrary, $R = 1$.

Example 1.6. The radius of convergence of the Neumann series (1.68) becomes more interesting in the context of

$$\frac{1}{1+x^2} = 1 - x^2 + x^4 + \cdots \tag{1.69}$$

This Lorentzian distribution is perfectly smooth when viewed on $x \in \mathbb{R}$. Why does its Taylor series expansion have $R = 1$ about the origin? Following the extension into the complex plane, $x \to z \in \mathbb{Z}$, the expansion in partial fractions

$$\frac{1}{1+z^2} = \frac{1}{2i}\left(\frac{1}{z-i} - \frac{1}{z+i}\right) \tag{1.70}$$

shows two singularities at $z = \pm i$ at a distance $R = |i| = 1$ from the origin. For a domain of convergence in the form of a disk around the origin, they spoil convergence at the distance to the origin $R = 1$.

Example 1.7. The Taylor series expansion of the exponential function

$$e^z = \sum_{n=0}^{\infty} \frac{1}{n!} z^n, \tag{1.71}$$

converges for all $z \in \mathbb{Z}$. Functions satisfying this property are called *entire*. That (1.71) has an infinite radius of convergence follows directly from (1.16), since e^r is well-defined for all r. The latter can be seen explicitly by noting that $n!$ grows much faster than z^n for n sufficiently large. By aforementioned Stirling's formula (1.56), we have for $n > N > er$, $r = |z|$,

$$\rho = \frac{er}{N} < 1 \tag{1.72}$$

and

$$\frac{r^n}{1 \cdot 2 \cdots n} < \left(\frac{er}{n}\right)^n < \left(\frac{er}{N}\right)^n = \rho^n. \tag{1.73}$$

The tails of the partial sums of (1.71) hereby satisfy

$$\left|\sum_{k=n}^{\infty} \frac{1}{k!} z^k\right| < \sum_{k=n}^{\infty} \frac{1}{k!} |z|^k < \sum_{k=n}^{\infty} \rho^k = \frac{\rho^n}{1-\rho}, \tag{1.74}$$

showing convergence to zero as n approaches infinity. By the same arguments on the radius of convergence of (1.64), we see that (1.71) is absolutely convergent. Since z was chosen arbitrarily, this property holds throughout the complex plane, i.e., the exponential function (1.71) is entire.

A key property of power series is that they are *unique*: if the left and right hand sides of

$$a_0 + a_1 z + a_2 z^2 + \cdots = b_0 + b_1 z + b_2 z^2 + \cdots \tag{1.75}$$

define convergent series in some common domain of convergence, then $a_n = b_n$ for all n. This can be slightly strengthened. It suffices for (1.75) to hold on an infinite sequence z_n with a finite accumulation point (a limit of a subsequence of z_n), e.g., $z_n = 1/n$ with limit $z = 0$. Thus, a sufficient condition for obtaining a power series in a_n is that (1.75) holds on such infinite number of points from some convergent series.

Successive Taylor series may provide a means to continuation, going beyond the domain of convergence of a single Taylor series expansion. Continuation along a curve γ in the complex plane consists of calculating Taylor series about different points z_k on γ, where the series about z_{k+1} obtains from the series about z_k proviso $|z_{k+1} - z_k|$ is less than the radius of convergence R_k of the latter. This idea has broad applications to numerical continuation, especially to nonlinear problems, e.g., strong shocks in compressible magnetohydrodynamics. In some cases, the problem at hand may be viewed as a continuation of a weakly nonlinear problem that is relatively tractable, analytically or numerically. If so, let z denote a control parameter that takes us from a relatively simple model with known solution(s) at $z = 0$ to the original problem at $z = 1$. Let γ from $z = 0$ to $z = 1$ be a path of continuation, taking us to solutions at $z = 1$ by successively calculating solutions at $z_k = k/N$ $(k = 0, 1, \ldots, N-1)$ for some suitable choice of N that ensures that a solution at z_{k+1} obtains from the solution at z_k. By way of example, solutions at z_{k+1} obtain by

Newton iterations with an initial guess given by the solution at z_k. Continuation may be successful upon choosing N sufficiently large. Even so, uniqueness at $z = 1$ is not guaranteed, as the result may depend on the choice of path. A further detailed discussion on this topic falls outside the present scope, however.

1.5 Minkowski Spacetime

According to the Euler's formula, the trigonometric functions cosine and sine are linear combinations of the exponential function with imaginary exponent. The exponential function being entire, we consider cosine and sine functions with imaginary arguments, giving the hyperbolic cosine and hyperbolic sine

$$\cosh z = \frac{1}{2}\left(e^z + e^{-z}\right), \quad \sinh z = \frac{1}{2}\left(e^z - e^{-z}\right), \tag{1.76}$$

satisfying

$$\cosh^2 z - \sinh^2 z = 1. \tag{1.77}$$

Their ratio is the hyperbolic tangent

$$\tanh z = \frac{\sinh z}{\cosh z}. \tag{1.78}$$

These hyperbolic functions allow for a succinct description of the causal structure in Minkowski spacetime by parameterizing velocity four-vectors of massive particles. These tangents to their world-lines are within the light cones, described as follows.

Putting the velocity of light c equal to 1, Minkowski spacetime has a *Lorentz invariant causal structure* described by light cones. In $1+1$ space-time, this is expressed by invariance of the signed distances defined by the Lorentz metric

$$ds^2 = dt^2 - dx^2, \tag{1.79}$$

associated with light-like and space-like separations $ds^2 > 0$ and $ds^2 < 0$, respectively.

A light cone is along null-directions

$$ds^2 = 0. \tag{1.80}$$

Here, we use the sign convention $(+-)$. The line-element (1.79) gives a signed measure of spacetime intervals such that

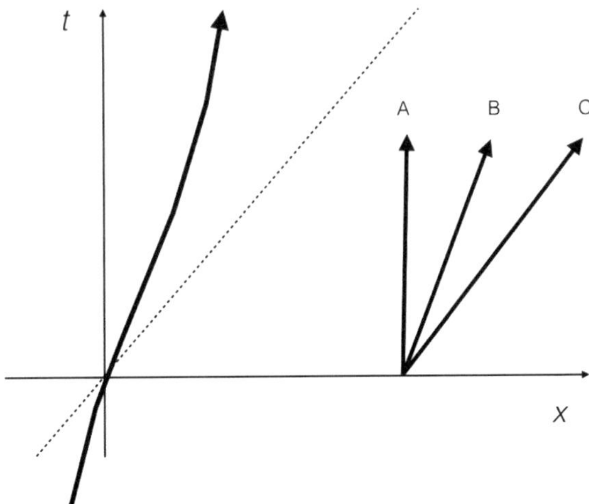

Fig. 1.5 Minkowski diagram in 1 + 1, showing a null-ray, the worldline of a massive particle inside and three worldlines with different velocities. Particle A is at rest in (t, x), whereas B and C are moving

$$\frac{\Delta x'}{\Delta t'} = \frac{\Delta x}{\Delta t} = 1 \tag{1.81}$$

in the transformation of spacetime intervals over light cones between frames $K = (t, x)$ and $K' = (t', x')$. By (1.80), (1.81) is preserved when

$$dt^2 - dx^2 = (dt')^2 - (dx')^2. \tag{1.82}$$

Massive particles move inside these light cones (Fig. 1.5). Their worldlines have tangents described by velocity four-vectors

$$u^b = \left(u^t, u^x\right) = \left(\frac{dt}{d\tau}, \frac{dx}{d\tau}\right), \tag{1.83}$$

where τ denotes their *eigentime*, i.e., the invariant length of their world-lines measured by the wrist watch of a comoving observer. According to (1.79),

$$1 = \left(\frac{ds}{d\tau}\right)^2 = \left(\frac{dt}{d\tau}\right)^2 - \left(\frac{dx}{d\tau}\right)^2. \tag{1.84}$$

In light of (1.77), we define the contravariant and covariant tangents

$$u^b = (\cosh\lambda, \sinh\lambda), \ u_b = (\cosh\lambda, -\sinh\lambda) \tag{1.85}$$

in terms of a *rapidity* λ, whereby (1.84) is identically satisfied:

$$u^c u_c = (u^t)^2 - (u^x)^2 = \cosh^2\lambda - \sinh^2\lambda \equiv 1. \tag{1.86}$$

While u^b describes the tangent to a worldline, $u_b = (e, P)$ describes the energy and momentum of a particle. Illustrative to the latter is the covariant gradient $\partial_a \phi = (\omega, k)$ of the phase $\phi = \omega t - kx$ of a electromagnetic plane wave with angular velocity ω and wave number k that, multiplied by Planck's constant, define its unit of energy and momentum.

The particle *three-velocity*

$$V \equiv \frac{dx}{dt} = \frac{dx/d\tau}{dt/d\tau} = \frac{u^x}{u^t} = \tanh\lambda. \tag{1.87}$$

satisfies $|V| < 1$ since $-1 < \tanh x < 1$. The time component u^t of the velocity four-vector is the *Lorentz factor*

$$u^t = \cosh\lambda = \frac{1}{\sqrt{1-\tanh^2\lambda}} = \frac{1}{\sqrt{1-V^2}} \equiv \Gamma. \tag{1.88}$$

Consider B with velocity four-vector $u^b = (\cosh\mu, \sinh\mu)$ and three-velocity $U = \tanh\mu$ in $K = (t, x)$. In B's frame $K' = (t', x')$, a particle C has a velocity four-vector $v^{b'} = (\cosh\lambda, \sinh\lambda)$ with three-velocity $V = \tanh\lambda$.

To determine C's velocity four-vector $v^b = (\cosh\alpha, \sinh\alpha)$ in K, we consider u^b of B to be a *Lorentz boost* by μ from rest in K, i.e.,

$$\begin{pmatrix} u^t \\ u^x \end{pmatrix} = \Lambda \begin{pmatrix} 1 \\ 0 \end{pmatrix} \tag{1.89}$$

with

$$\Lambda - \begin{pmatrix} \cosh\mu & \sinh\mu \\ \sinh\mu & \cosh\mu \end{pmatrix}. \tag{1.90}$$

Applying the same boost to C leaves invariant B's observation of C's velocity, whereby[7]

$$\begin{pmatrix} v^t \\ v^x \end{pmatrix} = \Lambda \begin{pmatrix} v^{t'} \\ v^{x'} \end{pmatrix}, \tag{1.91}$$

thus taking λ into $\alpha = \mu + \lambda$. Consequently,

$$\alpha = \lambda + \mu: \quad W = \tanh\alpha = \frac{U+V}{1+UV} \tag{1.92}$$

is C's three-velocity in the laboratory frame of A.

In $1+1$ Minkowski spacetime, therefore, three-velocities combine by linear superposition of rapidities. The Galilean transformation is the non-relativistic limit

[7]More on matrices in Chap. 3.

λ, $\mu << 1$, satisfying

$$W \simeq U + V. \tag{1.93}$$

The Lorentz transformation (1.91) also derives by analytic continuation of rotations on the unit circle S^1, described by

$$e^{i\phi}e^{i\psi} = e^{i\gamma} : \ \gamma = \phi + \psi. \tag{1.94}$$

In matrix notion, this is expressed by a rotation of position vectors

$$\begin{pmatrix} \cos\gamma \\ \sin\gamma \end{pmatrix} = \begin{pmatrix} \cos\phi & -\sin\phi \\ \sin\phi & \cos\phi \end{pmatrix} \begin{pmatrix} \cos\psi \\ \sin\psi \end{pmatrix}, \tag{1.95}$$

associated with the Euclidean line-element $dx^2 + dy^2$ of the two-dimensional plane. Taking $\phi = i\mu$, $\psi = i\lambda$ and $\gamma = i\alpha$, (1.95) becomes

$$\begin{pmatrix} \cosh\alpha \\ i\sinh\alpha \end{pmatrix} = \begin{pmatrix} \cosh\mu & -i\sinh\mu \\ i\sinh\mu & \cosh\mu \end{pmatrix} \begin{pmatrix} \cosh\lambda \\ i\sinh\lambda \end{pmatrix}, \tag{1.96}$$

identical to (1.91) along with the hyperbolic metric $dx^2 + dy^2 = -dt^2 + dx^2$ by $(x, y) = (x, it)$. In Minkowski spacetime, the Galilean transformation has become rule of addition of hyperbolic angles.

1.5.1 Rindler Observers

According to (1.83) and (1.85), we are at liberty to consider trajectories with time-dependent rapidity

$$\lambda(\tau) = a\tau, \tag{1.97}$$

where a constant a defines *Rindler observers* $\mathcal{O}$ with acceleration

$$a^b = \frac{du^b}{d\tau}, \ \ a^c a_c = a^2. \tag{1.98}$$

Their trajectories obtain as the integral of u^b,

$$x^b(\tau) = \frac{1}{a}\left(\sinh\lambda, \cosh\lambda\right). \tag{1.99}$$

Here, we use the convention $x^b(0) = (0, \xi)$, $\xi = 1/a$, i.e., the distance to the origin ξ is given by the reciprocal of acceleration.

In fact, at any point along the trajectory, x^b is a spacelike separation to the origin, normal to the trajectory with distance

$$\text{(a) } x^c u_c = 0, \quad \text{(b) } \sqrt{x^c x_c} = \frac{1}{a} = \xi. \tag{1.100}$$

Here, (a) can be seen by direct evaluation or by taking the time derivative of (b). Consequently, *$\mathcal{O}$ considers itself at constant distance to the light cone H with vertex at the origin (Fig. 1.6)* and *$\mathcal{O}$ can receive signals region IV but not from III or II.*

Example 1.8. Consider the world-line of a particle at rest at a distance $\mu\xi$ from the origin with $0 < \mu < 1$, shown in Fig. 1.6. According to (1.99), its intersection with the spacelike separation of $\mathcal{O}$ to the origin satisfies $\lambda\xi\cosh(a\tau) = \mu\xi$, i.e.,

$$\lambda = \frac{\mu}{\cosh(a\tau)}, \tag{1.101}$$

giving rise to the curved trajectory with horizon distance $\mu\xi/\cosh(a\tau)$ shown in the right panel of Fig. 1.6. While it appears with infinite coordinate length in $\mathcal{O}$'s coordinate system, it represents only the segment of finite (invariant) length

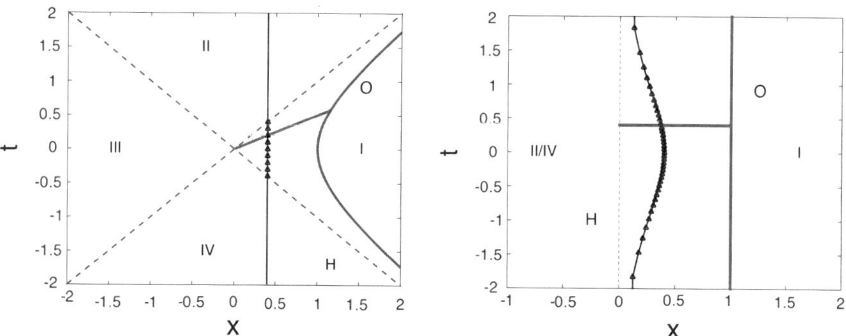

Fig. 1.6 (*Left*) The trajectory of a Rindler observer $\mathcal{O}$ at constant acceleration in Minkowski spacetime. At any point along its trajectory, the separation to the origin is spacelike with constant distance in the Minkowski line-element equal to the inverse of its acceleration. Show is further the world-line of a particle at rest at an intermediate distance to the origin, that temporarily visits region I. (*Right*) The same as observed in a comoving coordinate system of $\mathcal{O}$, wherein it manifestly has a constant distance to the event horizon H. To $\mathcal{O}$, the (curved) trajectory of the particle at rest appears to visit region I eternally

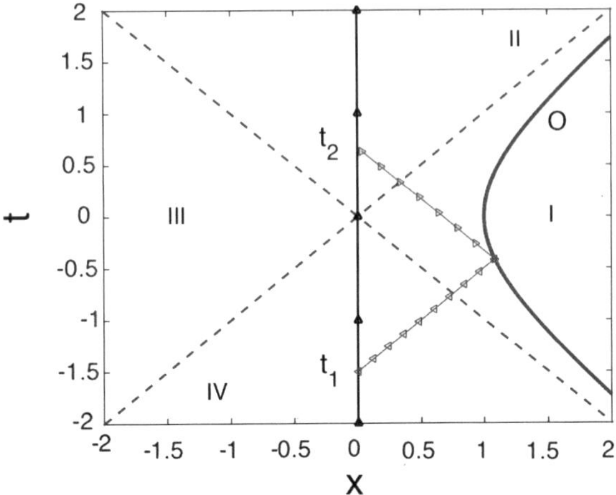

Fig. 1.7 Reflection of light from an inertial observer at the origin off a mirror at contant acceleration gives a reflection time $t_2 - t_1$ indicated by ray tracing along null-directions

covering the particle's visit in region I, in Minkowski coordinates $y^b = (t, x)$ satisfying

$$y^b = \lambda x^b = \frac{\mu}{a} (\tanh(a\tau), 1) \,. \tag{1.102}$$

The Minkowski time $y^0 = t$ is eigentime of the particle at rest. It has a corresponding stretched time τ of $\mathcal{O}$, satisying $\tau = a^{-1} \tanh^{-1}(at/\mu)$, indicated by the non-uniform distribution of markers along its world-line in $\mathcal{O}$'s frame of reference (Fig. 1.6).

Example 1.8. Consider the time delay in bouncing a signal off a moving mirror at constant acceleration a (Fig. 1.7). Emitted at t_1 from the world-line of an inertial, zero-velocity observer at the origin, it bounces at (t, x) in Minkowski coordinates, satisfying

$$(t_1 + x, x) = a^{-1} (\sinh(a\tau), \cosh(a\tau)) \tag{1.103}$$

by (1.99). The delay time $\Delta = t_2 - t_1$ upon return satisfies

$$a\Delta t = \mu + \frac{1}{\mu}, \quad \mu = |at_1| \,. \tag{1.104}$$

A bounce at $t = 0$ gives the familiar result $\Delta t = 2\xi$, $\xi = 1/a$.

The above can be slightly extended, upon fixing the acceleration a and allowing ξ to be free. This recovers (b) in (1.100) and (1.99) only for $\xi = 1/a$. The new set of orthogonal *Rindler coordinates* (ξ, τ)

$$t = \xi \sinh(a\tau), \quad x = \xi \cosh(a\tau) \tag{1.105}$$

cover region I shown in Fig. 1.6 with line-element (1.79) transformed to

$$ds^2 = \xi^2 d\tau^2 - d\xi^2. \tag{1.106}$$

A set of Rindler coordinates gives a *chart* of region I of Minkowski spacetime, as defined by the light cone with vertex at the origin. That (1.106) is flat can be seen directly by

$$(\xi, \tau) = (r, i\varphi), \tag{1.107}$$

taking it into the line-element $ds^2 = dr^2 + r^2 d\varphi^2$ of Euclidean space in polar coordinates.

Example 1.9. Rindler charts are often used to model the local structure of spacetime near the horizon of a black hole. A black hole of mass M is described by the Schwarzschild line-element in spherical coordinates (t, r, θ, φ)

$$ds^2 = \alpha^2 dt^2 - \frac{dr^2}{\alpha^2} + r^2 d\theta^2 + r^2 \sin\theta d\varphi^2 \tag{1.108}$$

with horizon at radius $r = 2M$ and redshift factor, measured by locally static observers,

$$\alpha \equiv \frac{\text{eigentime}}{\text{time-at-infinity}} = \sqrt{1 - \frac{2M}{r}}. \tag{1.109}$$

The redshift factor defines the energy-at-infinity of a test particle of unit rest-mass, giving rise to the surface gravity

$$\frac{d\alpha}{ds} = \frac{\alpha d\alpha}{dr} = \left.\frac{M}{r^2}\right|_{r=2M} = \frac{1}{4M}. \tag{1.110}$$

Deredshifted to the eigentime τ, a locally static observer hereby experiences a constant acceleration

$$a = \alpha^{-1}\frac{M}{r^2} \simeq \frac{1}{4M}\left(\frac{x}{2M}\right)^{-1/2} \tag{1.111}$$

at physical distance to the event horizon

$$\xi = \int_{2M}^{2M+x} \frac{dr}{\alpha} = 4M\left(\frac{x}{2M}\right)^{1/2}. \tag{1.112}$$

The local results (1.111–1.112) satisfy

$$a\xi = 1 \tag{1.113}$$

as in (b) of (1.100) of Rindler observers in Minkowski spacetime.

1.6 The Logarithm and Winding Number

The logarithm $\log z$ is defined as the inverse of e^z. If $w = e^z$, then

$$z = \log w = \log |w| + i \arg w. \tag{1.114}$$

The logarithm is multivalued, since $\arg w = \operatorname{Arg} w + 2\pi k$, i.e.,

$$z = \log w = \log |w| + i \operatorname{Arg} w + 2\pi i k \quad (k \in \mathbb{Z}). \tag{1.115}$$

In particular, (1.12) shows

$$\log 1 = 2\pi i k \quad (k \in \mathbb{Z}). \tag{1.116}$$

Example 1.10. Consider w on the unit circle, $w = e^{i\theta}$. In this event, $\log w = i(\theta + 2\pi k)$. The multiplicity $2\pi k$ in the argument has a topological interpretation in k denoting the number of times the unit circle is traversed in an identity map $w \to w$ by some function $\theta = \theta(\lambda)$ on $\lambda \in [0, 1]$ satisfying $\theta(1) - \theta(0) = 2\pi k$, i.e.:

$$w(1) = w(0): \quad e^{i\theta(1)} = e^{i\theta(0)}. \tag{1.117}$$

Thus, k is referred to as the *winding number* of $w(\lambda)$.

In Example 1.8, the multiplicity k is a winding number about origin. This suggests a generalization to closed loops, more general than S^1. Pick $w \in \mathbb{C}$ and let γ be a closed curve from and to w (w is a base point of γ):

$$\gamma:\ \xi=\xi(\lambda),\ \ \xi(0)=\xi(1)=\boldsymbol{w}\ \ (\lambda\epsilon[0,1]). \tag{1.118}$$

When γ does not pass through the origin, $\xi(\lambda)\neq 0$ for all $\lambda\epsilon[0,1]$. Start at $\lambda=0$ with a choice of the argument of $\boldsymbol{w}$, e.g., the principle value

$$\theta=\mathrm{Arg}\,\boldsymbol{w} \tag{1.119}$$

such that $-\pi\leq\theta<\pi$ as in (1.20). We take (1.119) as an initial value for $\boldsymbol{w}(0)$ in a (continuous) continuation $\boldsymbol{w}(\lambda)$, marching along γ to $\boldsymbol{w}(1)$. While $|\boldsymbol{w}(0)|=|\boldsymbol{w}(1)|$ is obvious, we encounter

$$\Delta\theta=2\pi k \tag{1.120}$$

as γ winds the origin $k\,\epsilon\,\mathbb{Z}$ times.

When considering closed loops γ, therefore, the logarithm is a natural function to calculate winding numbers in the complex plane.

1.7 Branch Cuts for log *Z*

With a finite radius of convergence R, the power series (1.67) defines functions that, as will be discussed in more detail in the next chapter, are analytic in $|z|<R$. By analytic continuation, they hereby represent Taylor series expansions of functions possibly extending into $|z|\geq R$. As briefly alluded to earlier, continuation over extended paths is not necessarily unique. Since Taylor series are unique, a Taylor series inevitably picks out a particular branch.

To illustrate this, consider $\log(1+x)$ as defined by the calculus of integration

$$f(x)=\log(1+x)=\int_0^x\frac{dx}{1+x} \tag{1.121}$$

and the Neumann series (1.68), we have

$$\log(1+x)=\int_0^x\left(1-x+x^2+\cdots\right)dx=x-\frac{1}{2}x^2+\frac{1}{3}x^3+\cdots \tag{1.122}$$

The right hand side introduces a function $f(z)$ by the power series

$$f(z)=\sum_{n=0}^{\infty}\frac{(-1)^n}{n+1}z^{n+1} \tag{1.123}$$

with radius of convergence 1. Since $f(x) = \log(1 + x)$ on $-1 < x < 2$, (1.121) defines the Taylor series of $\log(1 + z)$ about $z = 0$, of $f(x)$ with x extended to the complex plane.

Consider a closed loop γ in the domain $|z| < 1$. Then $\gamma_1 = \{1 + z \mid z\epsilon\gamma\}$ is a loop away from the origin: the winding number of γ_1 around the origin is zero. It follows that $\log(1 + z)$ is *univalent* on γ, after choosing a specific value of $\log(1 + z)$ for a given z on γ.

Since $\arg w$ is undefined at the origin $w = 0$, the origin is a *branch point* for $\log w$. In many applications, we wish to work with univalent functions. (In calculus of a real variable, a function commonly refers to a univalent map.) At the root is the multivalued argument of complex numbers, i.e., (1.12) and (1.116), that represent winding numbers of loops around the origin. The logarithm becomes univalent upon building a restriction in the complex plane, permitting loops with zero winding numbers only.[8] To this end, we cut the plane, by a branch cut connecting a branch point to infinity. For $\log z$, this branch point is the origin. The branch cut acts as a barrier, across which no curve shall pass. Any closed curve not crossing the branch cut has zero winding number around the origin. It fixes a univalent value of the logarithm, after choosing a value of $\log z$ for a given $z_0\epsilon\mathbb{C}\backslash\{0\}$.

With this in mind, we interpret $f(z)$ in (1.121) as the branch of $\log(1 + z)$ in the domain $|z| < 1$ with $f(2) = \log 2$, distinct from other branches $\log 2 + 2\pi i k$ $(k \neq 0)$. *A power series selects a univalent branch of a possibly multivalued function.*

Example 1.11. Consider $h(z) = (1 + z)^s$ on $|z| < 1$ with possibly complex valued s. With $h(z) = e^{s\log(1+z)}$, we write

$$h(z) = e^{s(\log|1+z|+i\text{Arg}\,(1+z)+2\pi ik)} = |1 + z|^s\, e^{is\text{Arg}(1+z)} e^{2\pi iks}. \quad (1.124)$$

With $k\,\epsilon\,\mathbb{Z}$, we see that $h(z)$ is multivalued whenever $e^{2\pi iks}$ is not identically equal to 1, i.e., when s is not integer. A univalent function obtains by fixing $h(1)$, e.g., $h(1) = 0$ corresponding to $k = 0$, and introducing a cut from the branch point $z = -1$ to infinity. The associated Taylor series obtains as

$$\begin{aligned} h(z) = e^{sf(z)} &= 1 + sf(z) + \tfrac{1}{2}(sf(z))^2 + \cdots \\ &= 1 + s\left(z - \tfrac{1}{2}z^2 + \cdots\right) + \tfrac{1}{2}s^2\left(z - \tfrac{1}{2}z^2 + \cdots\right)^2 + \cdots \quad (1.125) \\ &= 1 + sz + \tfrac{1}{2}s(s-1)z^2 + \tfrac{1}{6}s(s-1)(s-2)z^3 + \cdots \end{aligned}$$

[8]More accurately, a branch cut *preserves* univalence in extending a function to $f(z)$ by analytic continuation from a choice of $w_0 = f(z_0)$ at $z = z_0$. The branch cut defines a barrier for paths of continuation, preventing closed loops around a branch point.

1.8 Branch Cuts for $z^{\frac{1}{p}}$

With (1.12) and (1.116) in mind, the square root is the multivalued map

$$w = z^{\frac{1}{2}} = \left(e^{\log|z|+2\pi i k}\right)^{\frac{1}{2}} = |z|^{\frac{1}{2}}1^{\frac{1}{2}} \ \ (k \in \mathbb{Z}). \tag{1.126}$$

In particular, we have

$$z = 1: \ \ w \in \{-1, 1\}. \tag{1.127}$$

As before, we select a univalent branch by fixing w for a choice of z, e.g., $w = 1$ for $z = 1$, corresponding to $k = 0$ in (1.126), together with a branch cut from the origin to infinity (Fig. 1.8). Thus, a univalent square root obtains from a univalent $\log z$ by a choice of corresponding branch cut.

Consider $w = \sqrt{z^2 - 1} = \sqrt{(z-1)(z+1)}$. We have the pair of branch points $z = -1, 1$ corresponding to $1 - z^2 = 0$. To obtain a univalent function, we cut the complex plane such that (i) closed loops can wind around both but not one of them, or (ii) no closed loops can wind around $z = -1$ or $z = 1$. For (i), each winding along a loop γ enclosing both gives a phase change of $2 \times 2\pi$ in $z^2 - 1$ and hence a phase chance of 2π in w, illustrated in Fig. 1.9.

The examples above serve to show that *developing a Taylor series expansion begins with a selection of a univalent branch, comprising a choice of $f(z_0)$ and branch cuts of the complex plane associated with one or more branch points.* For instance, the quartic root of $z = re^{i\theta}$,

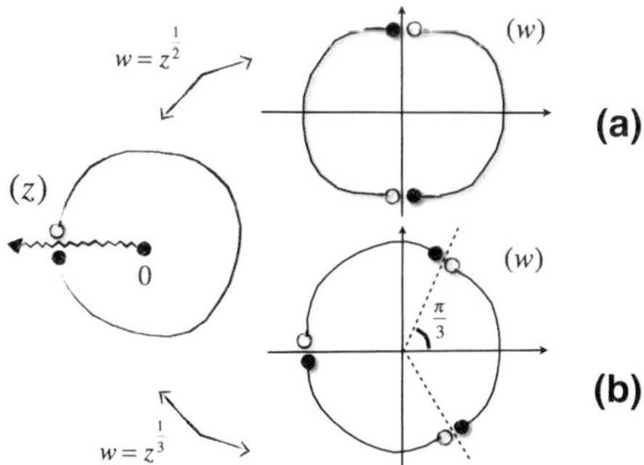

Fig. 1.8 **a** The square root $w = z^{\frac{1}{2}}$ maps circles around the origin into two half-circles corresponding to a different branch associated with $w(1) = 1$ and $w(1) = -1$, respectively. To obtain a univalent map with the property $w(1) = 1$, we cut the complex plane as shown. This univalent map is discontinuous across the *branch cut*: the wiggly line from $-\infty$ to the branch point $z = 0$. **b** Similar arguments apply to the three univalent branches of $w = z^{\frac{1}{3}}$

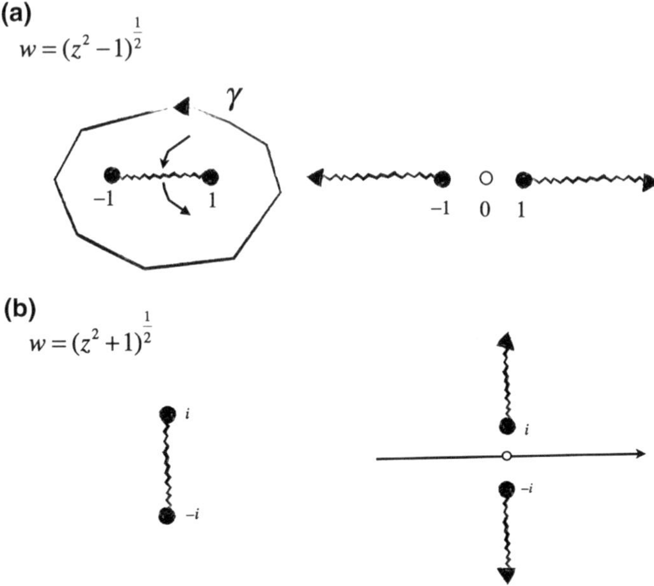

Fig. 1.9 **a** Branch cuts for a univalent function $w = (z^2 - 1)^{\frac{1}{2}}$, connecting the two branch points $z = -1, 1$ or connecting each to infinity. For the first, $\Delta\arg w = 2\pi$ as z traverses along a contour enclosing $-1, 1$. This branch is discontinuous across the branch cut $[-1, 1]$ shown. For the second, w is continuous around the origin. The first and second are automatically selected by the Taylor series expansions of w around infinity (defined as the Taylor series expansion of w as a function of $1/z$ about $z = 0$) and, respectively, the origin. **b** Similar arguments apply to $w = (z^2 + 1)^{\frac{1}{2}}$ with branch points $z = i, -i$

$$w = z^{\frac{1}{4}} = r^{\frac{1}{4}}\left[\cos\left(\frac{\theta}{4}\right) + i\sin\left(\frac{\theta}{4}\right)\right], \tag{1.128}$$

has one branch point at $z = 0$ and four solution branches associated with w at $z = 1$:

$$w(1) = 1, i, -1, \text{ or } -i. \tag{1.129}$$

Next, consider [1]

$$f(z) = \log\left(1 + \left(\frac{z+1}{z-1}\right)^{\frac{1}{4}}\right), \quad f(-1) = 0. \tag{1.130}$$

For a loop γ going around both $z = -1, 1$, the change in argument in the Möbius transformation

$$\xi = \frac{z+1}{z-1} \tag{1.131}$$

Table 1.1 Complex numbers

1. Complex numbers are invented to solve $z^2+1=0$ by the imaginary numbers $z=\pm i$, where $i=\sqrt{-1}$. Complex numbers $z=x+iy$ are conveniently expressed by their real and imaginary parts x and, respectively, y or in polar form $z=re^{i\theta}$ by their norm $|z|=r$ and argument θ, where θ is determined up to a multiplicity of 2π. The latter gives Euler's identity $e^{i\pi}+1=0$.

2. e^z with complex argument z is the extension of e^x as a function of x on the real line. Its inverse is the logarithm $\log z=\log r+i\arg z$, where $\arg z$ has multiplicity 2π. The n-th root of $1=e^{2\pi ik}$ is the set of points $e^{2\pi ik/n}$ on S^1 $(k=0,1,2,\cdots n-1)$.

3. $\log z$ has a branch point at the origin, where $\arg z$ is not defined. $\log z$ can be made univalent by introducing a branch cut in the complex plane from $z=0$ to infinity, restricting paths of continuation to those with winding number zero about the origin.

4. A series $\{z_n\}_{n=0}^{\infty}$ converges to z_* if the tails $\{z_n\}_{n=N}^{M}\subset B_\epsilon(z_*)$, where N is sufficiently large for a ball of given size ϵ.

5. Power series $\sum_{n=0}^{\infty}a_nz^n$ have a radius of convergence R, where R may be zero, finite or infinite. Convergence is absolute in $|z|<R$ and divergent in $|z|>R$. On $|z|=R$, the sum may converge or diverge. The exponential function $\sum_{n=0}^{\infty}(1/n!)z^n$ has $R=\infty$.

is zero. At $z=2$, fix $\xi^{\frac{1}{4}}=3^{\frac{1}{4}}>0$ with a branch cut along the interval $[-1,1]$ connecting the two branch points. This gives a univalent branch of $f(z)$ with $f(-1)=0$. It can easily be seen that the branch point $\xi^{\frac{1}{4}}=-1$ of $\log(1+\xi^{\frac{1}{4}})$ is never encountered.

Table 1.1 summarizes this discussion.

1.9 Exercises

1.1. Show by explicit evaluation that the real and imaginary part of e^z with $z=i\theta$ in (1.14) recover the Taylor series expressions for $\cos\theta$ and $\sin\theta$.

1.2. Expand (1.26) in a Taylor series up to third order, and show that the expansions on the left and right hand-side agree term by term.

1.3. Find the radius of convergence of the following power series

$$\text{(a)} \sum \frac{1}{n} z^n, \text{(b)} \sum z^n, \text{(c)} \sum \frac{1}{n!} z^n, \text{(d)} \, x - \frac{1}{3!}x^3 + \frac{1}{5!}x^5 + \cdots, \tag{1.132}$$

where summations are over the integers n.

1.4. Find the singularities of the functions

$$\text{(a)} \frac{1}{1+z^2}, \text{(b)} \frac{1}{\sin z}, \text{(c)} \frac{1}{z^2}, \text{(d)} \, e^{\frac{1}{z}}. \tag{1.133}$$

1.5. Identify the branch points and introduce suitable branch cuts in the complex plane for the univalent functions:

- $f(z) = \sqrt{1+z}$, $f(0) = 1$
- $f(z) = \sqrt{1+z^2}$, $f(0) = 1$
- $f(z) = \log z$, $f(1) = 0$
- $f(z) = z^{\frac{1}{3}}$, $f(1) = 1$

1.6. Extend the de Moivre formula for trigonometric functions to hyperbolic functions.

1.7. Construct a sequence $\{x_n\}_{n=0}^{\infty}$ which converges to zero, while x_{2n}/x_{2n+1} has no limit.

1.8. Consider a series $\sum_{n=0}^{\infty} a_n$ that is absolutely convergent, i.e., the partial sums $\sum_{n=0}^{N} |a_n|$ converge in the limit as N goes to infinity. Show that the order in which the terms are summed is immaterial. [*Hint*: Show that $\sum_A a_n$ is convergent for any permutation A of $\mathbb{N}$.]

1.9. Show that the alternating sequence $1 - \frac{1}{2} + \frac{1}{4} - \frac{1}{8} + \cdots$ is convergent, but is not absolutely convergent. What is the radius of convergence of $\sum 2^{-n} z^n$, summed over $n \geq 0$?

1.10. In Question 1.5, what happens to $f(z)$ when we cross a branch cut?

1.11. Generalize (1.26) to hyperbolic functions, following $\theta \to i\theta$. Show that any cubic equation can be transformed to either $4y^3 \pm 3y = x$ or $y^3 = x$. Use the resulting identifies to derive analytic solutions for $4y^3 \pm 3y = x$ for all x.

1.12. Show that the two convergence criteria (1.47) are equivalent. [*Hint:* Put $w_n = \log B_n$.]

1.13 Let s_n be a decreasing sequence bounded below. Show that s_n has a limit by modifying the proof of boundedness of Cauchy sequences, or show that if s_{n_k} is a convergent subsequence, then any other subsequence s_{n_l} is convergent with the same limit.

1.14. Consider $f(x) = \tanh(x)$ $(-\infty < x < \infty)$. Sketch the graph of $f(x)$ and give the Taylor series expansion of $f(x) = \tanh(x)$ about $x = 0$ to third order.

1.15. With the Minkowski metric

$$\eta_{ab} = \begin{pmatrix} 1 & 0 \\ 0 & -1 \end{pmatrix} \tag{1.134}$$

the Minkowski line-element (1.79) can be written as

$$ds^2 = \eta_{ab} dx^a dx^b, \tag{1.135}$$

where summation over the indices a and b runs over (t, x). Consider the Lorentz boosts

$$\Lambda^a{}_b = \begin{pmatrix} \cosh\lambda & \sinh\lambda \\ \sinh\lambda & \cosh\lambda \end{pmatrix}, \quad \Lambda_a{}^b = \begin{pmatrix} \cosh\lambda & -\sinh\lambda \\ -\sinh\lambda & \cosh\lambda \end{pmatrix}. \tag{1.136}$$

(a) Show that

$$\Lambda_a{}^b = \eta_{aa'}\eta^{bb'}\Lambda^{a'}{}_{b'}, \tag{1.137}$$

where η^{ab} is the inverse of η_{ab}, i.e., $\eta_{ab}\eta^{bc} = \delta_a^c$ in terms of the Kronecker delta function $\delta_a^b = 0$ $(a \neq b)$, $\delta_a^b = 1$ $(a = b)$. (b) Under a Lorentz boost, η_{ab} transforms as

$$\eta_{a'b'} = \eta_{ab}\Lambda_{a'}^{a}\Lambda_{b'}^{b}. \tag{1.138}$$

Show that $\eta_{a'b'}$ is again of the form (1.134). (c) Define

$$v_a = \eta_{ab} v^b. \tag{1.139}$$

Under a Lorentz boost, u^b transforms as

$$u^b = \Lambda^b{}_{a'} u^{a'}. \tag{1.140}$$

Show that the contraction

$$u^c v_c = u^t v^t - u^x v^x \tag{1.141}$$

is invariant under Lorentz boosts.[9]

1.16. Based on Fig. 1.6, argue that a Rindler observer attributes a constant surface gravity $a = 1/\xi$ to the event horizon H.

1.17. The contravariant vector u^b represents a directional derivative in the invariant expression $u^c \partial_c \varphi$, where $\varphi = \varphi(t, x)$ is a scalar field. Let φ denote the total phase in a wave $e^{i\varphi}$. Let u^b denote the velocity four-vector of an observer. Then $\omega = u^c \partial_c \varphi$ denotes the observed angular frequency. The energy and momentum of the wave are $(\hbar\omega, \hbar k)$, where $\hbar$ is Planck's constant and k denotes the wave number. In the laboratory frame, observers have $u^t = (1, 0)$, whereby $\omega = \partial_t \varphi$ and $k = \partial_x \varphi$. Show the invariant $\partial^c \partial_c \varphi = 0$ for a photon, i.e., a massless particle satisfying the dispersion relation $\omega = ck$, where c is the velocity of light.

1.18. Using (1.139), the contravariant four-momentum $p^b = mu^b$ of a particle of mass m has a covariant four-momentum $p_a = mu_a$. Following **1.12**, $E = p_t$ is the the energy of the particle measured in the laboratory frame in units with $c = 1$. Derive

$$E = m\sqrt{1 + \sinh^2 \lambda} \simeq m + \frac{1}{2} m V^2 \ (|\lambda| << 1). \tag{1.142}$$

Show that the three-velocity satisfies

$$V = \frac{u^x}{u^t} = \frac{dE}{dP}, \tag{1.143}$$

where $P = m \sinh \lambda$. Plot the *dispersion relations* $E(P)$ and $\omega(k)$ of a photon and explain their common asymptotic behavior for large momenta. $E_0 = mc^2$ at $P = 0$ is the rest mass energy of the particle. If m is 1 g, how long can it power a laptop?

1.19. Consider a monochromatic point source emitting that flash of photons in all directions. Show that this shell of photons is a spherical in any frame of reference. Does the light have the same color in all directions? Show that the shape of a sphere of matter at high Lorentz Lorentz factors is a pancake.

1.20. Factor (1.96) over the Lorentz factors Γ_i of $i = A, B, C$,

[9] In general relativity, Minkowski space describes the local causal structure in the tangent spaces of four-dimensional spacetime manifolds with metric g_{ab}, which may be curved in the presence of matter and fields. By general coordinate covariance, $u^c u_c = g_{ab} u^a u^b$ and $u^c v_c = g_{ab} u^a v^b$ are then invariant with respect to arbitrary coordinate transformations.

$$\Gamma_A \begin{pmatrix} 1 \\ W \end{pmatrix} = \Gamma_B \Gamma_C \begin{pmatrix} 1 & U \\ U & 1 \end{pmatrix} \begin{pmatrix} 1 \\ V \end{pmatrix}. \tag{1.144}$$

Consider A emitting a particle to a mirror B moving towards A with Lorentz factor Γ_B. Compute the boost in Lorentz factor of the particle in A's frame after bouncing off the mirror back towards A. Compute the limit in case the particle is a photon. [*Hint.* Transform the velocity four-vector of the particle to the frame of the mirror and, after bouncing, back to the laboratory frame.]

1.21. For the hyperbolic functions $\cosh\lambda$ and $\sinh\lambda$, establish identities equivalent to (1.26). Give an exact solution to the cubic equation $4x^3 + 2x + 2 = 0$.

1.22. Given (1.131), show that $\xi^{\frac{1}{4}}$ never equals -1.

Reference

1. Carrier, G.F., Crook, M., and Pearson, C.E., 1983, *Functions of a Complex Variable*, Hod Books, Ithaca, New York.

Chapter 2
Complex Function Theory

We next turn to functions of a complex variable, that are perhaps best known for solving integrals by contour deformation in the complex plane. It also provides a starting point to special functions and linear transform methods, some to be introduced in subsequent chapters. Complex functions also give a concise description of idealized two-dimensional fluids, in the approximation of incompressible and irrotational flows. These developments derive from *analyticity*. If a function $f(z)$ of a complex variable z is differentiable at some z, we say that $f(z)$ is analytic at z. Extended over an open region in the complex plane, this property has broad implications.

2.1 Analytic Functions

Complex function theory considers maps $w = f(z)$ given by functions

$$f(z) = u(x, y) + iv(x, y) \tag{2.1}$$

from the complex plane to itself, where we decompose $f(z)$ into its real and imaginary parts $u(x, y)$ and $v(x, y)$, respectively. We define the complex conjugate $\bar{z} = x - iy$, whereby

$$x = \mathrm{Re}\, z = \frac{1}{2}\,(z + \bar{z})\,, \quad y = \mathrm{Im}\, z = \frac{1}{2i}\,(z - \bar{z})\,. \tag{2.2}$$

Likewise, we have

$$\begin{aligned} u(x, y) &= \mathrm{Re} f(z) = \tfrac{1}{2}\left(f(z) + \overline{f(z)}\right), \\ v(x, y) &= \mathrm{Im} f(z) = \tfrac{1}{2i}\left(f(z) - \overline{f(z)}\right). \end{aligned} \tag{2.3}$$

M.H.P.M. van Putten, *Introduction to Methods of Approximation in Physics and Astronomy*, Undergraduate Lecture Notes in Physics,
DOI 10.1007/978-981-10-2932-5_2

Our focus will be on functions that are differentiable. We say that $f(z)$ is differentiable at $z = z_0$ if the limit

$$f'(z_0) = \lim_{z \to z_0} \frac{f(z) - f(z_0)}{z - z_0} \tag{2.4}$$

exists. It means that the limit gives a unique finite answer, regardless of the way z_0 is approached. Equivalently, we formulate (2.4) in terms of sequences $\{z_n\}_{n=0}^{\infty}$ with $z_n \to z_0$ in the limit as n approaches infinity, insisting that

$$\lim_{n \to \infty} \frac{f(z_n) - f(z_0)}{z_n - z_0} \tag{2.5}$$

exists for any such sequence approaching z_0. Of particular interest are the two alternatives of approaching z_0 along the x and y directions, i.e., $z = z_0 + h$ and $z = z_0 + ih$, respectively. The limit (2.4) exists iff

$$A(h) = \frac{f(z_0 + h) - f(z_0)}{h}, \quad B(h) = \frac{f(z_0 + ih) - f(z_0)}{ih} \tag{2.6}$$

approach finite limits which are the same as h approaches zero. In the notation of (2.1),

$$A_0 \equiv \lim_{h \to 0} A(h), \quad B_0 \equiv \lim_{h \to 0} B(h) \tag{2.7}$$

gives

$$A_0 = u_x(x_0, y_0) + iv_x(x_0, y_0), \quad B_0 = v_y(x_0, y_0) - iu_y(x_0, y_0). \tag{2.8}$$

where we used $1/i = -i$. Equating $A_0 = B_0$, we obtain the *Cauchy-Riemann relations*

$$u_x = v_y, \quad u_y = -v_x \tag{2.9}$$

at the point $z = z_0$ in the complex plane.

If (2.9) holds true throughout some region (an open domain) D in the complex plane, we say that $f(z)$ is *analytic* in D. (A more restricted focus would be analyticity along a curve.)

Example 2.1. Consider $f(z) = z$ with $u(x, y) = x$ and $v(x, y) = y$. The Cauchy-Riemann relations (2.9) are satisfied by inspection, so $f(z)$ is analytic for all z, i.e., $f(z)$ is entire. For $f(z) = z^2$, we have $u(x, y) = x^2 - y^2$ and $v(x, y) = 2xy$, and hence

$$u_x = 2x, \quad v_y = 2x, \quad u_y = -2y, \quad v_x = 2y. \tag{2.10}$$

It follows that (2.9) holds for all z, i.e., also $f(z) = z^2$ is entire. Summarizing, we have

$$f(z) = z \to f'(z) = 1, \quad f(z) = z^2 \to f'(z) = 2z. \tag{2.11}$$

The latter can also be seen from

$$f'(z) = \lim_{\xi \to z} \frac{\xi^2 - z^2}{\xi - z} = \lim_{\xi \to z} \frac{(\xi - z)(\xi + z)}{\xi - z} = 2z. \tag{2.12}$$

Analyticity is a strong condition. Functions of a complex variable that are not analytic are easily found. For instance, $f(z) = z\bar{z} = x^2 + y^2$ is *not* analytic. Since $f(z)$ is constant along circles concentric at the origin, its tangential derivative is zero, while its normal derivative is $2|z|$. The requirement that the derivative of $f(z)$ be the same irrespective of the direction is hereby not satisfied and the Cauchy-Riemann relations (2.9) do not hold. Also, writing $f(z) = u(x, y) + iv(x, y)$ into its real and imaginary parts shows

$$\Delta u(x, y) = 4, \quad \Delta v(x, y) = 0. \tag{2.13}$$

Its real part is not harmonic, and hence $f(z)$ is not analytic.

When functions $f(z)$ and $g(z)$ are analytic in a common domain, then the following holds:

$$\begin{aligned} (f(z)g(z))' &= f'(z)g(z) + f(z)g'(z), \\ (f(z)/g(z))' &= \left[f'(z)g(z) - f(z)g'(z)\right]/g(z)^2 \end{aligned} \tag{2.14}$$

and if $g(z)$ is analytic on a domain D and $f(z)$ is analytic on the image $D' = g(D)$, then

$$[f(g(z))]' = f'(g(z))g'(z). \tag{2.15}$$

These results may be seen as *analytic continuations of existing identities on the real line.* By elementary considerations, we have

$$\begin{aligned} z^n &\to nz^{n-1} \quad (n \in \mathbb{Z}), \\ e^{az} &\to ae^{az} \quad (a, z \in \mathbb{C}) \\ \sin z, \ \cos z, \ \sinh z, \ \cosh z &\to \cos z, \ -\sin z, \ \cosh z, \ \sinh z \\ \arctan z &\to \tfrac{1}{1+z^2} \quad (z \neq \pm i). \end{aligned} \tag{2.16}$$

2.2 Cauchy's Integral Formula

The strength of the condition of analyticity comes about mostly importantly in *Cauchy's Theorem*: if $f(z)$ is analytic in a region D,[1] then

$$\int_{\gamma} f(z)dz = 0 \tag{2.17}$$

for any closed contour γ in D.[2] Here, the integral is defined according to

$$\int_{\gamma}(u+iv)(dx+idy) = \int_{\gamma}(udx - vdy) + i\int_{\gamma}(vdx+udy). \tag{2.18}$$

If, by analyticity, the Cauchy-Riemann relations (2.9) hold true throughout some open domain containing γ, then by Green's theorem

$$\begin{aligned} \int_{\gamma}(udx - vdy) &= -\int_{I(\gamma)}(u_y + v_x)dxdy = 0, \\ \int_{\gamma}(vdx + udy) &= \int_{I(\gamma)}(u_x - v_y)dxdy = 0, \end{aligned} \tag{2.19}$$

giving (2.17).

A fundamental proof of Cauchy's Theorem is due to Goursat, assuming only the existence of $f'(z)$, i.e., (2.9) with no a priori condition on continuity. Goursat's proof applies to regions D which are simply connected (no punctures or holes), as illustrated in Figs. 2.1 and 2.2. In particular, the unit disk is simply connected, but the unit disk with the origin removed is not.

As a consequence of (2.17), the integral

$$F(z) = \int_{z_0}^{z} f(z)dz \tag{2.20}$$

is a well-defined for z and z_0 in D. Any two paths $\gamma_{1,2}$ connecting them give rise to a loop γ by following γ_1 and traversing γ_2 in reverse, so that by (2.17)

$$0 = \int_{\gamma} f(z)dz = \int_{\gamma_1} f(z)dz - \int_{\gamma_2} f(z)dz, \tag{2.21}$$

whereby (2.20) is *path independent*. By the fundamental theorem of calculus, $F'(z) = f(z)$, whereby (2.9) applies to $F(z) = U(x, y) + iV(x, y)$.

[1] Here, D is an open subset of the complex plane which is simply connected.

[2] A closed contour γ divides the complex plane into a region $I(\gamma)$ within γ and an unbounded region outside, according to the Jordan curve theorem.

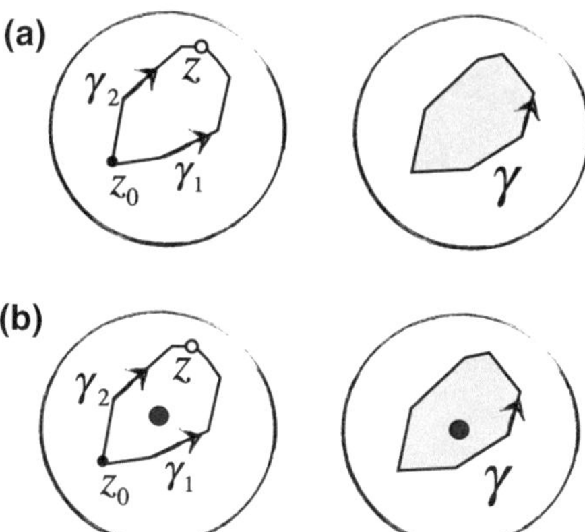

Fig. 2.1 **a** The unit disk $|z| < 1$ is a simply connected domain: any two points z_0 and z can be connected by a path γ_1 within, and choosing another such path γ_2 obtains a closed loop γ as shown, whose interior $I(\gamma)$ lies within $|z| < 1$ also). **b** The punctured disk obtained by removing the origin fails to have this property. The interior of the closed loop γ shown contains the origin, which is not part of the punctured disk. The punctured disk is not simply connected

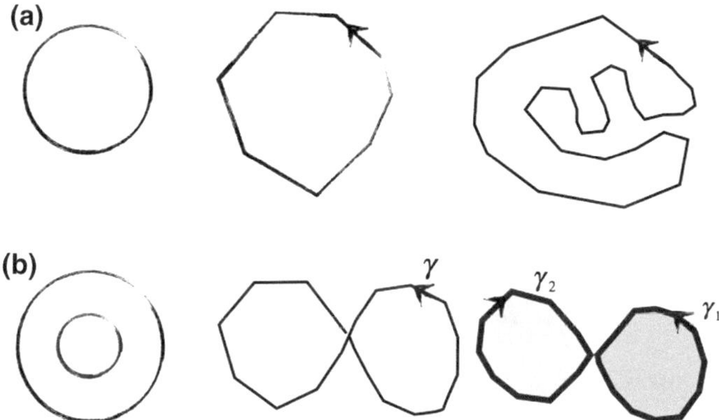

Fig. 2.2 **a** Simply connected domains such as the unit disk can be deformed by continuous deformations. Preserving no self-intersecting boundaries, their boundaries remain simple contours. **b** An annulus obtains by making a hole in the unit disk. Like the punctured disk, it is not simply connected. The figure eight exemplifies a self-intersection curve, whose interior forms two disjoint simply connected domains. There are no paths connecting points z_0 in one and z in the other domain. The figure eight is not a simple contour

Example 2.2. The function $f(z) = 1/(z - a)$ is analytic everywhere in the complex plane except at the isolated pole $z = a$. Let γ be a contour whose interior $I(\gamma)$ does not contain a. The orientation of γ is taken to go around a once in a counter clock-wise direction, as defined by tracking the change in

$$\frac{1}{2\pi i}\int_\gamma \frac{d\xi}{\xi - a} = \begin{cases} 1 \\ 0 \\ \frac{1}{2} \end{cases}$$

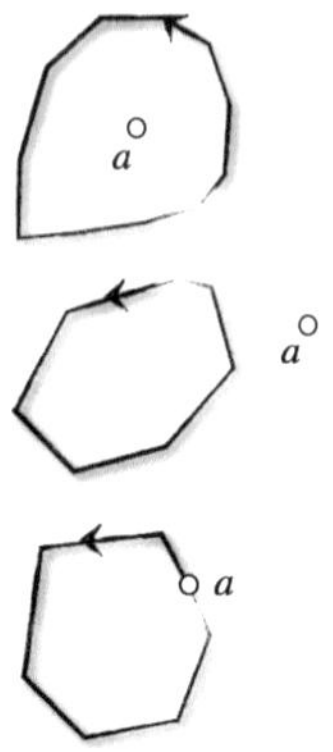

Fig. 2.3 Contour integration of $1/(z-a)$ as shown equals 1 or 0 if a is in or out of $I(\gamma)$. The integral equals $\frac{1}{2}$ if a is on γ, when taking the principle value of the integral, provided a is on a smooth section of γ. If a is located at a corner of γ, e.g., a polygon, the result is $\alpha/(2\pi)$, where α denotes the interior angle at that corner

argument of $\log(z-a)$ as z traverses γ from a point $z_0 \in \gamma$ back to itself. Then $f(z)$ is analytic everywhere in $I(\gamma)$. By (2.9), we have

$$\frac{1}{2\pi i}\int_\gamma f(z)dz = 0. \tag{2.22}$$

Next, let $a \in I(\gamma)$, so that (2.9) does not apply. Specializing to a circle C: $|z-a| = \rho$ (with counter-clockwise orientation and winding number 1), we put $z = a + \rho e^{i\theta}$, so that $dz = i\rho e^{i\theta}d\theta$. As a result,

$$\frac{1}{2\pi i}\int_\gamma \frac{dz}{z-a} = \frac{1}{2\pi i}\int_0^{2\pi} id\theta = 1. \tag{2.23}$$

The above generalizes to γ traversing a N times, by which

$$\frac{1}{2\pi i}\int_\gamma \frac{dz}{z-a} = N \tag{2.24}$$

measures the winding number of γ around $z = a$. For simple contours (no self-crossing), (2.22) and (2.24) are summarized in Fig. 2.3.

For a more general γ, consider *contour deformation* following Fig. 2.4. Here, we give γ a detour branching from $z_0 \epsilon \gamma$ to the circle $C: |z-a| = \rho$, with ρ sufficiently such that C lies in $I(\gamma)$, and back to z_0. A single path P connecting C with z_0 may be used twice (acting as a bridge), in going from z_0 to C and back. The new contour Γ consisting of γ, P and C circumvents the singularity $z = a$, whereby $f(z)$ is analytic in $I(\Gamma)$. It follows that

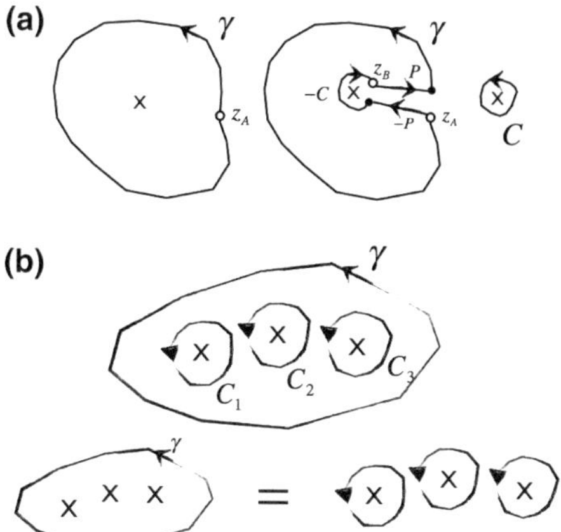

Fig. 2.4 **a** z_A on a simple contour γ can be connected to a contour $-C$ within via a bridge P between z_A and z_B on $-C$. Here, $-C$ refers to C with opposite orientation. The interior of this new contour does not contain X shown. Integration of $f(z)$ that is analytic in a domain containing γ, except possibly at X, then vanishes by Cauchy's theorem. Since integration over the bridge forth and back produces zero, the integral of $f(z)$ over γ equals the integral of $f(z)$ over C. This procedure is exemplifies *contour deformation*. **b** The same applies to multiple isolated singularities. Integration of $f(z)$ over γ is then equivalent to the sum of the integrations of $f(z)$ over contours around each singularity individually. The result gives the *residue theorem*: the net result of integration is defined by the coefficients A_i in $A_i/(z - z_i)$ in the expansion of $f(z)$ in partial fractions up to an arbitrary function which is analytic throughout $I(\gamma)$

$$0 = \int_\Gamma f(z)dz = \int_\gamma f(z)dz - \int_C f(z)dz \tag{2.25}$$

if we define the orientation of C also to be counter clock wise. Here, we used the fact that integration over P forth and back between z_0 and C produces zero. The result is a deformation of γ into C:

$$\frac{1}{2\pi i}\int_\gamma f(z)dz = \frac{1}{2\pi i}\int_C f(z)dz. \tag{2.26}$$

The arguments from Example 2.2 give *Cauchy's integral formula*

$$f(z) = \frac{1}{2\pi i}\int_\gamma \frac{f(\xi)}{\xi - z}d\xi \tag{2.27}$$

with the same conditions on γ as before. In fact, for $f(z)$ analytic in D containing γ,

$$f(\xi) = f(z) + (f'(z) + A(\xi, z))(\xi - z) \tag{2.28}$$

where $A(\xi) \to 0$ as $\xi \to z$, by the existence of the derivative of $f(\xi)$ at $\xi = z$. The result (2.27) obtains directly following a contour deformation of γ to a circle C with radius ρ as above, when we let ρ approach zero.

By Cauchy's integral formula, the n-th derivative satisfies

$$\frac{1}{n!} f^{(n)}(z) = \frac{1}{2\pi i} \int_\gamma \frac{f(\xi)}{(\xi - z)^{n+1}} d\xi, \tag{2.29}$$

showing that *if $f(z)$ is analytic, then $f(z)$ is infinitely differentiable.* An important corollary is that analyticity of a function in a domain D implies the existence of a power series with a finite radius of convergence. That is, $f(z)$ has a *Taylor series in z about $z_0 \in D$*,

$$f(z) = \sum_{n=0}^{\infty} a_n (z - z_0)^n, \tag{2.30}$$

where the a_n (that depend on the choice of z_0) are defined by (2.29).

2.3 Evaluation of a Real Integral

As an application of the above, consider the integral

$$K = \int_{-\infty}^{\infty} \frac{dx}{1+x^2} = \lim_{R\to\infty} K_R, \quad K_R = \int_{-R}^{R} \frac{dx}{1+x^2}. \tag{2.31}$$

First, we consider the analytic extension

$$f(x) = \frac{1}{1+x^2} \to f(z) = \frac{1}{1+z^2} \tag{2.32}$$

by taking x into the complex plane. Here, we note the poles $z = \pm i$ and the associated partial fraction expansion,

$$f(z) = \frac{A_1}{z-i} + \frac{A_2}{z+i} = \frac{1}{2i}\left(\frac{1}{z-i} - \frac{1}{z+i}\right). \tag{2.33}$$

Next, we consider a contour given by $[-R, R]$ and the semi-circle $C: \ z = Re^{i\theta}$, $0 \le \theta \le \pi$ (Fig. 2.5). It encloses the singularity $z = i$ but not $z = -i$. By (2.24) and (2.22) (see also Fig. 2.3), we have

$$\frac{1}{2\pi i}\left[\int_{-R}^{R} f(x)dx + \int_C f(z)dz\right] = A_1 = \frac{1}{2i}. \tag{2.34}$$

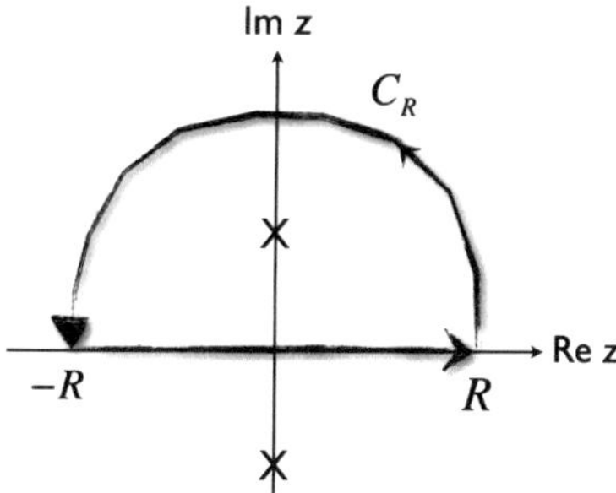

Fig. 2.5 Shown is a contour comprising $[-R, R]$ and the semi-circle C_R in the upper half plane with counter clockwise orientation, to evaluate the integral $A = \int dx/(1+x^2)$ over the real line in terms of $A_R = \int dx/(1+x^2)$ over $[-R, R]$, following an extension of x to complex values z and taking R to infinity. As a function of z, $f(z) = 1/(1+z^2)$ has two poles at $z = i, -i$. The contour shown can be deformed to a circle around $z = i$, to give $2\pi i$ times the residue $1/(2i)$ of $f(z)$ at $z = i$, while the integral over C_R approaches zero in the limit as R approaches infinity. As a result, $A = \pi$

Here,

$$\left|\int_C f(z)dz\right| \leq \int_C |f(z)|\,|dz| \leq \int_0^{\pi} \frac{1}{R^2-1} R d\theta = O\left(\frac{1}{R}\right). \tag{2.35}$$

Taking R to infinity in (2.34) hereby gives

$$K = 2\pi i \times \frac{1}{2i} = \pi. \tag{2.36}$$

2.4 Residue Theorem

Consider contour integration of the function

$$f(z) = \frac{1}{z^2(z-\frac{1}{2})} \tag{2.37}$$

in light of its isolated singularities at $z_1 = 0$ and $z_2 = \frac{1}{2}$. The singularity at $z = 0$ is of second order. and the singularity at $z_2 = \frac{1}{2}$ is a simple pole. Integration over a contour around either one of them will pick up contributions according to their *residues* according to either one of the following.

1. By partial fractions, we have the expansion

$$f(z) = \frac{1}{z} \times \frac{1}{z(z-\frac{1}{2})} = \frac{2}{z} \times \left(\frac{1}{z-\frac{1}{2}} - \frac{1}{z} \right) = -\frac{2}{z^2} - \frac{4}{z} + \frac{4}{z-\frac{1}{2}} \tag{2.38}$$

2. In the disk $D : |z| < \frac{1}{2}$, we have, alternatively, the expansion

$$f(z) = \frac{1}{z^2(z-\frac{1}{2})} = -\frac{2}{z^2} \times \frac{1}{1-2z} = -\frac{2}{z^2}\left[1 + 2z + 4z^2 g(z)\right], \tag{2.39}$$

where $g(z) = 1 + 2z + 4z^2 + \cdots$ is analytic in D. We thus have the *Laurent expansion*

$$f(z) = \frac{1}{z^2(z-\frac{1}{2})} = -\frac{2}{z^2} - \frac{4}{z} - 8g(z). \tag{2.40}$$

3. We compute

$$\lim_{z\to 0} \frac{d}{dz}\left[z^2 f(z)\right] = -4. \tag{2.41}$$

According to (1), integration of $f(z)$ over an anti-clockwise contour γ gives

$$\frac{1}{2\pi i}\int_\gamma f(z)dz = -4,\ 4,\ 0 \tag{2.42}$$

according to whether γ encloses (a) $z_1 = 0$, (b) $z_2 = \frac{1}{2}$ or (c) both $z = z_1$ and $z = z_2$. Here, we use the fact that (cf. Fig. 2.3)

$$\frac{1}{2\pi i}\int_\gamma \frac{dz}{z^n} = \delta_{n0}, \tag{2.43}$$

where $\delta_{ij} = 1\ (i = j)$ and $\delta_{ij} = 0\ (i \neq j)$ denotes the Kronecker delta symbol and γ is an anti-clock wise oriented contour around the origin (see Exercise 2.11).

According to (2–3), the result for (a) obtains by noting that the remainder $8u(z)$ is analytic in D, which is immaterial by Cauchy's theorem. Only the term A_{-1}/z matters in view of the identity (2.43). For a contour in D, it thereby follows that

$$\frac{1}{2\pi i}\int_\gamma f(z)dz = A_{-1} = -4. \tag{2.44}$$

The same arguments can be readily extended to functions with a number of isolated singularities (Fig. 2.4):

$$\frac{1}{2\pi i}\int_\gamma f(z)dz = \sum_{k=1}^{N} \mathrm{Res}_{z=z_k} f(z) \tag{2.45}$$

in terms of

$$\mathrm{Res}_{z_k} f(z) = A_k, \quad f(z) = \frac{A_k}{z - z_k} + g_{(k)}(z) \tag{2.46}$$

where $g_{(k)}(z)$ denotes a remainder which is analytic within γ, except perhaps at $z = z_m$, $m \neq k$.

To further illustrate Cauchy's integral formula (2.27), we calculate the following contour integrals. The first two are over C, given by a circle of radius $\rho > 2$ around the origin, described by $z = \rho e^{i\theta}$, $0 \le \theta \le 2\pi$.

Example 2.3. Consider the contour integral

$$I = \frac{1}{2\pi i}\int_C \frac{f(z)}{z^2 + 2z + 2}dz = \frac{1}{2\pi i}\int_C \frac{f(z)}{(z - z_0)(z - z_1)}dz, \tag{2.47}$$

where $z_0 = -1 + i$ and $z_1 = -1 - i$ denote the zeros of the polynomial $p(z) = z^2 + 2z + 2$. Proceeding by partial fractions, we have

$$I = \frac{1}{2\pi i}\frac{1}{z_0 - z_1}\int_C e^z \left[\frac{1}{z - z_0} - \frac{1}{z - z_1}\right] dz. \tag{2.48}$$

By Cauchy's integral formula (2.27)

$$I = \frac{1}{2\pi i}\frac{1}{2i}\left[\int_C \frac{e^z}{z - z_0}dz - \int_C \frac{e^z}{z - z_1}dz\right] = e^{-1}\frac{1}{2i}(e^i - e^{-i}) = e^{-1}\sin(1). \tag{2.49}$$

Example 2.4. By (2.29), the integral

$$I = \frac{1}{2\pi i}\int_C \frac{e^{2z}}{(z + 1)^4}dz \tag{2.50}$$

is $1/3!$ times the third derivative of

$$g(w) = \frac{1}{2\pi i} \int_C \frac{e^{2z}}{z - w} dz = e^{2w} \tag{2.51}$$

evaluated at $w = -1$. Therefore,

$$I = \frac{1}{6} g^{(3)}(-1) = \frac{4}{3} e^{-2}. \tag{2.52}$$

Example 2.5. Let C_1 denote the unit circle $|z| = 1$. By (2.27)

$$I = \int_{C_1} \frac{e^{kz}}{z} dz = 2\pi i \tag{2.53}$$

is readily evaluated. With $dz = ie^{i\theta} d\theta$, the left hand side can be expanded in real and imaginary parts

$$I = \int_0^{2\pi} e^{k(\cos\theta + i\sin\theta)} i d\theta = i \int_0^{2\pi} \left[e^{k\cos\theta} (\cos(k\sin\theta) + i\sin(k\sin\theta)) \right] d\theta \tag{2.54}$$

so that

$$\int_0^{2\pi} e^{k\cos\theta} \cos(k\sin\theta) d\theta = 2\pi, \quad \int_0^{2\pi} e^{k\cos\theta} \sin(k\sin\theta) d\theta = 0. \tag{2.55}$$

These relations hold for all k.

Example 2.6. The Poisson integral

$$I = \int_{\mathbb{R}} e^{-x^2} \cos(2bx) dx = e^{-b^2} \sqrt{\pi} \tag{2.56}$$

is an exact result, illustrating that a smooth function, here e^{-x^2}, integrated against the oscillatory function $\cos(2bx)$ goes to zero as the frequency k

approaches infinity.[3] To see this, we first consider the finite integral over the finite interval $\Lambda: \; -L \leq x \leq L$,

$$I_L = \int_\Lambda e^{-x^2}\cos(2bx)dx = \frac{1}{2}e^{-b^2}\int_\Lambda \left[e^{-(x-ib)^2} + e^{-(x+ib)^2}\right]. \quad (2.57)$$

To treat the integration of $e^{-(x+ib)^2}$ on the right hand side of (2.57), let Γ_L denote the segment $z = x + ib$ with $x \in \Lambda$ and complete it to a clock-wise oriented closed contour γ containing Λ as follows,

$$\int_{\Gamma_L} e^{-z^2}dx - i\int_L^{L+ib} e^{-z^2}dy - \int_\Lambda e^{-x^2}dx + i\int_{-L}^{-L+ib} e^{-z^2}dy = 0, \quad (2.58)$$

where the sum vanishes since e^{-z^2} is entire. On the sides at $x = \pm L$ of γ, we have

$$\left|e^{-(L\pm iy)^2}\right| = \left|e^{-(L^2-y^2)}e^{\pm 2iyL}\right| = e^{-L^2-y^2} \quad (2.59)$$

and so

$$\left|\int_L^{L+ib} e^{-z^2}dz\right| \leq \int_L^{L+ib} e^{-(L^2-b^2)}dy \leq be^{L^2-b^2} \to 0 \quad (2.60)$$

in the limit as L approaches infinity. Therefore,

$$\lim_{L\to\infty}\int_{\Gamma_L} e^{-z^2}dx = \int_{\mathbb{R}} e^{-x^2}dx = \sqrt{\pi}. \quad (2.61)$$

Since (2.61) is independent of b, the result of integration of $e^{-(x-ib)^2}$ on the right hand side of (2.57) will be the same, thus showing (2.56).

Example 2.7. Compute $f'(1)$ of

$$f(z) = \int_{|\xi|=3} \frac{\xi^2 + 2\xi + 2}{\xi - z}d\xi. \quad (2.62)$$

By partial fractions, there exist constants A, B and C such that

[3] Known as the Riemann-Lebesgue theorem in Fourier transforms, see e.g., van Putten, M.H.P.M., 1998, SIAM Rev., 40(2), 333.

$$f'(z) = \int_{|\xi|=3} \left[\frac{A}{(\xi - z)^2} + \frac{B}{\xi - z} + C \right] d\xi = 2\pi i B(z). \tag{2.63}$$

We can extract B from the numerator $p = \xi^2 + 2\xi + 2 = C(\xi - z)^2 + B(\xi - z) + A$ in (2.62) as the residue

$$B = \text{Res}_{\xi=z} \left[\frac{\xi^2 + 2\xi + 2}{(\xi - z)^2} \right] = \lim_{\xi \to z} p'(\xi) = 2z + 2. \tag{2.64}$$

Evaluated at $z = 1$, we conclude $f'(1) = 8\pi i$.

2.5 Morera's Theorem

Morera's theorem[4] states that given a continuous function $f(z)$ in some domain D, if

$$\int_C f(z)dz = 0 \tag{2.65}$$

for every closed curve C in D, then $f(z)$ is analytic (holomorphic) in D. Morera's theorem is hereby a converse to Cauchy's theorem of Sect. 2.2. It follows that

$$F(z) = \int_{z_0}^{z} f(\xi)d\xi \tag{2.66}$$

is a well-defined primitive of $f(z)$, independent of the path of integration from z_0 to z. Thus, $F'(z) = f(z)$ and hence $F(z)$ is analytic in D. For complex functions, this implies that $F(z)$ is infinitely differentiable and in particular, $F''(z)$ exists, showing that $f(z)$ is analytic in D.

Write $f(z) = u + iv$ in its real and imaginary parts as before. If we assume that u and v are differentiable, then

$$\int_C f(z)dz = \int_C (u + iv)(dx + idy) = \int_C (udx - vdy) + i \int_C (udx + vdy). \tag{2.67}$$

implies by Green's theorem as before in (2.19)

[4]Giacinto Morera 1856–1909.

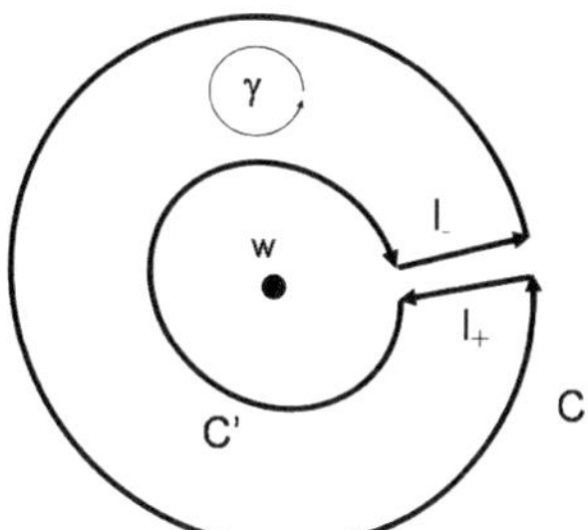

Fig. 2.6 Shown is an oriented path γ formed out of an outer contour C, a counter-oriented inner contour C' and a bridge I that is traversed twice, from C to C' over I_+ and back over I_-. The inner contour C' encircles a point w

$$\int_C f(z)dz = -\int_{I(C)} (u_x + v_y)dA + i\int_{I(C)} (u_x - v_y)dA. \tag{2.68}$$

Implied by (2.65) is that the Cauchy-Riemann relations hold

$$u_x = v_y, \quad u_y = -v_x, \tag{2.69}$$

previously derived as conditions for the derivative of $f(z)$ to be the same regardless of direction of differentiation,

$$\frac{df(z)}{dx} = u_x + iv_x, \quad \frac{df(z)}{idy} = -iu_y + v_x. \tag{2.70}$$

Let $f(z)$ satisfy Morera's condition (2.65) and consider

$$g(z) = \frac{f(z)}{z - w} \tag{2.71}$$

for some w in D. Then $g(z)$ satisfies Morera's condition (2.65) in the punctured domain D' with w removed. Consider integration over the contour γ shown in Fig. 2.6

$$0 = \int g(z)dz = \int_C + \int_{I_+} + \int_{C'} + \int_{I_-} = \int_C + \int_{C'}, \tag{2.72}$$

where $-C'$ is the clockwise oriented curve obtained by reversing orientation of the counter-clockwise oriented C', so that

$$\int_C g(z)dz = \int_{-C'} g(z)dz. \tag{2.73}$$

Note that no specific shape of C or C' has been used; C can be deformed to any other C' (with the same orientation), provided that in a continuous deformation the region traced out enclosed between C and C' is within D'. Allow C' to become a small circle around w, i.e., $z - w = \epsilon e^{i\theta}$. If $f(z)$ is continuous at w, then $f(z) = f(w) + \eta(z, w)$ with $\eta(z, w)$ approaching zero in the limit as $z \to w$. By (2.73) and $dz = ie^{i\theta}d\theta$, we have

$$\int_{-C'} g(z) = \int_0^{2\pi} \frac{f(w+\epsilon e^{i\theta})}{\epsilon e^{i\theta}} i e^{i\theta} d\theta = i \int_0^{2\pi} f(w+\epsilon e^{i\theta}) d\theta \to 2\pi f(w) \quad (2.74)$$

in the limit as $\epsilon \to 0$. Thus, (2.73) gives (2.27) once more. An immediate consequence is that $f(z)$ is infinitely times differentiable, and the n-th derivative of $f(z)$ satisfies (2.29). *A continuous complex function satisfying Morera's theorem is infinitely times differentiable.* Finally, consider the circle C: $z - w = \rho e^{i\theta}$. A similar calculation to (2.74) shows the *Mean Value Theorem*

$$f(w) = \frac{1}{2\pi} \int_0^{2\pi} f(w + \rho e^{i\theta}) d\theta. \quad (2.75)$$

Viewed as a mean value over a boundary, it immediately follows that the real and imaginary parts u and v of $f(w) = u + iv$ are bounded by the extrema of u and v on C. This so-called *minmax theorem* holds true for boundaries of arbitrary shape.

2.6 Liouville's Theorem

If $f(z)$ is entire *and* $f(z)$ is bounded, i.e., $f(z) \le M$ for all $z \in \mathbb{C}$, then $f(z)$ reduces to a constant. This is *Liouville's theorem*, and it follows readily from Cauchy's integral formula (2.27),

$$f'(z) = \frac{1}{2\pi i} \int_C \frac{f(\xi)}{(\xi - z)^2} dz \quad (2.76)$$

where C is a simple closed contour whose interior contains z. We may choose C to be a circle $\xi - z = Re^{i\theta}$ of radius R. Taking the modulus gives

$$\left| f'(z) \right| \le \frac{1}{2\pi} \int_C \frac{|f(\xi)|}{R^2} R\, d\theta = \frac{M}{R}. \quad (2.77)$$

Since R was arbitrary, we may take it to infinity, and hence $|f'(z)| = 0$. It follows that $f(z)$ is a constant.

Liouville's theorem can be used to show that every polynomial

$$p_n(z) = a_0 + a_1 z + a_2 z^2 + \cdots a_n z^n \quad (a_n \ne 0) \quad (2.78)$$

of degree n has exactly n roots, $p_n(z_i) = 0, i = 1, 2, \cdots, n$, including multiplicities. To see this, let

$$h(z) = \frac{1}{p_n(z)}. \quad (2.79)$$

If $p_n(z)$ has no zeros anywhere in $\mathbb{C}$, then $h(z)$ is entire and reduces to a constant by Liouville's theorem. Then $p_n(z) = 1/h(z)$ is a constant, which contradicts our assumption that $a_n \neq 0$. It follows that $p_n(z)$ has at least one zero, say, z_1. Next, consider $h_1(z) = (z - z_1)/p_n(z)$. Repeating, we encounter additional zeros. This procedure terminates when the constant $h_n(z) = a_n$ is reached, thus retrieving n zeros.

As the examples show, we often encounter polynomials (2.78) with real coefficients. In this event, *roots come in pairs of complex conjugates*. If $p_n(z) = 0$, then

$$0 = \overline{p_n(z)} = p_n(\bar{z}) \tag{2.80}$$

so $\bar{z}$ is also a root. If the root z is real, then this produces no new root, but if it is complex, then conjugation produces a genuine second root.

2.7 Poisson Kernel

On a circle C of radius a, Cauchy's integral formula for a function $f(z)$ reduces to an analytic extension of by a real kernel into the region in C. To see this, consider $z = re^{i\theta}$ inside of C and a point

$$z^* = \frac{a^2}{\bar{z}} \tag{2.81}$$

symmetric with respect to C. Since z^* is outside, we have

$$f(z) = \frac{1}{2\pi i} \int_C f(\xi) \left[\frac{1}{\xi - z} - \frac{1}{\xi - z^*} \right] d\xi. \tag{2.82}$$

Since $a^2 = \xi\bar{\xi}$, (2.82) reduces to

$$f(z) = \frac{1}{2\pi} \int_0^{2\pi} f\left(ae^{i\alpha}\right) \frac{a^2 - r^2}{a^2 + r^2 - 2ra\cos(\alpha - \theta)} d\alpha. \tag{2.83}$$

Since the Poisson kernel $(a^2 - r^2)/(a^2 + r^2 - 2ar\cos(\alpha - \theta))$ is real, (2.83) defines analytic continuation of the real and imaginary parts of $f(z)$ separately, i.e., giving analytic continations of harmonic functions on the disk. This result is closely connected to the Fourier transform.[5] To see this, specialize to $a = 1$ and let $u(r, \theta)$ denote the real part of $f(z)$ with Fourier coefficients

$$C_n = \frac{1}{2\pi} \int_0^{2\pi} u_0(\alpha) e^{-in\alpha} d\alpha \quad (n \in \mathbb{Z}) \tag{2.84}$$

[5] More on this in Chap. 6.

of data $u_0(\alpha) = u(a,\alpha)$ on S^1. Since $u(r,\theta)$ is real, $C_{-n} = \bar{C}_n$, and hence

$$u(r,\theta) = C_0 + \sum_{n\geq 1} r^n \left(C_n e^{in\theta} + \bar{C}_n e^{-in\theta}\right) \quad (r \leq 1) \tag{2.85}$$

defines the harmonic extension of $u_0(\alpha)$ into $r \leq 1$. Using (2.84), we have

$$u(r,\theta) = \frac{1}{2\pi}\int_0^{2\pi} u_0(\alpha)\left[1 + \sum_{n\geq 1}\left(r^n e^{-in(\alpha-\theta)} + r^n e^{in(\alpha-\theta)}\right)\right] d\alpha. \tag{2.86}$$

With $\sum_{n\geq 1} z^n = z/(1-z)$, $z = re^{-i(\alpha-\theta)}$, $r < 1$, the kernel in (2.86) evaluates to

$$1 + \frac{z}{1-z} + \frac{\bar{z}}{1-\bar{z}} = \frac{1-z\bar{z}}{1+z\bar{z}-z-\bar{z}} = \frac{1-r^2}{1+r^2-2r\cos(\alpha-\theta)}, \tag{2.87}$$

thus recovering the Poisson kernel in (2.83).

2.8 Flux and Circulation

The equations of motion of fluids derive from conservation laws of mass, energy and momentum. Solutions critically depend on the Reynolds number, a dimensionless ratio of convective to diffusive momentum transport,[6] due to randow walks of molecules or atoms that make up the fluid. For large Reynolds numbers, flows are generally complex and inherently time-dependent, that rarely permit analytical solutions. Even so, some key aspects of conservation of mass and momentum are amenable to analytical solutions, particularly for *solenoidal* (incompressible) and *irrotational* (vanishing vorticity) flows. In two dimensions, these properties are described by a *complex velocity potential*

$$w(z) = \phi(x,y) - i\psi(x,y) \tag{2.88}$$

representing a flow velocity

$$\mathbf{u} = u\mathbf{i} + v\mathbf{j} = \phi_x\mathbf{i} + \phi_y\mathbf{j}, \tag{2.89}$$

satisfying

$$u_x + v_y = 0, \quad u_y - v_x = 0. \tag{2.90}$$

[6] An excellent introduction to classical fluid dynamics is [1].

by virtue of the Cauchy-Riemann relations (2.9) *with a change in sign in* v. The power of complex function theory can hereby be introduced to describe this class of two-dimensional flows.

Consider a smooth curve Γ connecting two points P and Q. We denote the unit tangent vector by τ and the unit normal by $\mathbf{n}$. In two dimensions, we can fix $\mathbf{n}$ uniquely by defining it to be a rotation over $-\pi/2$ of τ. With this convention, $\mathbf{n}$ is the outer normal for a simply positively connected curve, i.e.,

$$\boldsymbol{\tau} = \tau_x \mathbf{i} + \tau_y \mathbf{j}, \quad \mathbf{n} = \tau_y \mathbf{i} - \tau_x \mathbf{j}. \tag{2.91}$$

Thus, the tangent and normal are related by a clockwise rotation over $\pi/2$,

$$\mathbf{n} = \begin{pmatrix} 0 & 1 \\ -1 & 0 \end{pmatrix} \boldsymbol{\tau}. \tag{2.92}$$

For a smooth flow, consider the path integrals

$$I_A = \int_\Gamma \mathbf{u} \cdot \mathbf{n} \, ds, \quad I_B = \int_\Gamma \mathbf{u} \cdot \boldsymbol{\tau} \, ds \tag{2.93}$$

over a curve Γ from P to Q. In general, these two integrals are path dependent. Consider two alternative paths Γ_1 and Γ_2 as shown in Fig. 2.7. Then Γ_1 as shown followed by Γ_2 in the opposite direction produces a closed loop γ with positive orientation. By Green's theorem, we have

$$\begin{aligned} \mathcal{A} &= \int_\gamma \mathbf{u} \cdot \mathbf{n} \, ds = \int_{I(\gamma)} (u_x + v_y) dA, \\ \mathcal{B} &= \int_\gamma \mathbf{u} \cdot \boldsymbol{\tau} \, ds = \int_{I(\gamma)} (v_x - u_y) dA, \end{aligned} \tag{2.94}$$

where $I(\gamma)$ denotes the interior of γ. The first integral in (2.94) vanishes when the flow is solenoidal, and the second vanishes when the flow is irrotational. If these two conditions are satisfied, then the integrals (2.93) are *path independent*. They are hereby well-defined as functions of P and Q without specifying the path connecting them. Exceptions may still arise, when $I(\gamma)$ is not simply connected, e.g., when $I(\gamma)$

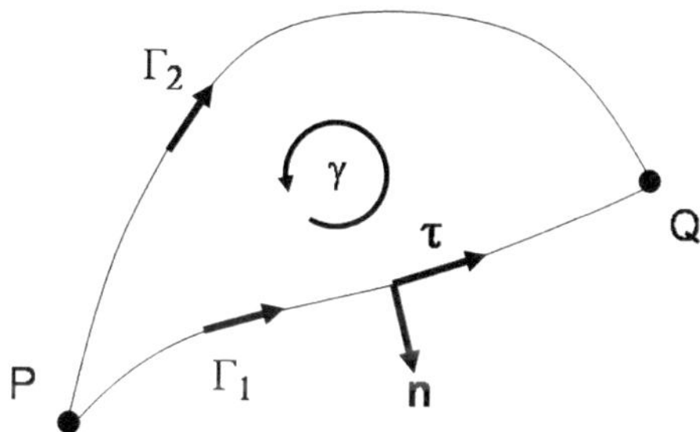

Fig. 2.7 Two paths Γ_1 and Γ_2 from P to Q form a closed loop γ. The loop γ is positively oriented, when formed buy traversing along Γ_1 and traversing Γ_2 in the opposite direction

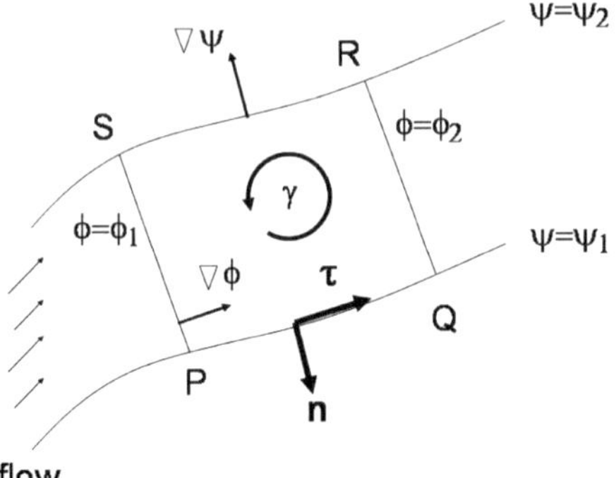

Fig. 2.8 In solenoidal and irrotational flows, streamlines are level curves of the stream function ψ, which are orthogonal to level curves of the flow potential ϕ. The separation between two streamlines forms a duct of *flux*, that is conserved as measured across level curves of the flow potential at $P - S$ and $Q - R$. Circulation obtains from integration over γ, which vanishes for irrotational flow as generated by a flow potential ϕ

is an annulus. In this event, the winding number of γ becomes relevant. We will not consider these possibilities here.

For a flow in a simply connected region, we now consider the *stream function* and the *flow potential* given by integration over a path from A to B,

$$\psi(B) = \psi(A) - \int_A^B \mathbf{u} \cdot \mathbf{n}\, ds, \quad \varphi(B) = \varphi(A) + \int_A^B \mathbf{u} \cdot \boldsymbol{\tau}\, ds. \tag{2.95}$$

As illustrated in Fig. 2.8, ψ is constant along the streamlines of $\mathbf{u}$. Consider a closed loop γ as indicated. Integration of $\mathbf{u} \cdot \mathbf{n}$ over γ receives contributions from segments $Q - R$ and $S - P$ only, and so

$$\Delta\psi = [\psi]_Q^R = [\psi]_P^S. \tag{2.96}$$

It expresses a conserved flux passing through a level curve of ϕ between P to S as well as between Q to R, i.e., the flux between two streamlines $\psi_1 = \psi(Q)$ and $\psi_2 = \psi(R)$. Integration of $\mathbf{u} \cdot \tau$ over γ expresses the *circulation* of the flow over γ. It vanishes based on the assumption of irrotational flow. When the segments $Q - R$ and $S - P$ are orthogonal ($\mathbf{u} \cdot \tau = 0$) to the flow, integration receives contributions from segments $P - Q$ and $R - P$ only, and so

$$\Delta\phi = [\phi]_P^Q = [\phi]_S^R \tag{2.97}$$

measures the strength of the flow along the streamlines.

With ψ and ϕ defined by (2.95) with (2.96), we have $u = \phi_x = \psi_y, \quad v = \phi_y = -\psi_x$, that is, the Cauchy-Riemann relations for the real and imaginary part of an analytic function

$$w(z) = \phi + i\psi, \quad w'(z) = u - iv, \quad \phi(z) = \mathrm{Re}\, w(z), \quad \psi(z) = \mathrm{Im}\, w(z) \tag{2.98}$$

as a function of $z = x + iy$.

2.9 Examples of Potential Flows

A uniform flow with unit velocity has a complex velocity potential

$$w(z) = z, \tag{2.99}$$

whose streamlines of constant $\psi(x, y) = \mathrm{Im}\, w(z)$ are parallel to the x-axis. A potential $w(z) = e^{-i\alpha}z$ describes a uniform flow at an inclination angle α to the x-axis.

In a potential flow past a solid body, the surface of the body is a streamline, along which $\psi(x, y) = \mathrm{Im}\, w(z)$ is constant. If the same flow is uniform at infinity, then $w(z) \sim z$ at large z, perhaps up to a complex constant that defines the direction of the flow. The potential

$$w(z) = z + \frac{1}{z} \tag{2.100}$$

hereby describes a flow past a cylinder of unit radius, since $w(z) = \cos\theta$ and hence $\mathrm{Im}\, w(z) = 0$ on $z = e^{i\theta}$. The potential (2.100) has two stagnation points at $z = \pm 1$, where the velocity

$$w'(z) = \frac{z^2 - 1}{z^2} \tag{2.101}$$

vanishes. One can envision sliding the location of these stagnation points over the cylinder, by modifying the zeros of $w'(z)$ to, e.g., $z_1 = e^{-i\alpha}$ and $z_2 = e^{i(\pi+\alpha)}$, changing the numerator in (2.101) to

$$(z - z_1)(z - z_2) = z^2 - 2i \sin\alpha z - 1. \tag{2.102}$$

The associated velocity potential

$$w(z) = z + \frac{1}{z} + 2i \sin\alpha \ln z, \tag{2.103}$$

where the logarithm adds ciruclation to the flow past a cylinder.

The flow potentials

$$w_1(z) = \frac{Q}{2\pi}\ln z, \quad w_2(z) = \frac{\Gamma}{2\pi i}\ln z \tag{2.104}$$

describe point sources of flow with streamlines emanating radially from and, respectively, concentric about the origin. The total flux and circulation integrals (2.93) over $S^1 : z = e^{i\theta}$ give

$$\int_C w_1'(z)(-idz) = Q, \quad \int_C w_2'(z)dz = \Gamma, \tag{2.105}$$

where we used the correspondences $\mathbf{n}ds = -idz$ and $\boldsymbol{\tau}\, ds = dz$. With $\Gamma = -4\pi \sin\alpha$, the potential (2.103) obtains in standard form

$$w(z) = z + \frac{1}{z} + \frac{\Gamma}{2\pi i}\ln z. \tag{2.106}$$

As a map, $\zeta = z + 1/z$ in (2.101) is known as the Joukowski transformation. This map is conformal (having non-zero derivative) away from $z = \pm 1$. It generates Joukowski airfoil profiles as images of circles C with radius $a > 1$ that pass through $z = 1$. These images of C are smooth except for a cusp at $\zeta(1) = 2$, where the map fails to be conformal. The cusp sets location of the corresponding stagnation point on C, to fix circulation and to avoid flow separation and the shedding of vorticies. Let $w(z)$ denote a complex flow potential about a C, such as (2.103). Then $W(\zeta) \equiv w(z(\zeta)))$ is a flow potential past the Joukowski airfoil with velocity

$$W'(\zeta) = w'(z)\left(\frac{d\zeta}{dz}\right)^{-1} = \frac{z^2 w'(z)}{z^2 - 1}. \tag{2.107}$$

Table 2.1 summarizes this discussion.

2.10 Exercises

2.1. Show by explicit evaluation that the real and imaginary part of e^z with $z = i\varphi$ in the defining Taylor series of the exponential function recovers the Taylor series expressions for $\cos\varphi$ and $\sin\varphi$.

2.2. Show that an analytic function $f(z)$ is a conformal map whenever $f'(z) \neq 0$, that is, if α is the angle between the tangents of two curves intersecting at $z = z_0$, then α is also the angle between the tangents to the images of these two curves under $f(z)$.

Table 2.1 Complex function theory

1. Cauchy's integral formula gives a representation of functions $f(z)$ analytic in some domain D according to $f(z) = (1/2\pi i)\int_\gamma f(\xi)/(\xi - z)d\xi$, where γ is a simple counterclockwise oriented contour in D that encloses $z \in D$.

2. The winding number of a contour γ about z satisfies $N = (1/2\pi i)\int_\gamma d\xi/(\xi - z) \in \mathbb{Z}$. $N = 0$, $1/2$ or 1 depending on whether, respectively, z is outside, on or inside γ.

3. Contour integrals of $f(z)$ are determined by residues of $f(z)$ at isolated singularities $z = z_0$, given by the coefficient A_{-1} in expansion $f(z) = \cdots + A_{-1}/(z - z_0) + \cdots$.

4. Analytic functions describe the complex velocity potential $w(z) = \phi + i\psi$ of irrotational solenoidal flows in two dimensions, whereby $w'(z) = u - iv$. The real part ϕ is the velocity potential $\mathbf{u} = \nabla\phi$ and the imaginary part ψ is the stream function.

5. Streamlines are level curves of ψ along which the gradient in ϕ expresses the flow velocity.

2.3. Verify by explicit calculation that the Cauchy-Riemann relations for $f(z) = z^n$ for the three cases $n = 0, 1, 2$.

2.4. Prove (2.16).

2.5. We say that $f(z)$ is analytic at infinity if $g(w) = f(1/w)$ is analytic at $w = 0$. Consider

$$f(z) = \frac{1}{z^2 + 1}. \tag{2.108}$$

Show that $f(z)$ is analytic at infinity. What is the radius of convergence of the Taylor series of $g(w)$?

2.6. Obtain the Laurant series expansion of

$$f(z) = \frac{1}{(z^2 + 1)(4 - z^2)} \tag{2.109}$$

valid in the annulus $1 < |z| < 2$.

2.7. Derive (2.43) by contour deformation to $z = e^{i\theta}$.

2.8. Calculate the integrals

$$(a) \int_\gamma \frac{dz}{z^n}, \quad (b) \int_\gamma \log(1+z)dz \tag{2.110}$$

where n is an integer and γ is a small loop around the origin.

2.9. Obtain the partial fractions of

$$\text{(a)} \ \frac{1}{z^2-1}, \ \text{(b)} \ \frac{2z}{z^2+1}, \ \text{(c)} \ \frac{1}{z^2+1}, \ \text{(d)} \ \frac{z-2}{z^2+z}. \tag{2.111}$$

2.10. Obtain the partial fractions of

$$\text{(a)} \ \frac{1}{z^2(z-\frac{1}{2})(z^2+4)} \quad \text{(b)} \ \frac{1}{z^n(z+1)} \ (n = 1, 2, 3) \tag{2.112}$$

2.11. Obtain the integrals

$$\text{(a)} \int_{1+i}^{3-2i} \sin z dz, \ \text{(b)} \int_\gamma z^n dz \ (n \in \mathbb{N}), \ \text{(c)} \int_\gamma \log(1+z)dz \tag{2.113}$$

where γ is a contour in the unit disk.

2.12. Let γ be the unit circle $|z| = 1$ with counter clockwise orientation. Compute the following complex integrals

1. $\int_\gamma \frac{dz}{(z-a)^n}$ for $a = 0, 2$ and $n = 1, 2$.
2. $\int_\gamma \frac{dz}{\cosh^2 z}$
3. $\int_\gamma \frac{\sin(\pi z^2)}{(z-1/2)(z-2)} dz$.

2.13. Obtain the integral

$$\frac{1}{2\pi i} \int_\gamma \frac{\sin z}{2z+i} \tag{2.114}$$

where γ is a contour which encloses $z = -\frac{1}{2i}$.

2.14. Find all the *branch points*, where the following function does *not* satisfy the

Cauchy-Riemann relations:

$$f(z) = \sqrt{e^z + 1}. \tag{2.115}$$

2.15. Prove the minmax theorem based on (2.75).

2.16. Consider the functions

$$f(z) = \frac{2}{z} + 3 + 4z, \quad g(z) = \frac{1}{z^2} + f(z) \tag{2.116}$$

and the residues defined by the contour integrals

$$\text{Res}_{z=0}\, f(z) = \frac{1}{2\pi i}\int_\gamma f(z)dz, \quad \text{Res}_{z=0}\, g(z) = \frac{1}{2\pi}\int_\gamma g(z)dz \tag{2.117}$$

over a contour γ that encloses the origin $z = 0$. Show that

$$\text{Res}_{z=0}\, f(z) = \lim_{z\to 0} zf(z), \quad \text{Res}_{z=0}\, g(z) = \lim_{z\to 0}\frac{d}{dz}\left[z^2 g(z)\right]. \tag{2.118}$$

2.17. The Möbius transformation $w(z) = (z - i)/(z + i)$ maps the real line onto the unit circle. Cauchy's integral formula on the unit circle hereby transforms to one on the real line. Follow steps similar to those in deriving the Poisson kernel (2.83), now with the point $z^* = \bar{z}$ symmetric with respect to the real line, to obtain

$$u(x, y) = \frac{y}{\pi}\int_{-\infty}^{\infty}\frac{u_0(x')}{(x' - x)^2 + y^2}dx'. \tag{2.119}$$

2.18. Sketch the streamlines including the direction of flow of the following complex velocity potentials

$$\text{(a)}\ \ w(z) = z^k\ \left(k = \frac{1}{2}, 2, 3\right), \quad w(z) = \frac{1}{z}. \tag{2.120}$$

$$\text{(b)}\ \ w(z) = \frac{1}{2\pi}\ln(z^2 - 1), \quad w(z) = z + \frac{1}{2\pi}\ln z. \tag{2.121}$$

2.19. Sketch the field lines of the complex velocity potential of the dipole

$$w(z) = \frac{1}{2\pi\epsilon}\left[\ln(z + \epsilon) - \ln(z - \epsilon)\right] \tag{2.122}$$

and determine the limit as ϵ approaches zero.

2.20. Show that

$$f(z) = (1+z)^{\frac{1}{z}} \tag{2.123}$$

has a removable singularity at the origin by deriving a Taylor series expansion $f(z) = e\Sigma_{m=0}^{\infty} c_m z^m$ about $z = 0$. Note that the radius of convergence is 1 in view of the singularity at $z = -1$. Match the behavior of $f(z)$ as z approaches -1 from the right to the Taylor series expansion to show that the (rational) coefficients c_m approach $e^{-1}(-1)^m$. [*Hint.* Use $e^{z^{-1}\log(1+z)} = e^{1-\frac{1}{2}z+\frac{1}{3}z^2-\frac{1}{4}z^3+\cdots}$.]

2.21. Consider an asymptotically flat black hole spacetime. Viewed as an analytic function of radius $z \in \mathbb{C}$, the metric satisfies $\eta_{ab} + O(z^{-1})$, where η_{ab} denotes the Minkowski metric. Use Liouville's theorem to argue that the metric must have at least one singularity in $\mathbb{C}$ [2].

References

1. Batchelor, G.K., *An Introduction to Fluid Dynamics* (Cambridge University Press, Cambridge, 1990)
2. van Putten, M.H.P.M., PNAS, 2006, 103, 519

Chapter 3
Vectors and Linear Algebra

3.1 Introduction

Linear algebra is the language describing systems in finite (or countably infinite) dimensions, where *dimension* represents the number of variables at hand. This appears naturally in systems with more than one degrees of freedom or in approximate descriptions of complex systems in a discrete set of variables. Linear algebra also gives the basic framework of quantum mechanics, describing observables in terms of eigenvalues. In two and more dimensions, much of linear algebra is illustrated by vectors and their linear transformations.

Angular momentum $\mathbf{J}$ is a vector that appears in numerous problems. Like energy and linear momentum, total angular momentum is a conserved quantity. For freely rotating rigid bodies, angular momentum is in proportion to angular velocity

$$\mathbf{\Omega} = \Omega\,\mathbf{n}. \tag{3.1}$$

The length Ω denotes the rate of rotation and the direction denotes its orientation. *According to Mach's principle*, angular velocity is commonly defined relative to the distant stars, where most of the mass is. For periodic motion with period P, the angular velocity satisfies

$$\Omega = \frac{2\pi}{P}. \tag{3.2}$$

For motion at a separation $\mathbf{r}$ about to a given axis, the instantaneous velocity is a tangent

$$\mathbf{v} = \frac{d\mathbf{r}}{dt} = \mathbf{\Omega} \times \mathbf{r}. \tag{3.3}$$

The associated linear momentum is the vector

M.H.P.M. van Putten, *Introduction to Methods of Approximation in Physics and Astronomy*, Undergraduate Lecture Notes in Physics,
DOI 10.1007/978-981-10-2932-5_3

$$\mathbf{p} = \frac{d}{dt} m\mathbf{r} = m\mathbf{v}, \tag{3.4}$$

when the mass m is time-independent. The associated angular momentum of a particle with linear momentum $\mathbf{p}$ is

$$\mathbf{J} = \mathbf{r} \times \mathbf{p}. \tag{3.5}$$

Thus, $\mathbf{J}$ is a vector formed out of $\mathbf{r}$ and $\mathbf{p}$. Transformations of $\mathbf{J}$ follow the rules for vectors, e.g., when considering translation or rotation of a coordinate system.

Maps in linear algebra are represented by $n \times m$ matrices, from a linear vector space of dimension m to a linear vector space of dimension n. These vector spaces are often over the real or complex numbers, e.g., $\mathbb{R}^n$ or $\mathbb{C}^n$. As such, matrices are comprised of row and column vectors of length m and, respectively, n. An $n \times m$ matrix is said to be of dimension $n \times m$.

Consider, for instance, a 2×2 matrix

$$C = \begin{pmatrix} 1 & 2 \\ 2 & 0 \end{pmatrix}. \tag{3.6}$$

Its rows $\mathbf{r}_{(i)}$ and columns $\mathbf{c}_{(i)}$ $(i = 1, 2)$ can be schematically indicated as

$$C = \begin{pmatrix} \text{———} \\ \text{———} \end{pmatrix} = \begin{pmatrix} \mathbf{r}_1^T \\ \mathbf{r}_2^T \end{pmatrix}, \quad C = \begin{pmatrix} | & | \\ | & | \end{pmatrix} = (\mathbf{c}_1 \; \mathbf{c}_2) \tag{3.7}$$

with

$$\mathbf{r}_1^T = (1 \; 2), \;\; \mathbf{r}_2^T = (2 \; 0), \;\; \mathbf{c}_1 = \begin{pmatrix} 1 \\ 2 \end{pmatrix}, \;\; \mathbf{c}_2 = \begin{pmatrix} 2 \\ 0 \end{pmatrix}, \tag{3.8}$$

where we explicitly include the *transpose* T to denote row vectors according to

$$\begin{pmatrix} x \\ y \end{pmatrix}^T = (x \; y). \tag{3.9}$$

These 2×2 matrices describe various transformations in the two-dimensional plane, such as reflections, rotations and coordinate permutations. Key properties are eigenvalues and the associated eigenvectors, much of which depends on their determinants and symmetry properties.

3.2 Inner and Outer Products

Two linearly independent vectors **a** and **b** span a parallelogram. The *projection* of **a** onto **b** defines the inner product (further Sect. 3.5) $\mathbf{a}\cdot\mathbf{b} = |\mathbf{a}||\mathbf{b}|\cos\theta$ with

$$\cos\theta = \angle(\mathbf{a},\mathbf{b}) = \frac{\mathbf{a}\cdot\mathbf{b}}{|\mathbf{a}||\mathbf{b}|} \tag{3.10}$$

denoting the cosine of the angle between the two, where $|\mathbf{a}|$ refers to the length of **a** satisfying $|\mathbf{a}| = \sqrt{\mathbf{a}\cdot\mathbf{a}}$. Referenced to a Cartesian coordinate system with basis vectors $\{\mathbf{i},\mathbf{j},\mathbf{k}\}$, expressions obtain in component form, In three dimensions, we have

$$\mathbf{a} = a_1\mathbf{i} + a_2\mathbf{j} + a_3\mathbf{k},\quad \mathbf{b} = b_1\mathbf{i} + b_2\mathbf{j} + b_3\mathbf{k} \tag{3.11}$$

and so

$$\mathbf{a}\cdot\mathbf{b} = a_1 b_a + a_2 b_2 + a_3 b_3. \tag{3.12}$$

The outer product represents the area element, in area and orientation, of the parallellogram, represented by the normal vector

$$(a_2b_3 - a_3b_2)\mathbf{i} + (a_3b_1 - a_1b_3)\mathbf{j} + (a_1b_2 - a_2b_1)\mathbf{k} = \begin{pmatrix} a_2b_3 - a_3b_2 \\ a_3b_1 - a_1b_3 \\ a_1b_2 - a_2b_1 \end{pmatrix}, \tag{3.13}$$

where we used the right handed rule in the direction of movement of a corkscrew turned from **a** to **b**. Its length equals the area of the parallelogram

$$|\mathbf{a}\times\mathbf{b}| = |\mathbf{a}||\mathbf{b}|\sin\theta. \tag{3.14}$$

3.3 Angular Momentum Vector

In circular motion, angular momentum **J** is a vector with the same orientation as the angular velocity $\boldsymbol{\Omega}$. By the vector identity

$$\mathbf{a}\times(\mathbf{b}\times\mathbf{c}) = \mathbf{b}(\mathbf{a}\cdot\mathbf{c}) - \mathbf{c}(\mathbf{a}\cdot\mathbf{b}) \tag{3.15}$$

between vectors **a**, **b**, **c**, circular motion gives the *specific angular momentum* (angular momentum per unit mass)

$$\mathbf{j} = \mathbf{r}\times\mathbf{v} = \mathbf{r}\times(\boldsymbol{\Omega}\times\mathbf{r}) = r^2\boldsymbol{\Omega} = r^2\Omega\mathbf{n}, \tag{3.16}$$

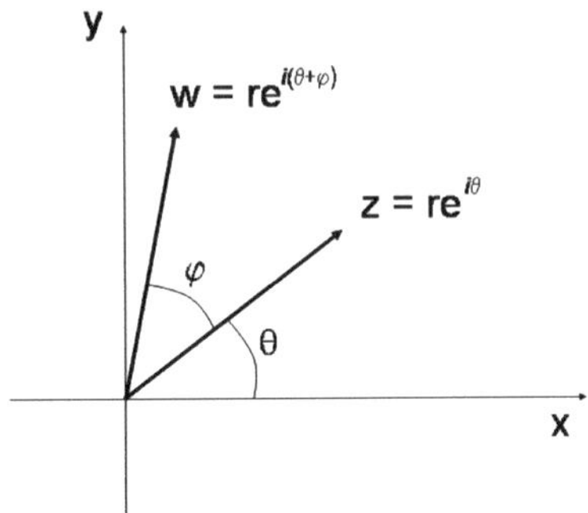

Fig. 3.1 Rotation in the (x, y)-plane over an angle φ obtains by multiplication by a 2×2 matrix $R(\varphi)$ of vectors $\mathbf{z} = x\mathbf{i}_x + y\mathbf{i}_y$. In the complex plane, it corresponds to multiplication by $e^{i\varphi}$

since $\mathbf{r} \cdot \mathbf{r} = r^2$ and $\mathbf{r} \cdot \boldsymbol{\Omega} = 0$. With (3.4, 3.5), our model problem of circular motion, therefore, implies

$$\mathbf{j} = r^2 \frac{2\pi}{P} = 2\frac{\pi r^2}{P} = 2\frac{dA}{dt}\mathbf{n}, \tag{3.17}$$

that is, $\mathbf{j}$ represents twice the rate-of-change of surface area traced out by the radius $\mathbf{r}$ in the orbital motion. Based on (3.4–3.17), this is a geometrical identify, not restricted to circular motion, familiar as Kepler's third law in planetary motion.

3.3.1 Rotations

If $z = re^{i\theta}$, then

$$w = ze^{i\varphi} = re^{i(\theta+\varphi)} = r\left(\cos(\theta + \varphi) + i\sin(\theta + \varphi)\right), \tag{3.18}$$

as illustrated in Fig. 3.1. That is

$$w = re^{i\theta}e^{i\psi} = r\left[\cos\theta\cos\phi - \sin\theta\sin\phi + i(\sin\theta\cos\phi + \cos\theta\sin\theta)\right]. \tag{3.19}$$

Applied to the basic vectors $\{\mathbf{i}_x, \mathbf{i}_y\}$, we have

$$\mathbf{i}_x' = \mathbf{i}_x\cos\theta + \mathbf{i}_y\sin\theta, \quad \mathbf{i}_y' = -\mathbf{i}_x\sin\theta + \mathbf{i}_y\cos\theta. \tag{3.20}$$

Example 3.1. Consider a basis $\{\mathbf{i}, \mathbf{j}\}$ rotating along with a point (x, y) over the unit circle S^1. That is, $\mathbf{i}$ points to (x, y) with local tangent $\mathbf{j}$ to S^1 with counter-clockwise orientation. Moving along S^1 at constant angular velocity ω, $\theta = \omega t$ as a function time t and (3.20) implies

$$\frac{d\mathbf{i}}{dt} = \omega\left[-\mathbf{i}\sin\theta + \mathbf{j}\cos\theta\right], \quad \frac{d\mathbf{j}}{dt} = \omega\left[-\mathbf{i}\cos\theta - \mathbf{j}\sin\theta\right], \tag{3.21}$$

and so

$$\frac{d\mathbf{i}}{dt} = \omega\mathbf{j}, \quad \frac{d\mathbf{j}}{dt} = -\omega\mathbf{i}. \tag{3.22}$$

3.3.2 *Angular Momentum and Mach's Principle*

Following (3.5) and (3.23), circular particle motion satisfies

$$\mathbf{J} = I\,\mathbf{n}, \quad I = m\Omega r^2, \tag{3.23}$$

where I denotes the moment of inertia about $\mathbf{n}$.[1] Evidently, (3.23) implies that $J = 0$ whenever $\Omega = 0$ and visa-versa. In an astronomical context, we may follow Mach[2] and define the angular velocity as the rate of change of angles *measured relative to the distant stars*. Does (3.23) hold in general?

It turns out that angular momentum is sensitive to matter in the universe anywhere. While (3.23) holds true to great precision under ordinary circumstances when Ω is defined relative to the distant stars, deviations appear in the proximity of massive rotating bodies. This can be detected in tracking the orientation $\mathbf{n}$ of a freely suspended gyroscope relative to a distant star. Recently, the NASA satellite Gravity Probe B[3] did just that, and measured an angular velocity in $\mathbf{n}$ at a minute rate of

$$\omega = -39\,\text{mas yr}^{-1} = -6\times 10^{-16}\,\text{rad s}^{-1}. \tag{3.24}$$

It agrees within a 20% window of uncertainty with the *frame-dragging* angular velocity of space-time around the earth, induced by Earth's angular momentum according to the theory of general relativity. According to the exact solution of rotating black holes in general relativity [3], (3.24) is the frame-dragging angular velocity at about 5 million Schwarzschild radii around a maximally spinning black hole with the same angular as the Earth (and 27 times its mass).

Though small, (3.24) defines a key result in our views on the relation between rotation and angular momentum, that comes out non-trivially in curved space-time predicted by the theory of general relativity. In particular, it changes our perception

---

[1]Formally I_{nn}, since I is generally a two-index tensor.

[2]Ernst Mach (1838–1916).

[3]Everitt et al. [1] a local measurement on the Riemann tensor. LAGEOS II detected frame dragging in the orientation of polar orbits [2].

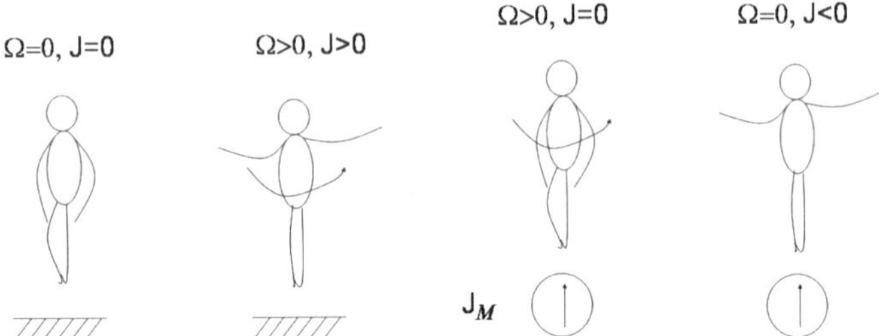

Fig. 3.2 (*Left*) In flat space-time, the ballerina effect a correspondence between zero angular velocity Ω relative to the distant stars and zero angular momentum J (Mach's principle). (*Right.*) In curved space-time, the ballerina effect is different. Here, $J = (\Omega - \omega)I$, where I denotes the moment of inertia and ω is the frame-dragging angular velocity along the angular momentum J_M of a massive object nearby. As a result, $\Omega = \omega$ for $J = 0$ and $J < 0$ when $\Omega = 0$. Mach's principle is to be generalized include all matter, including massive objects in a local neighborhood

of the ballerina effect (Fig. 3.2). In reality, a ballerina standing *still* with respect to the distant stars experiences a slight lifting of her arms *up*, due to her non-zero angular momentum imparted by frame-dragging around Earth.[4] In a twist to the original formulation of Mach's principle, she would experience co-rotation with an angular velocity (3.24) for her arms to be *down* in a fully relaxed state.

Frame dragging (3.24) induced by the angular momentum of the Earth is manifest also in energetic spin-spin interactions.[5] In response, particles with angular momentum J_p about the spin axis of the Earth experience a potential energy [4]

$$E = \omega J_p, \tag{3.25}$$

that represents a line-integral of *Papapetrou forces* [5] mediated by ω. The energy (3.25) is notoriously small for J_p of classical objects. However, for charged particles like electrons or protons in magnetic fields around black holes, E can be huge and reach energies on the scale of Ultra High Energy Cosmic Rays (UHECRs). Measurement of (3.25) around the Earth awaits future satellite experiments.

3.3.3 *Energy and Torque*

Angular momentum $\mathbf{J} = J\mathbf{n}$ can be changed by application of a torque, defined as

$$\mathbf{T} = \frac{d}{dt}\mathbf{J} = \mathbf{n}\frac{d}{dt}J + J\frac{d}{dt}\mathbf{n}. \tag{3.26}$$

[4] A related result affects her weight by Papapetrou forces [5], here expressed in Eq. (3.25).

[5] The complete set of frame dragging induced interactions is described by the Riemann tensor.

The dimension of torque is energy, as follows from $[J]$=g cm^2 s^{-1} (mass times rate of change of area). Because angular momentum is a vector, (3.26) shows the appearance of a torque already when changing its orientation, even when keeping its magnitude constant. In this case, (3.26) may be due to a rotation, i.e.,

$$\Delta \mathbf{T} = J\,(\mathcal{R} - I)\,\mathbf{n} \tag{3.27}$$

where $\mathcal{R}$ is a rotation matrix. (More on matrices in Sect. 3.5) For a rotation over an angle φ about the x-axis, for example, we have (Sect. 3.4.2)

$$\mathcal{R} = \begin{pmatrix} 1 & 0 & 0 \\ 0 & \cos\varphi & -\sin\varphi \\ 0 & -\sin\varphi & \cos\varphi \end{pmatrix} \tag{3.28}$$

Feymnan [6] gives an illustrative set-up that can be performed using a bicycle wheel attached freely to a rod. In this event, $\mathbf{n}$ is along the y-axis when the rod is initially held horizontally. Attempting to rotate the rod along the x-axis in an effort to move the wheel overhead is described by (3.27), see Fig. 3.3. By (3.28), it introduces a component of $\Delta\mathbf{T}$ along the z-axis. The person performing the rotation will experience a tendency to start rotating in the opposite direction to the angular momentum of the wheel, by conservation of total angular momentum *in all three dimensions* (in each of the three components x, y and z), i.e.,

$$\mathbf{J}_{wheel} + \mathbf{J}_{person} = \mathbf{0}. \tag{3.29}$$

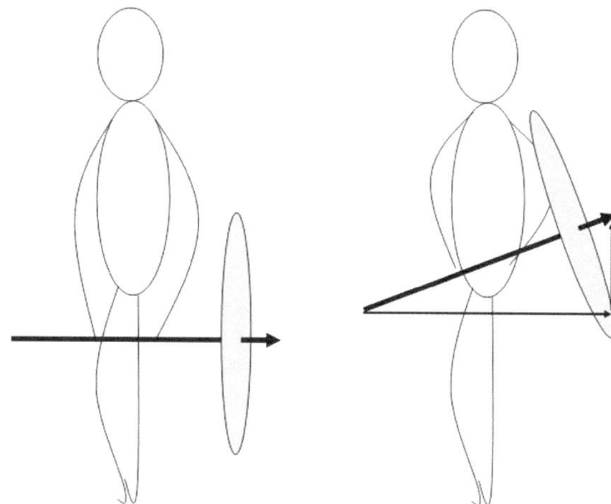

Fig. 3.3 Changing the orientation $\mathbf{n}$ of the angular momentum of a spinning wheel by a rotation introduces a component in an orthogonal direction, here along the vertical direction. Since angular momentum is conserved, a corresponding negative amount of angular momentum along the vertical direction is imparted by the person holding the wheel. The person will experience a counter-torque along the *vertical axis*

Since *power* is a scalar of dimension energy s^{-1}, the power delivered to or extracted from a rotating object is given by the inner product of torque and angular velocity, i.e.,

$$P = \mathbf{\Omega} \cdot \mathbf{T}. \tag{3.30}$$

For our circular motion, we have $\mathbf{T} = \frac{d}{dt}\mathbf{J} = I\frac{d}{dt}\mathbf{\Omega}$, and hence

$$P = \frac{d}{dt}\left(\frac{1}{2}\Omega^2\right) \tag{3.31}$$

It follows that the rotational energy in case of $\mathbf{J} = I\mathbf{\Omega}$ satisfies

$$E_{rot} = \frac{1}{2}\Omega^2 I = \frac{1}{2}\mathbf{\Omega} \cdot \mathbf{J}. \tag{3.32}$$

Although (3.32) applies to non-relativistic mechanics such as spinning tops, somewhat remarkably it gives a fairly good approximation also to the rotational energy $E_{rot} = k\,\mathbf{\Omega} \cdot \mathbf{J}$, $k^{-1} = 2\cos^2(\lambda/4)$, of a rotating black hole with non-dimensional angular momentum $\sin\lambda$, since

$$\frac{1}{2} \leq k \leq 0.5858. \tag{3.33}$$

To exemplify angular momentum conservation, consider the problem of the Moon's migration, in absorbing angular momentum in the Earth's spin due to a gravitational tidal torque.

Example 3.2. Some 4.52 Gyr ago, the Earth's spin period at birth was $P = 5.4\,\mathrm{h}$ before the Moon was born. The Earth's normalized angular velocity

$$A_0 = \left(\frac{\Omega}{\Omega_b}\right)_\oplus \tag{3.34}$$

then (4.52 Gyr ago) was very similar to the same for Jupiter today, where

$$\Omega = \frac{2\pi}{P},\quad \Omega_b = \sqrt{\frac{GM}{R^3}} \tag{3.35}$$

denote the actual and, respectively, break-up angular velocity for a planet of mass M and radius R, and G is Newton's constant. Some data:

$$\begin{aligned} &\text{Earth:}\quad M_\oplus = 5.97 \times 10^{27}\,\mathrm{g},\ R_\oplus = 6000\,\mathrm{km},\quad P_\oplus = 24\,\mathrm{h}, \\ &\text{Jupiter: } M \simeq 320 M_\oplus\ , R \simeq 11 R_\oplus,\ \ P \simeq 0.5 P_\oplus. \end{aligned} \tag{3.36}$$

The above follows from the following.

- For the Earth's $\Omega_{\oplus,b}$ and today's value $\Omega_\oplus = 2\pi/P_\oplus$, we have

$$A_1 = \left(\frac{\Omega}{\Omega_b}\right)_\oplus . \tag{3.37}$$

- The change $P_\oplus$ to 5.4 h from 24 h today satisfies the scaling

$$\left(\frac{\Omega}{\Omega_b}\right)_\oplus \propto P_\oplus^{-1}. \tag{3.38}$$

- Consequently, the spin angular velocity relative to break up at birth satisfies

$$A_0 = \left(\frac{5.4\,\text{h}}{24\,\text{h}}\right)^{-1} A_1. \tag{3.39}$$

that may be compared to the same ratio of Jupiter today.

With Newton's constant $G = 6.67 \times 10^{-8}$ g^{-1} cm^3 s^{-2} (recall that $G\rho$ has dimension angular velocity squared, i.e., s^{-2}.), we have by explicit calculation

$$\Omega_\oplus = 7.27 \times 10^{-5}\,\text{rad s}^{-1}, \;\; \Omega_{\oplus,b} = 1.36 \times 10^{-3}\,\text{rad s}^{-1} \tag{3.40}$$

and hence the ratio

$$A_1 = 0.0536. \tag{3.41}$$

By aforementioned scaling with $P_\oplus$, we have

$$A_0 = \left(\frac{24\,\text{h}}{5.4\,\text{h}}\right) A_1 = 4.44 A_0 \simeq 0.2380 \tag{3.42}$$

Repeating the above for Jupiter,

$$B_1 = \left(\frac{\Omega}{\Omega_b}\right)_J = 0.2185, \tag{3.43}$$

that is, our A_0 4.52 Gyr ago and Jupiter's B_1 today are very similar. As a consequence, we expect the weather of the Earth at birth to be very similar to that of Jupiter today, essentially a permanent storm by exceedingly large Coriolis forces. Recall that Coriolis forces scale with $\Omega_\oplus^2 \propto P_\oplus^{-2}$. They were initially some 20 times stronger than they are now. Thanks, in part, to spin down by the Moon, we can enjoy today's clement climate [7].

3.3.4 Coriolis Forces

Conservation of angular momentum gives rise to apparent forces when moving things around by external forces that leave the angular momentum invariant, as in the absence of any frictional forces. The specific angular momentum in the presence of an angular velocity ω is

$$j = \omega\sigma^2 : \omega = \frac{j}{\sigma^2}, \tag{3.44}$$

where σ denotes the distance to the axis of rotation. Moving a fluid element along the radial direction changes ω, as when the ballerina moves stretched arms inwards, according to $\delta\omega = -2j\sigma^{-3}\delta\sigma$. It comes with a change in azimuthal velocity $\delta v_\varphi = \sigma\delta\omega$ seen in a corotating frame, satisfying

$$\delta v_\varphi = -2\omega\delta\sigma. \tag{3.45}$$

In vector form, (3.45) is

$$\frac{d}{dt}\mathbf{v}_\varphi = -2\boldsymbol{\omega} \times \mathbf{v}_\sigma. \tag{3.46}$$

This result is commonly expressed in terms of the Coriolis force

$$F_c = m\frac{d}{dt}\mathbf{v}_\varphi = 2m\mathbf{v} \times \boldsymbol{\omega} \tag{3.47}$$

Coriolis forces are particularly relevant when working in a rotating frame of reference. In particular, all of us terrestrial inhabitants living with the rotating frame fixed to Earth's surface. Air moving to a different latitude is subject to (3.47), since it changes the distance σ to the Earth's axis of rotation, which is approximately polar. Let Ω denote the absolute angular velocity of the Earth (relative to the distant stars), and express the angular velocity of the air $\omega' = \omega - \Omega$ relative to it, as measured in this rotating frame. Since $\delta\omega' = \delta\omega$, moving air from, say, in the direction of the equator produces a retrograde azimuthal velocity (rotation at an angular velocity $\omega < \Omega$). Moving it a constant angular velocity towards the equator produces a curved trajectory in response to the (retrograde) constant Coriolis force (3.47). This may give rise to large scale circulation patterns in combination with pressure gradients.

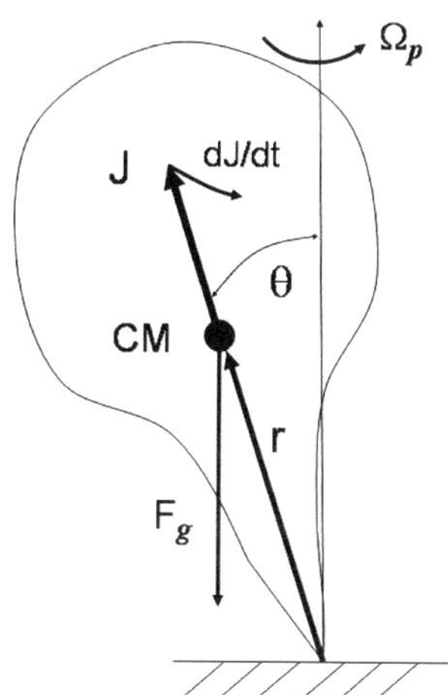

Fig. 3.4 Precession of the spinning top causes a velocity $\dot{\mathbf{n}}$ in the orientation $\mathbf{n}$ of the angular momentum, such that $d\mathbf{J}/dt = J\dot{\mathbf{n}}$ absorbs the torque due to the gravitational force $\mathbf{F}_g$ applied at its center of mass CM. In the idealized friction-free set-up, this process involves no exchange of energy or dissipation

3.3.5 *Spinning Top*

The motion of a spinning top tilted at at angle θ exemplifies the interaction of angular momentum as a vector with a torque, $\mathbf{T}$, applied continuously by the Earth's gravitational force $\mathbf{F}_g$ as illustrated in Fig. 3.4. In general, we have the relations

$$\mathbf{T} = \frac{d}{dt}\mathbf{J} = \mathbf{r} \times \frac{d}{dt}\mathbf{p} = \mathbf{r} \times \mathbf{F}_g. \tag{3.48}$$

For a top that spins with no friction, the magnitude of its angular momentum vector is conserved. By (3.26, 3.27), the top precesses at an angular velocity $\mathbf{\Omega}_p$ about the z-axis, $\Omega_p = d\phi/dt$, satisfying

$$\frac{d}{dt}\mathbf{J} = J\frac{d}{dt}\mathbf{n} = J\mathbf{\Omega}_p \times \mathbf{n} = \mathbf{\Omega}_p \times \mathbf{J}. \tag{3.49}$$

By (3.48), $T = \Omega_p J \sin\theta = rW \sin\theta$, and hence the angular velocity of precession about the vertical axis satisfies

$$\Omega_p J = rW, \tag{3.50}$$

where W denotes the weight of the top and r the distance of its center of mass away from its pivot on the table.

Example 3.3. Illustrative for some vector calculations is a more explicit calculation of the precession frequency (3.50). To this end, Fig. 3.5 shows a massive ring of radius R spinning at an angular velocity ω, whereby it attains an angular momentum per unit mass $J = \omega R^2$. Suppose it is mounted to one end of a rod,

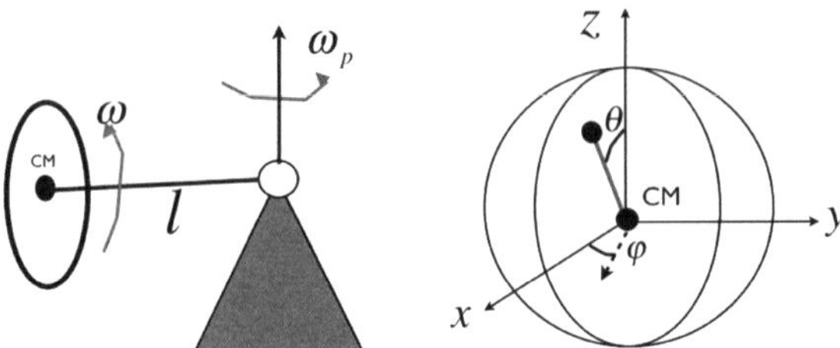

Fig. 3.5 Shown is a ring of radius R rotating at an angular velocity ω about a *horizontal axis* of length l, supported by a pivot that allow rotation at a precession angular velocity ω_p about the *vertical axis*. Upon translation of the center of mass (CM) to the origin of a spherical coordinate system, mass elements on the ring move over the surface of a sphere of radius R, parameterized by a poloidal and azimuthal angle θ and, respectively, φ, wherein $d\theta/dt = \omega$ and $d\varphi/dt = \omega_p$

that is suspended at a pivot at the other end. An approximately horizontal rod hereby precesses with an angular velocity ω_p about the vertical axis without dropping to a vertical position, satisfying (3.49). This result is invariant under linear translation of the CM. Precession is entirely due to the motion of mass-elements about the ring's CM, allowing us to place the CM at the origin of a spherical coordinate system (r, θ, φ), as if the CM where placed at the pivot. With $\boldsymbol{\omega} = \omega \mathbf{i}_z$, the outer product $\omega \times \mathbf{r}$ is the rotational velocity $\mathbf{v}_\phi$ of the end point of a vector $\mathbf{r}$ and that $\mathbf{v}_\phi = \omega\sigma$, where $\sigma = b \sin\theta$ is the distance to the axis of rotation. A mass element $\delta m = (M/2\pi)\delta\theta$ in the ring herein assumes an angular momentum $\delta\mathbf{J} = \mathbf{r} \times \delta\mathbf{p} = \delta m \mathbf{r} \times \mathbf{v}$ with position vector

$$\mathbf{r} = b \begin{pmatrix} \sin\theta\cos\varphi \\ \sin\theta\sin\varphi \\ \cos\theta \end{pmatrix}, \quad \omega = \frac{d\theta}{dt}, \quad \omega_p = \frac{d\varphi}{dt}, \tag{3.51}$$

and associated velocity $\mathbf{v} = d\mathbf{r}/dt$,

$$\mathbf{v} = b \begin{pmatrix} \cos\theta\cos\varphi \\ \cos\theta\sin\varphi \\ -\sin\theta \end{pmatrix} \omega + b \begin{pmatrix} -\sin\theta\sin\varphi \\ \sin\theta\cos\varphi \\ 0 \end{pmatrix} \omega_p, \tag{3.52}$$

and acceleration $\mathbf{a} = d\mathbf{v}/dt$,

$$\mathbf{a} = -b \begin{pmatrix} \sin\theta\cos\varphi \\ \sin\theta\sin\varphi \\ \cos\theta \end{pmatrix} \omega^2 - b \begin{pmatrix} \sin\theta\cos\varphi \\ \sin\theta\sin\varphi \\ 0 \end{pmatrix} \omega_p^2 \tag{3.53}$$

$$+2b \begin{pmatrix} -\cos\theta\sin\varphi \\ \cos\theta\cos\varphi \\ 0 \end{pmatrix} \omega\omega_p. \tag{3.54}$$

Its inertia introduces a torque

$$\delta T = \frac{d\delta \mathbf{J}}{dt} = \delta m \left(\frac{d}{dt}\mathbf{r} \times \mathbf{v} + \mathbf{r} \times \frac{d}{dt}\mathbf{v} \right) = \delta m \mathbf{r} \times \mathbf{a} \tag{3.55}$$

that evaluates to

$$\delta \mathbf{T} = \delta m b \left(\omega_p^2 \mathbf{r} \times \mathbf{i}_z + 2\omega\omega_p \mathbf{r} \times \begin{pmatrix} -\cos\theta\sin\varphi \\ \cos\theta\cos\varphi \\ 0 \end{pmatrix} \right) \tag{3.56}$$

To finalize, we integrate (3.56) over all mass elements δm. Making use of the following averages over the *fast angle* θ,

$$\langle \mathbf{r} \times \mathbf{i}_z \rangle = \frac{1}{2\pi}\int_0^{2\pi} \mathbf{r} \times \mathbf{i}_z d\theta = \frac{b}{2\pi}\int_0^{2\pi} \begin{pmatrix} \sin\theta\sin\varphi \\ -\sin\theta\cos\varphi \\ 0 \end{pmatrix} d\theta = \mathbf{0} \tag{3.57}$$

and

$$\left\langle \mathbf{r} \times \begin{pmatrix} -\cos\theta\sin\varphi \\ \cos\theta\cos\varphi \\ 0 \end{pmatrix} \right\rangle = \frac{1}{2\pi}\int_0^{2\pi} \mathbf{r} \times \begin{pmatrix} -\cos\theta\sin\varphi \\ \cos\theta\cos\varphi \\ 0 \end{pmatrix} d\theta = \tag{3.58}$$

$$\frac{b}{2\pi}\int_0^{2\pi} \begin{pmatrix} -\cos^2\theta\cos\varphi \\ -\cos^2\theta\sin\varphi \\ 2\sin\theta\cos\theta \end{pmatrix} d\theta = -\frac{b}{2}\begin{pmatrix} \cos\varphi \\ \sin\varphi \\ 0 \end{pmatrix}, \tag{3.59}$$

we arrive at a total inertial torque $\mathbf{T} = \int_0^{2\pi} \delta\mathbf{T}$,

$$\mathbf{T} = \frac{M}{2\pi}\int_0^{2\pi} \mathbf{r} \times \mathbf{a}\, d\theta = Mb\left(\omega_p^2 \langle \mathbf{r} \times \mathbf{i}_z \rangle + 2\omega\omega_p \left\langle \mathbf{r} \times \begin{pmatrix} -\cos\theta\sin\varphi \\ \cos\theta\cos\varphi \\ 0 \end{pmatrix} \right\rangle \right). \tag{3.60}$$

With $J = I\omega$ expressed in the moment of inertia $I = Mb^2$, the latter reduces to $T = \omega\omega_p Mb^2 = \omega_p J$, i.e., our vector identity (3.49).

In Fig. 3.5, if the bar holding the rotating wheel is initially suspended horizontally at the pivot with zero angular momentum about the z-axis, then the onset of precession ω_p—balancing inertial to gravitational torque $gM\sigma$—produces a finite angular momentum $J_z = M\sigma^2\omega_p$ about the z-axis (upwards, say), where $\sigma = l\cos\alpha$ is the arm length to the z-axis, now at a dip angle α. Since the total angular momentum about the z-axis remains zero, $J_z = J\sin\theta$ (pointing downwards). Given $\omega_p J = Mg\sigma$, it follows that (cf. Exercise 3.3)

$$\tan\alpha = \left(\omega_p/\Omega\right)^2, \tag{3.61}$$

where $\Omega = \sqrt{g/\sigma}$.

3.4 Elementary Transformations in the Plane

In the two-dimensional plane with Cartesian coordinates (x, y), transformations describe a map

$$\mathbf{z} = x\mathbf{i}_x + y\mathbf{i}_y \rightarrow \mathbf{w} = x'\mathbf{i}_x + y'\mathbf{i}_y. \tag{3.62}$$

When linear, such map is a matrix multiplication $\mathbf{w} = C\mathbf{z}$,

$$\mathbf{z} = \begin{pmatrix} x \\ y \end{pmatrix} = x\begin{pmatrix} 1 \\ 0 \end{pmatrix} + y\begin{pmatrix} 0 \\ 1 \end{pmatrix} \tag{3.63}$$

with

$$C\mathbf{z} = \begin{pmatrix} \text{———} \\ \text{———} \end{pmatrix}\mathbf{z} = \begin{pmatrix} \mathbf{r}_1^T\mathbf{z} \\ \mathbf{r}_2^T\mathbf{z} \end{pmatrix} \tag{3.64}$$

and

$$\mathbf{r}^T = (a\ b): \ \ \mathbf{r}^T\mathbf{z} = ax + by. \tag{3.65}$$

Equivalently, we have

$$C\mathbf{z} = \begin{pmatrix} | & | \\ | & | \end{pmatrix}\left\{x\begin{pmatrix} 1 \\ 0 \end{pmatrix} + y\begin{pmatrix} 0 \\ 1 \end{pmatrix}\right\} = x\mathbf{c}_1 + y\mathbf{c}_2. \tag{3.66}$$

These two views (3.64–3.66) explicitly bring about linearity in the row and column vectors of C.

When working in the two-dimensional plane, we note that (3.62) is equivalent to a map of complex numbers $z = x + iy \rightarrow w = x' + iy'$, that is occasionally useful when working with conformal transformations $w = w(z)$ $(w'(z) \neq 0)$.

3.4.1 *Reflection Matrix*

Figure 3.6 illustrates reflections in the two-dimensional plane about the x-axis, the y-axis and through the origin, $\mathcal{O} = (0, 0)$. Reflection about the x-axis is described by

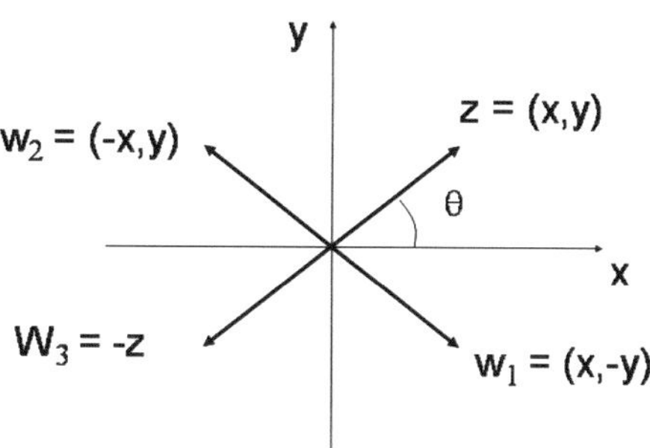

Fig. 3.6 Reflections in the (x, y)-plane about the x-axis, the y-axis and through the origin, take $z = (x, y)$ to, respectively, $w_1 = (x, -y)$, $w_2 = (-x, y)$ and $w_3 = -z$. Each transformation is described by a 2×2 matrix acting on the vector $\mathbf{z} = x\mathbf{i}_x + y\mathbf{i}_y$

$$\mathbf{z} = x\mathbf{i}_x + y\mathbf{i}_y \to \mathbf{w} = x\mathbf{i}_x - y\mathbf{i}_y. \tag{3.67}$$

The same transformation can be written as a matrix equation for the equations $x' = x$ and $y' = -y$ as follows

$$\begin{pmatrix} x' \\ y' \end{pmatrix} = \begin{pmatrix} 1 & 0 \\ 0 & -1 \end{pmatrix} \begin{pmatrix} x \\ y \end{pmatrix}. \tag{3.68}$$

Reflection about the y-axis is described by

$$\mathbf{z} = x\mathbf{i}_x + y\mathbf{i}_y \to \mathbf{w} = -x\mathbf{i}_x + y\mathbf{i}_y. \tag{3.69}$$

The same transformation can be written as a matrix equation for $x' = -x$ and $y' = y$ as follows

$$\begin{pmatrix} x' \\ y' \end{pmatrix} = \begin{pmatrix} -1 & 0 \\ 0 & 1 \end{pmatrix} \begin{pmatrix} x \\ y \end{pmatrix}. \tag{3.70}$$

As mentioned above, (3.67–3.69) are equivalent to taking $z \in \mathbb{C}$ into, respectively,

$$w_1 = \bar{z} = x - iy, \quad w_2 = -\bar{z} = -x + iy. \tag{3.71}$$

Reflection about the origin is described by

$$\mathbf{z} = x\mathbf{i}_x + y\mathbf{i}_y \to \mathbf{w} = -x\mathbf{i}_x - y\mathbf{i}_y, \quad w_3 = -z, \tag{3.72}$$

The same transformation can be written as a matrix equation for the equations $x' = -x$ and $y' = -y$ as follows

$$\begin{pmatrix} x' \\ y' \end{pmatrix} = \begin{pmatrix} -1 & 0 \\ 0 & -1 \end{pmatrix} \begin{pmatrix} x \\ y \end{pmatrix}. \tag{3.73}$$

The *identity matrix* is the defined by the transformation which leaves $\mathbf{z}$ the same, i.e.,

$$I = \begin{pmatrix} 1 & 0 \\ 0 & 1 \end{pmatrix}. \tag{3.74}$$

3.4.2 Rotation Matrix

The above can be extended to continuous transformations such as rotations. The rotation matrix can be derived from the multiplication of complex numbers following (3.18) and (3.62). With $\mathbf{z} = r\cos\theta\mathbf{i}_x + r\sin\theta\mathbf{i}_y$, we have

$$\begin{pmatrix} x' \\ y' \end{pmatrix} = R(\varphi)\begin{pmatrix} x \\ y \end{pmatrix}, \tag{3.75}$$

in terms of the rotation matrix

$$R(\varphi) = \begin{pmatrix} \cos\varphi & -\sin\varphi \\ \sin\varphi & \cos\varphi \end{pmatrix}. \tag{3.76}$$

Evidently, it satisfies

$$\det R = 1, \quad R(-\varphi) = R^{-1}(\varphi) = R^T(\varphi), \tag{3.77}$$

where R^{-1} refers to the inverse of R, R^T refers to the *transpose* and

$$\det\begin{pmatrix} a_{11} & a_{12} \\ a_{21} & a_{22} \end{pmatrix} = a_{11}a_{22} - a_{12}a_{21} \tag{3.78}$$

defines the determinant of a 2×2 matrix.

For what follows, we shall generalize (3.9) to matrices. For a square $n \times n$ matrix A, the transpose obtains by interchanging the off-diagonal components a_{ij} $(i \neq j)$ about the *principle diagonal* containing the a_{ii}. Schematically, if L refers to the *upper diagonal elements* and U refers to the *lower diagonal elements*, then

$$A = \begin{pmatrix} a_{11} & & U \\ & a_{22} & \\ & & \cdots & \\ L & & a_{nn} \end{pmatrix} \rightarrow A^T = \begin{pmatrix} a_{11} & & L \\ & a_{22} & \\ & & \cdots & \\ U & & a_{nn} \end{pmatrix}. \tag{3.79}$$

The rotation matrix $R(\varphi)$ in (3.76) is anti-symmetric in its off-diagonal elements, i.e., $U = -L$. A square matrix is said to be anti-symmetric, if $U = -L$ and the elements on the principle diagonal are zero. Since the diagonal elements in (3.76) are non-zero, $R(\varphi)$ is not an anti-symmetric matrix.

Example 3.4. A symmetric matrix, satisfying $U = L$ as defined in (3.79), is the Lorentz boost

$$\Lambda(\mu) = \begin{pmatrix} \cosh\mu & \sinh\mu \\ \sinh\mu & \cosh\mu \end{pmatrix}, \tag{3.80}$$

that appears in the transformation of four-momenta in Malinowski space. Both $R(\varphi)$ and $\Lambda(\mu)$ have determinant one,

$$\det R(\varphi) = \cos^2\varphi + \sin^2\varphi = 1, \quad \det\Lambda(\mu) = \cosh^2\mu - \sinh^2\mu = 1. \tag{3.81}$$

3.5 Matrix Algebra

Multiplication of two matrices A of dimension $p \times m$ and B of dimension $m \times q$ produces a new matrix $C = AB$ of dimension $p \times q$. Each entry of C is the inner product of a row from A and a column from B. Schematically, the product C of two 2×2 matrices is

$$C = \begin{pmatrix} \text{———} \\ \text{———} \end{pmatrix} \begin{pmatrix} | & | \\ | & | \end{pmatrix} = \begin{pmatrix} c_{11} & c_{12} \\ c_{21} & c_{22} \end{pmatrix}, \tag{3.82}$$

upon considering A in terms of its rows and B in terms of its columns. The entries of C satisfy

$$c_{ij} = (a_{i1}\ a_{i2}) \begin{pmatrix} b_{1j} \\ b_{2j} \end{pmatrix} = a_{i1}b_{1j} + a_{i2}b_{2j}. \tag{3.83}$$

The product $D = BA$ of the same 2×2 matrices satisfies

$$D = \begin{pmatrix} \text{———} \\ \text{———} \end{pmatrix} \begin{pmatrix} | & | \\ | & | \end{pmatrix} = \begin{pmatrix} d_{11} & d_{12} \\ d_{21} & d_{22} \end{pmatrix} \tag{3.84}$$

upon considering B in terms of its rows and A in terms of its columns, so that

$$d_{ij} = (b_{i1}\ b_{i2}) \begin{pmatrix} a_{j1} \\ a_{j2} \end{pmatrix} = b_{i1}a_{1j} + b_{i2}a_{2j}. \tag{3.85}$$

It is easy to see that in general $D \neq C$, i.e., matrix multiplication does not commute,

$$[A, B] = AB - BA \neq 0, \tag{3.86}$$

where the notation $[\cdot, \cdot]$ refers to the commutator.

Example 3.3. To illustrate, consider the two matrices

$$A = \begin{pmatrix} 0 & -1 \\ 1 & 0 \end{pmatrix}, \quad B = \begin{pmatrix} 0 & 1 \\ 1 & 0 \end{pmatrix}. \tag{3.87}$$

The commutation $[A, B]$ then evaluates to

$$\begin{pmatrix} 0 & -1 \\ 1 & 0 \end{pmatrix}\begin{pmatrix} 0 & 1 \\ 1 & 0 \end{pmatrix} - \begin{pmatrix} 0 & 1 \\ 1 & 0 \end{pmatrix}\begin{pmatrix} 0 & -1 \\ 1 & 0 \end{pmatrix} = 2\begin{pmatrix} -1 & 0 \\ 0 & 1 \end{pmatrix}. \tag{3.88}$$

3.6 Eigenvalue Problems

Eigenvalue problems are defined by the equation

$$A\mathbf{a} = \lambda\mathbf{a}, \tag{3.89}$$

where $\mathbf{a}$ refers to an eigenvector associated with the eigenvalue λ. Equivalently, $\mathbf{a}$ is in the *null-space* (is a right null-vector) of $A - \lambda I$:

$$(A - \lambda I)\,\mathbf{a} = \mathbf{0}. \tag{3.90}$$

For (3.90) to have a non-trivial solution $\mathbf{a}$, we must have

$$\det(A - \lambda I) = 0. \tag{3.91}$$

3.6.1 Eigenvalues of $R(\varphi)$

Let us explore (3.90, 3.91) for the rotation matrix $R(\varphi)$,

$$0 = |R - \lambda I| = (\cos\varphi - \lambda)^2 + \sin^2\varphi : \ \lambda_\pm = \cos\varphi + i\sin\varphi = e^{\pm i\varphi}. \tag{3.92}$$

The eigenvalues are S^1. It is a consequence of the fact that rotation is unitary (see Sect. 3.7). Also, the eigenvalues satisfy[6]

---

[6]The product of the eigenvalues equals the determinant of the matrix, as follows from, e.g., the Jordan decomposition theorem. The same theorem shows that the *trace* of a matrix given by the sum of the elements on the principle diagonal equals the sum of the eigenvalues.

$$\lambda_1 \lambda_2 = |R| = 1. \tag{3.93}$$

The associated eigenvectors

$$\mathbf{a} = \begin{pmatrix} \alpha_1 \\ \alpha_2 \end{pmatrix} \tag{3.94}$$

satisfy (3.90). To be definite, (3.90) defines two homogeneous equations in the two unknown coefficients (α_1, α_2),

$$\begin{cases} \alpha_1 \cos\varphi - \alpha_2 \sin\varphi - \lambda\alpha_1 = 0, \\ \alpha_1 \sin\varphi + \alpha_2 \cos\varphi - \lambda\alpha_2 = 0. \end{cases} \tag{3.95}$$

For the eigenvalues satisfying (3.93), these two equations are *linearly dependent*. It suffices to take one of them, to solve for α_1 and α_2,

$$\alpha_1(\cos\varphi - \lambda) - \alpha_2 \sin\varphi = 0 : \ \alpha_1 = -i\alpha_2, \ \ \alpha_1 = i\alpha_2 \tag{3.96}$$

for $\lambda = e^{i\varphi}$ and, respectively, $\lambda = e^{-i\varphi}$. We thus arrive at the eigenvector-eigenvalue pairs

$$\left\{ e^{i\varphi}, \begin{pmatrix} 1 \\ -i \end{pmatrix} \right\}, \left\{ e^{-i\varphi}, \begin{pmatrix} 1 \\ i \end{pmatrix} \right\}. \tag{3.97}$$

These two pairs are complex conjugates. This is no surprise since $R(\varphi)$ is a real matrix, whose determinant $|R - \lambda I|$ defines a quadratic polynomial in λ. With real coefficients, its roots are either both real or a pair of complex conjugates.

3.6.2 Eigenvalues of a Real-Symmetric Matrix

The matrix

$$A = \begin{pmatrix} 2 & 1 \\ 1 & 0 \end{pmatrix} \tag{3.98}$$

is real-symmetric with eigenvalues-eigenvectors $(\lambda_\pm, \mathbf{x}_\pm)$

$$\left\{ 1 + \sqrt{2}, \begin{pmatrix} 1 + \sqrt{2} \\ 1 \end{pmatrix} \right\}, \left\{ 1 - \sqrt{2}, \begin{pmatrix} 1 - \sqrt{2} \\ 1 \end{pmatrix} \right\}. \tag{3.99}$$

It is readily seen that $\mathbf{x}_\pm$ are *orthogonal*:

$$\mathbf{x}_+^T \mathbf{x}_- = 0. \tag{3.100}$$

We can normalize the eigenvectors to

$$\mathbf{e}_+ = \frac{1}{\sqrt{4+2\sqrt{2}}}\begin{pmatrix}1+\sqrt{2}\\1\end{pmatrix}, \quad \mathbf{e}_- = \frac{1}{\sqrt{4-2\sqrt{2}}}\begin{pmatrix}1-\sqrt{2}\\1\end{pmatrix}, \tag{3.101}$$

so that $(\mathbf{e}_+, \mathbf{e}_-)$ forms a new *orthonormal* basis set complementary to $(\mathbf{i}, \mathbf{j})$ along the x- and y-axis. Hence, we have the general decompositions

$$\mathbf{x} = x\mathbf{i} + y\mathbf{j} = a\mathbf{e}_+ + b\mathbf{e}_+, \tag{3.102}$$

where $x = \mathbf{i}\cdot\mathbf{x}$ and $y = \mathbf{j}\cdot\mathbf{y}$. The coefficients a and b can be read off using multiplication by $\mathbf{e}_\pm$:

$$a = x\,\mathbf{i}\cdot\mathbf{e}_+ + y\,\mathbf{j}\cdot\mathbf{e}_+, \quad b = x\,\mathbf{i}\cdot\mathbf{e}_- + y\,\mathbf{j}\cdot\mathbf{e}_-. \tag{3.103}$$

Note that (3.89) defines the eigenvectors as invariant subspaces. We now arrive at a new look at A as an operator on $\mathbf{x}$ in terms of multiplications by eigenvectors along the directions given by the associated eigenvectors,

$$A\mathbf{x} = a\,\lambda_+\mathbf{e}_+ + b\,\lambda_-\mathbf{e}_-. \tag{3.104}$$

3.6.3 Hermitian Matrices

Let † denote the *Hermitian conjugate*,[7] defined as the complex conjugate of the transpose of a matrix element, a column or row vector or a matrix. We define the scalar product of two vectors $\mathbf{a}$ and $\mathbf{b}$ in an n-dimensional vector space by

$$\mathbf{a}^\dagger\mathbf{b} = \bar{a}_1 b_1 + \bar{a}_2 b_2 + \cdots \bar{a}_n b_n. \tag{3.105}$$

Real-symmetric matrices generalize to complex valued matrices with the same properties of having real eigenvalues and mutually orthogonal eigenvectors associated with different eigenvalues according to (3.116) and, respectively, (3.119). Following the steps of the previous section, these are the self-adjoint or Hermitian matrices satisfying

$$H^\dagger = H, \tag{3.106}$$

defined by transformation of the entries $H^\dagger_{ij} = \bar{H}_{ji}$. Note that applying $\dagger$ twice is an identity operation, i.e., $(A^\dagger)^\dagger = A$ for any $n \times m$ matrix A. Hence, if H is an $n \times n$ matrix, we have

[7] Also referred to as the Hermitian transpose or the conjugate transpose.

$$H = \begin{pmatrix} a_{11} & & L^\dagger \\ & a_{22} & \\ & & \cdots \\ L & & a_{nn} \end{pmatrix} \qquad (3.107)$$

with real diagonal elements a_{ii} $(i = 1, 2, \cdots n)$.

Example 3.5. For instance, the rotation matrix $R(i\mu)$ with imaginary angle $\varphi = i\mu$,

$$H = \begin{pmatrix} \cosh\mu & -i\sinh\mu \\ i\sinh\mu & \cosh\mu \end{pmatrix}, \qquad (3.108)$$

is Hermitian. Since $|R(\varphi)| = 1$ for all φ, we have $|H| = 1$ by analytic continuation, which also follows by inspection,

$$|H| = \cosh^2\mu - \sinh^2\mu = 1. \qquad (3.109)$$

The eigenvalue-eigenvectors obtain by analytic continuation of (3.97), i.e.,

$$\left\{e^{-\mu}, \begin{pmatrix} 1 \\ -i \end{pmatrix}\right\}, \left\{e^{\mu}, \begin{pmatrix} 1 \\ i \end{pmatrix}\right\}. \qquad (3.110)$$

According to (3.105), the scalar product between the two eigenvectors satisfies

$$\begin{pmatrix} 1 \\ -i \end{pmatrix}^\dagger \begin{pmatrix} 1 \\ i \end{pmatrix} = (1\ i)\begin{pmatrix} 1 \\ i \end{pmatrix} = 1 + i^2 = 0. \qquad (3.111)$$

This result of Example 3.5 is expected, since (3.117–3.119) continues to hold upon replacing T by $\dagger$, i.e.,

$$\lambda_1 = \lambda_2 \text{ or } \mathbf{a}_1^\dagger \mathbf{a}_2 = 0. \qquad (3.112)$$

For a Hermitian matrix, the *eigenvectors of distinct eigenvalues are mutually orthogonal*, where orthogonality is defined according to the inner product (3.105).

Very similar properties of the eigenvalue problem (3.89) appear in the real-symmetric matrix $\Lambda(\mu)$ of (3.80). Again, we will find that the eigenvalues are real and distinct, whose accompanying eigenvectors are mutually orthogonal. These properties hold true for all real-symmetric matrices, as shown by the following.

Consider an eigenvalue-eigenvector pair $(\lambda, \mathbf{a})$ to a Hermition matrix A. Then

$$\mathbf{a}^\dagger A\mathbf{a} = \lambda \mathbf{a}^\dagger \mathbf{a}. \qquad (3.113)$$

Here, $\mathbf{a}^\dagger \mathbf{a}$ is real, obtained from the summation of the squared norms of the entries of $\mathbf{a}$. For (3.94), for example, we have

$$\mathbf{a}^\dagger \mathbf{a} = \overline{\alpha}_1 \alpha_1 + \overline{\alpha}_2 \alpha_2 \geq 0. \tag{3.114}$$

The transpose of the left hand side of (3.113) satisfies

$$\lambda \mathbf{a}^\dagger \mathbf{a} = \mathbf{a}^\dagger A \mathbf{a} = \left(\mathbf{a}^\dagger A \mathbf{a}\right)^T = \mathbf{a}^T A^T \bar{\mathbf{a}} = \mathbf{a}^T \overline{A^\dagger \mathbf{a}} = \mathbf{a}^T \overline{A\mathbf{a}} = \overline{\lambda} \mathbf{a}^T \bar{\mathbf{a}}. \tag{3.115}$$

and hence $\lambda \mathbf{a}^\dagger \mathbf{a} = \overline{\lambda}\ \overline{\mathbf{a}^\dagger \mathbf{a}} = \overline{\lambda} \mathbf{a}^\dagger \mathbf{a}$. It follows that the *eigenvalues of a Hermitian matrix are real:*

$$\bar{\lambda} = \lambda, \tag{3.116}$$

since $\mathbf{a}^T \bar{\mathbf{a}} \equiv \mathbf{a}^\dagger \mathbf{a}$.

Following similar arguments, consider

$$A\mathbf{a}_2 = \lambda_2 \mathbf{a} : \ \mathbf{a}_1^\dagger A \mathbf{a}_2 = \lambda_2 \mathbf{a}_1^\dagger \mathbf{a}_2. \tag{3.117}$$

For a Hermitian A, we have

$$\left(\mathbf{a}_1^\dagger A \mathbf{a}_2\right)^\dagger = \mathbf{a}_2^\dagger A^\dagger \mathbf{a}_1 = \mathbf{a}_2^\dagger A \mathbf{a}_1 = \lambda_1 \mathbf{a}_2^\dagger \mathbf{a}_1. \tag{3.118}$$

By (3.117, 3.118), we have $\lambda_2 \mathbf{a}_1^\dagger \mathbf{a}_2 = \lambda_1 \mathbf{a}_2^\dagger \mathbf{a}_1$. Since $\mathbf{a}_1^\dagger \mathbf{a}_2 = \mathbf{a}_2^\dagger \mathbf{a}_1$, it follows that

$$\lambda_1 = \lambda_2 \text{ or } \mathbf{a}_1^\dagger \mathbf{a}_2 = 0. \tag{3.119}$$

For a Hermitian matrix, the *eigenvectors of distinct eigenvalues are mutually orthogonal.*

Let us now turn to the example matrix $\Lambda(\mu)$ in (3.89). Its eigenvalues are defined by (3.91) with $A = \Lambda$, that is,

$$0 = |\Lambda - \lambda I| = (\cosh \mu - \lambda)^2 - \sinh^2 \mu, \tag{3.120}$$

whereby

$$\lambda_\pm = \cosh \mu + \sinh \mu = \begin{cases} e^\mu \\ e^{-\mu} \end{cases}. \tag{3.121}$$

Similar to (3.92), we note

$$\lambda_1 \lambda_2 = |\Lambda| = 1. \tag{3.122}$$

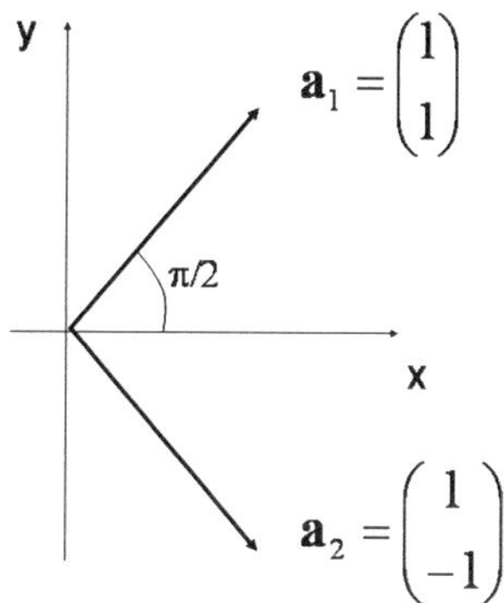

Fig. 3.7 The matrix Λ in (3.80) is real-symmetric. With two distinct eigenvalues, its eigenvectors are orthogonal. As shown, a second eigenvector $\mathbf{a}_2 = \mathbf{i}_x - \mathbf{i}_y$ hereby follows immediately from orthogonality to the first $\mathbf{a}_1 = \mathbf{i}_x + \mathbf{i}_y$

The equation for the eigenvectors (3.95) in terms of (α_1, α_2) are again a linearly dependent system of equations when λ assumes one of the eigenvalues (3.121). Considering the first of (3.95) with $\lambda = e^{\mu}$,

$$\alpha_1(\cosh \mu - e^{\mu}) + \alpha_2 \sinh \mu = 0 : \ \alpha_1 = \alpha_2, \tag{3.123}$$

we obtain the eigenvalue-eigenvector pair

$$\left\{ e^{\mu}, \begin{pmatrix} 1 \\ 1 \end{pmatrix} \right\}. \tag{3.124}$$

According to (3.119), the eigenvector associated with $\lambda = e^{-\mu}$ is orthogonal to that of (3.124). Since we are working in two dimensions, the second eigenvalue-eigenvector pair is therefore

$$\left\{ e^{-\mu}, \begin{pmatrix} 1 \\ -1 \end{pmatrix} \right\} \tag{3.125}$$

as illustrated in Fig. 3.7. The same obtains by solving (3.123) with e^{μ} replaced by $e^{-\mu}$.

3.7 Unitary Matrices and Invariants

The reflections and rotations shown in Fig. 3.6 *preserve norm and angles*. If $\mathbf{a}$ and $\mathbf{b}$ are two real vectors and $\mathbf{a}'$ and $\mathbf{b}'$ are their images, e.g.,

$$\mathbf{a}' = R(\varphi)\mathbf{a}, \ \ \mathbf{b}' = R(\varphi)\mathbf{b} \tag{3.126}$$

then the inner product

$$\rho = \mathbf{a}^T \mathbf{b} \tag{3.127}$$

is preserved, since

$$(\mathbf{a}')^T \mathbf{b}' = (R(\varphi)\mathbf{a})^T\, R(\varphi)\mathbf{b} = \mathbf{a}^T R(\varphi)^T R(\varphi)\mathbf{b} = \mathbf{a}^T \mathbf{b} \tag{3.128}$$

by the property of unitarity

$$R(\varphi)^T R(\varphi) = R(-\varphi)R(\varphi) = I. \tag{3.129}$$

In particular, $|\mathbf{a}'|^2 = (\mathbf{a}')^T\mathbf{a}' = |\mathbf{a}|^2$ and, likewise, $|\mathbf{b}'|^2 = |\mathbf{a}|^2$, showing that their norms are preserved. If θ and θ' refer to the angle between $(\mathbf{a}, \mathbf{b})$ and, respectively, $(\mathbf{a}', \mathbf{b}')$, then

$$|\mathbf{a}||\mathbf{b}|\cos\theta' = |\mathbf{a}'||\mathbf{b}'|\cos\theta' = (\mathbf{a}')^T\mathbf{b}' = \mathbf{a}^T\mathbf{b} = |\mathbf{a}||\mathbf{b}|\cos\theta, \tag{3.130}$$

which shows that $\cos\theta' = \cos\theta$. Since the norms and angles (between two vectors) are *invariant* under rotations, we say that $R(\phi)$ is *unitary*, defined by the property (3.129).

Generalized to complex valued matrices, we say that A is unitary if

$$A^\dagger A = I, \tag{3.131}$$

by which A is norm and angle preserving following (3.126–3.130) with $\dagger$ replacing T. In a unitary matrix, therefore, the columns and rows form orthonormal sets. This is evident by inspection in the rotation matrix $R(\varphi)$: its row

$$\mathbf{r}_1 = (\cos\varphi\ -\sin\varphi),\ \mathbf{r}_2 = (\sin\varphi\ \ \cos\varphi) \tag{3.132}$$

and column vectors

$$\mathbf{c}_1 = \begin{pmatrix} \cos\varphi \\ \sin\varphi \end{pmatrix},\ \mathbf{c}_2 = \begin{pmatrix} -\sin\varphi \\ \cos\varphi \end{pmatrix} \tag{3.133}$$

satisfy

$$\mathbf{r}_i\mathbf{r}_j^T = \delta_{ij},\ \ \mathbf{c}_i^T\mathbf{c}_j = \delta_{ij}, \tag{3.134}$$

where δ_{ij} denotes the Kronecker delta symbol ($\delta_{ij} = 1\ (i = j)$, $\delta_{ij} = 0\ (i \neq j)$).

The eigenvalues of a unitary matrix (3.131) are on the unit circle, as follows from

$$\mathbf{a}^\dagger\mathbf{a} = \mathbf{a}^\dagger A^\dagger A\mathbf{a} = (\lambda\mathbf{a})^\dagger(\lambda\mathbf{a}) = |\lambda|^2\mathbf{a}^\dagger\mathbf{a}, \tag{3.135}$$

where $\mathbf{a}$ denotes an eigenvector of the eigenvalue λ.

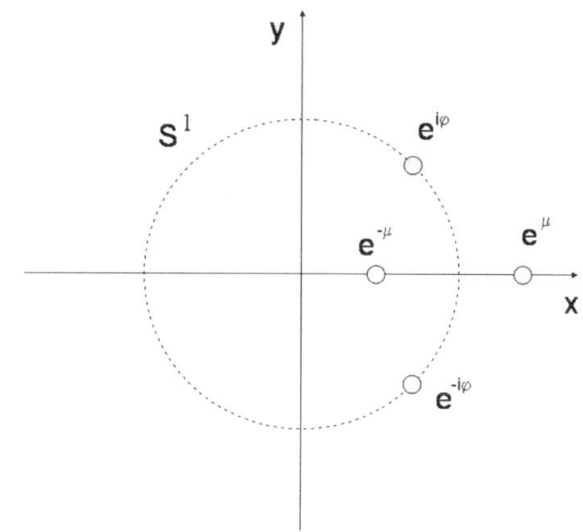

Fig. 3.8 Show are the eigenvalues $\lambda_\pm = e^{\pm i\varphi}$ of the rotation matrix $R(\varphi)$ on the unit circle S^1. Since R has a complete orthonormal set of eigenvectors, it is unitary. Shown are also $\lambda_\pm = e^{\pm\mu}$ of Λ. Away from S^1, Λ is not unitary

The $n \times n$ unitary matrices are $U(n)$. $U(n)$ is a group in that (i) $C = AB$ is in $U(n)$ for any $A, B \in U(n)$, (ii) every $A \in U(n)$ has an inverse $A^{-1} \in U(n)$ and (hence) (iii) $U(n)$ contains the identity matrix I. (Specifically, $AI = IA = A$.) For any two matrices A and B, we have

$$\det AB = \det A \det \mathrm{B} = \det BA, \tag{3.136}$$

that we state here without proof. (It may be seen from the fact that the determinant equals the product of eigenvalues.) According to (3.131), unitary matrices hereby satisfy

$$|\det A| = 1. \tag{3.137}$$

Elements of $U(n)$ have a complete set of orthonormal eigenvectors with eigenvalues on the unit circle (Fig. 3.8). The *special* unitary group $SU(n) \subset U(n)$ have unit determinant,

$$\det A = 1, \tag{3.138}$$

exemplified by the rotation matrices $R(\varphi)$ in (3.129).

In contrast, Hermitian matrices have a complete set of orthonormal eigenvectors with eigenvalues on the real axis. A matrix can be both unitary and Hermitian only if its eigenvalues are ± 1. Examples are $n \times n$ Householder matrices representing reflections across the plane normal to $\mathbf{u}$ in $\mathbb{R}^n$,

$$H = I - 2\mathbf{u}\mathbf{u}^\dagger, \ \mathbf{u}^\dagger\mathbf{u} = 1. \tag{3.139}$$

3.8 Hermitian Structure of Minkowski Spacetime

A Hermitian $n \times n$ matrix A on $\mathbb{C}^n$ introduces a metric structure through an inner product defined by their real eigenvalues λ_i,

$$(\mathbf{a}, \mathbf{b}) = \mathbf{a}^\dagger A \mathbf{b} = \sum_{i=1}^{n} \lambda_i \mathbf{a}^\dagger \mathbf{b}. \tag{3.140}$$

If all eigenvalues are positive, this metric structure introduces a norm equivalent to the Euclidean norm on $\mathbf{R}^{2n}$,

$$|a|_* = \sqrt{(\mathbf{a}, \mathbf{a})} = \sqrt{\sum_{i=1}^{n} \lambda_i \, |\mathbf{a}|^2}. \tag{3.141}$$

The Lorentz metric of Sect. 1.5 is an example of a real-symmetric matrix on $\mathbb{R}^4$ with signature $(1, -1, -1, -1)$, referring to one positive and three negative eigenvalues. The metric structure it introduces follows (3.140) with A given by η_{ab}, that is referred to as hyperbolic rather than Euclidean. We next strengthen this association to Hermitian matrices with some interesting consequences.

By dimension, we are at liberty to introduce complex combinations of the real 3+1 space-time components of a vector in terms of two complex-valued component vectors. Embedding the latter into a 2×2 Hermitian matrix, (a) the Lorentz metric obtains by the determinant of the matrix and (b) Lorentz transformations correspond to unitary transformations by unimodular matrices from SL(2,$\mathbb{C}$). Remarkably, the unimodular matrices giving a unitary transformation are effectively square roots of Lorentz transformations of four-vectors. For rotations on the unit sphere,[8] it gives a double cover of the rotations on the unit sphere S^2.[9]

The causal structure of Minkowski spacetime, defined by the Lorentz metric, is given geometrically by Lorentz invariant light cones. The generators of light cones are light rays. Light rays are integral curves of null-vectors with length zero. This refers to the fact that the change in total phase along a light ray is zero by definition—light rays define the propagation of wave fronts carrying constant total phase of electromagnetic radiation. They carry information on direction, but not distance. Projection of light rays onto the celestial sphere defines a one-to-one map of directions onto S^2. *Light rays hereby have two degrees of freedom,*[10] and are either future- or past-oriented with opposite signs of their angular velocity in the propagation of an electromagnetic wave. Since the dimension of Minkowski space is four, this suggests a formulation in two null-vectors.

[8]The celestial sphere in the language of cosmology.

[9]Commonly referred to as SO(3), described by rotation matrices with determinant +1.

[10]Photons carry an additional degree of freedom in polarization.

Expressed in terms of complex variables, a 2×2 formulation is realized by *spinors* of $\left(\epsilon_{AB}, \mathbb{C}^2\right)$, where ϵ_{AB} refers to the metric spinor as follows.

Given a four-vector $k^b = (k^t, k^x, k^y, k^z)$, consider the Hermitian matrix

$$K = \begin{pmatrix} k^t + k^z & k^x - ik^y \\ k^x + ik^y & k^t - k^z \end{pmatrix} = k^t\sigma_t + k^x\sigma_x + k^y\sigma_y + k^z\sigma_z, \tag{3.142}$$

expanded in terms of the Hermitian Pauli spin matrices[11]

$$\sigma_t = I,\ \sigma_x = \begin{pmatrix} 0 & 1 \\ 1 & 0 \end{pmatrix},\ \sigma_y = \begin{pmatrix} 0 & -i \\ i & 0 \end{pmatrix},\ \sigma_z = \begin{pmatrix} 1 & 0 \\ 0 & -1 \end{pmatrix} \tag{3.143}$$

with respective eigenvalues $\lambda = \pm 1$, $\lambda = \pm i$, $\lambda = \pm i$ and $\lambda = \pm i$. The Pauli spin matrices embed the basis vectors of Minkowski space,

$$\begin{aligned} \sigma_t &= K\left\{\begin{pmatrix} 1 \\ 0 \\ 0 \\ 0 \end{pmatrix}\right\},\quad \sigma_x = K\left\{\begin{pmatrix} 0 \\ 1 \\ 0 \\ 0 \end{pmatrix}\right\}, \\ \sigma_y &= K\left\{\begin{pmatrix} 0 \\ 0 \\ 1 \\ 0 \end{pmatrix}\right\},\quad \sigma_z = K\left\{\begin{pmatrix} 0 \\ 0 \\ 0 \\ 1 \end{pmatrix}\right\}. \end{aligned} \tag{3.144}$$

Notice that the σ_i $(i = x, y, z)$ are trace-free. From the determinant of Z,

$$\det K = \left(k^t\right)^2 - \left(k^x\right)^2 - \left(k^y\right)^2 - \left(k^z\right)^2, \tag{3.145}$$

the length of k^b in Minkowski space satisfies

$$s^2 = \det K, \tag{3.146}$$

incorporating the line-element $s^2 = \eta_{ab}k^a k^b$ with Minkowski metric

$$\eta_{ab} = \begin{pmatrix} 1 & 0 & 0 & 0 \\ 0 & -1 & 0 & 0 \\ 0 & 0 & -1 & 0 \\ 0 & 0 & 0 & -1 \end{pmatrix}. \tag{3.147}$$

Here, we use the Einstein summation convention of summing over all index values $a = t, x, y, z$ in combinations of covariant and contravariant indices. In Exercise 1.11, we noticed that η_{ab} reduced to 1+1 is invariant under Lorentz boosts. It is

[11]Wolfgang Pauli 1900–1958.

not difficult to ascertain that η_{ab} is invariant under general Lorentz transformations including rotations. As such, η_{ab} is a Lorentz invariant tensor.

Consider an element $L \epsilon$ SL(2,$\mathbb{C}$), mentioned above. These unimodular elements have $8-2=6$ degrees of freedom. Then

$$K \to L^\dagger K L \tag{3.148}$$

preserves (3.146), since

$$\det L^\dagger K L = \det K. \tag{3.149}$$

Notice that (3.149) holds true also for $L = -I$, showing that the sign of L is not determined by a given Lorentz transformation of k^b. Even so, a given L from SL(2,$\mathbb{C}$) in (3.148) defines a Lorentz transformation of k^b.

Example 3.6. Consider a Lorentz boost with rapidity μ of $k^b = (1,0,0,0)^T$ to $(\cosh\mu, 0, 0, \sinh\mu)^T$ along the z-axis,

$$\begin{pmatrix} 1 & 0 \\ 0 & 1 \end{pmatrix} \to \begin{pmatrix} \cosh\mu - \sinh\mu & 0 \\ 0 & \cosh\mu + \sinh\mu \end{pmatrix} = \begin{pmatrix} e^{-\mu} & 0 \\ 0 & e^{\mu} \end{pmatrix}. \tag{3.150}$$

It obtains by a boost

$$L(\mu) = \begin{pmatrix} e^{-\frac{1}{2}\mu} & 0 \\ 0 & -e^{\frac{1}{2}\mu} \end{pmatrix}, \tag{3.151}$$

whereas a rotation of $k^b = (0,1,0,0)$ to $(0, \cos\theta, \sin\theta, 0)$ about the z-axis,

$$\begin{pmatrix} 0 & 1 \\ 1 & 0 \end{pmatrix} \to \begin{pmatrix} 0 & \cos\theta + i\sin\theta \\ \cos\theta - i\sin\theta & 0 \end{pmatrix} = \begin{pmatrix} 0 & e^{i\theta} \\ e^{-i\theta} & 0 \end{pmatrix}, \tag{3.152}$$

obtains by a rotation

$$L(\theta) = \frac{1}{\sqrt{2}} \begin{pmatrix} e^{i\frac{1}{2}\theta} & ie^{-i\frac{1}{2}\theta} \\ ie^{i\frac{1}{2}\theta} & e^{-i\frac{1}{2}\theta} \end{pmatrix}. \tag{3.153}$$

Viewed by continuation starting from the identity matrix, (3.153) goes at the heart of spinors to be introduced below: a rotation over 2π in physical space gives rise to a change in sign in $L(\theta)$. A continuing rotation over 4π restores the original sign. The $L(\theta)$ in (3.153) are elements of SU(2), since $L^\dagger L = I$. Accordingly, SU(2) is a two-fold cover of SO(3).

Light cones are described by null-rays k^b, satisfying

$$s^2 = \det K = 0. \tag{3.154}$$

Their embedding (3.142) is therefore in rank-one matrices[12] (3.142) of the form

$$Z = \begin{pmatrix} \xi\bar{\xi} & \eta\bar{\xi} \\ \bar{\eta}\xi & \eta\bar{\eta} \end{pmatrix} = \begin{pmatrix} \bar{\xi} \\ \bar{\eta} \end{pmatrix} \left(\xi\ \eta\right), \tag{3.155}$$

whose determinant is identically equal to zero. Here, right hand side expresses the *spinor* and its transpose indicated by a primed index

$$\kappa^A = \left(\xi\ \eta\right), \quad \kappa^{A'} = \begin{pmatrix} \xi \\ \eta \end{pmatrix} \quad (\xi, \eta \in \mathbb{C}) \tag{3.156}$$

following the convention, to using unprimed and primed indices for a row and, respectively, column vector notation. Accordingly, we write

$$Z^{AA'} = \kappa^A \bar{\kappa}^{A'}, \tag{3.157}$$

where $\bar{\kappa}^{A'}$ is the Hermitian transpose of κ^A.

Let κ denote the row vector κ^A in (3.156). Then (3.148) implies a corresponding Lorentz transformation of a spinor κ,

$$Z \to (\kappa L)^\dagger (\kappa L): \quad \kappa \to \kappa L. \tag{3.158}$$

Rotation over 2π in real space now has a corresponding sign change in the spinor.

Now write the determinant of $K = K^{AA'}$ in (3.142) as

$$\det K = K^{11}K^{22} - K^{21}K^{12} \equiv \epsilon_{AB}\epsilon_{A'B'}K^{AA'}K^{BB'} \tag{3.159}$$

in terms of the anti-symmetric metric spinor $\epsilon_{AB} = -\epsilon_{BA}$, $\epsilon_{A'B'} = -\epsilon_{B'A'}$ with $\epsilon_{01} = \epsilon_{0'1'} = 1$, i.e.,

$$\epsilon_{AB} = \begin{pmatrix} 0 & 1 \\ -1 & 0 \end{pmatrix}, \quad \epsilon_{A'B'} = \begin{pmatrix} 0 & 1 \\ -1 & 0 \end{pmatrix}. \tag{3.160}$$

Then

$$\epsilon_{AB}\epsilon_{A'B'}K^{AA'}K^{BB'} = g_{ab}k^ak^b: \quad \epsilon_{AB}\epsilon_{A'B'} = g_{ab} \tag{3.161}$$

with implicit reference to the basis elements (3.170) to be discussed below. In (3.159), note that the incomplete contraction $\epsilon_{A'B'}K^{AA'}K^{BB'}$ is an antisymmetric tensor in

[12]The image space is comprised of multiples of one vector $\left(\xi\ \eta\right)^\dagger$.

our two-dimensional spinor space $\mathbb{C}^2$. Since the antisymmetric elements of $L(2,\mathbb{C})$ are spanned by ϵ_{AB},

$$\epsilon_{A'B'}K^{AA'}K^{BB'} = \frac{1}{2}\epsilon^{AB}\det K, \tag{3.162}$$

taking into account (3.159) and $\epsilon_{AB}\epsilon^{AB} = 2$.

The metric spinor allows lowering and raising indices

$$\kappa_A = \kappa^B\epsilon_{BA}, \quad \kappa^B = \epsilon^{BA}\kappa_A. \tag{3.163}$$

Lowering and raising is by multiplication from the left and, respectively, right. The same rules apply to A'. Since the metric spinor is skew symmetric, we automatically have that spinors are null,

$$\kappa^A\kappa_A = \kappa^A\kappa^B\epsilon_{BA} = \kappa^2\kappa^1 - \kappa^1\kappa^2 = 0. \tag{3.164}$$

In practical terms, the spinor κ^A is a square root of a null-vector k^b in $Z^{AA'}$.

Consider two null-vectors k^b and l^b represented by spinors o^A and ι^A. Then

$$k^c l_c = \left(o^A\bar{o}^{A'}\right)(\iota_A\bar{\iota}_{A'}) = \left(o^A\iota_A\right)\left(o^A\iota_A\right)^\dagger = \left|o^A\iota_A\right|^2 \geq 0, \tag{3.165}$$

whereby $k^c l_c \geq 0$, i.e., k^b and l^b share the same direction in time, e.g., are future-oriented. Choosing two distinct null-vectors, we may insist

$$k^c l_c = 1: \;\; o^A\iota_A = 1. \tag{3.166}$$

As members of $\mathbb{C}^2$, choosing such pair as a basis gives

$$\epsilon_{AB} = o_A\iota_B - \iota_A o_B = \begin{pmatrix} 0 & 1 \\ -1 & 0 \end{pmatrix}. \tag{3.167}$$

To be explicit, consider

$$o^A = \begin{pmatrix} 1 & 0 \end{pmatrix}, \;\; \iota^A = \begin{pmatrix} 0 & 1 \end{pmatrix}. \tag{3.168}$$

Then $\iota_A = \epsilon_{AB}\iota^B = \begin{pmatrix} 1 & 0 \end{pmatrix}$, whereby (3.166) is satisfied, and

$$\begin{aligned} o^A\bar{o}^{A'} &= \begin{pmatrix} 1 & 0 \\ 0 & 0 \end{pmatrix}, \iota^A\bar{\iota}^{A'} = \begin{pmatrix} 0 & 0 \\ 0 & 1 \end{pmatrix}, \\ o^A\bar{\iota}^{A'} &= \begin{pmatrix} 0 & 1 \\ 0 & 0 \end{pmatrix}, \iota^A\bar{o}^{A'} = \begin{pmatrix} 0 & 0 \\ 1 & 0 \end{pmatrix}. \end{aligned} \tag{3.169}$$

The result identifies the Pauli spin-matrices with metric spin-tensors

$$\begin{aligned}\sigma_t^{AA'} &= o^A\bar{o}^{A'} + \iota^A\bar{\iota}^{A'}, \quad \sigma_x^{AA'} = o^A\bar{\iota}^{A'} + \iota^A\bar{o}^{A'},\\ \sigma_y^{AA'} &= i\left(o^A\bar{\iota}^{A'} - \iota^A\bar{o}^{A'}\right), \ \sigma_z^{AA'} = o^A\bar{o}^{A'} - \iota^A\bar{\iota}^{A'}.\end{aligned} \tag{3.170}$$

In the notation of linear algebra, note that $\bar{o}^{A'} = (o^A)^\dagger$, etc. It recovers Pauli matrices in (3.142) as a basis of four-vectors in 3+1 Minkowski space. Note that (3.170) also introduces an algebraic map of a complex second-rank spinor, e.g., $\phi_{AA'}$, to four-vectors with possibly complex valued components.

3.9 Eigenvectors of Hermitian Matrices

For $n \times n$ Hermitian matrix A, $A^\dagger = A$, let $(\lambda, \mathbf{a})$ denote one of its eigenvalue-eigenvector pairs. The latter always exists by virtue of eigenvalue solutions to (3.91). Let

$$\hat{\mathbf{a}} = \frac{\mathbf{a}}{\sqrt{\mathbf{a}^\dagger\mathbf{a}}} \tag{3.171}$$

denote the normalized eigenvector, satisfying $\hat{\mathbf{a}}^\dagger\mathbf{a} = 1$. For instance, we have

$$\mathbf{a} = \begin{pmatrix}1\\1\end{pmatrix} \rightarrow \hat{\mathbf{a}} = \frac{1}{\sqrt{2}}\begin{pmatrix}1\\1\end{pmatrix}. \tag{3.172}$$

Let $\mathbf{u}$ be any vector. We may decompose it orthogonally as

$$\mathbf{u} = \mathbf{u}_{||} + \mathbf{u}_\perp. \tag{3.173}$$

Here, $\mathbf{u}_{||}$ and $\mathbf{u}_\perp$ are parallel and orthogonal to $\mathbf{a}$, obtained from the projection operator

$$P = I - \hat{\mathbf{a}}\hat{\mathbf{a}}^\dagger : \ \mathbf{u}_\perp = P\mathbf{u}, \ \mathbf{u}_{||} = (I - P)\mathbf{u}. \tag{3.174}$$

Geometrically, the *image space* of PA consists of all vectors orthogonal to $u_{||}$,

$$\mathrm{Im}PA = \left(u_{||}\right)^\perp. \tag{3.175}$$

We also note that $\mathbf{u}_\perp$ is in the plane with normal $\mathbf{a}$. The expansion (3.173) hereby satisfies

$$\mathbf{u} = (I - P)\mathbf{u} + P\mathbf{u}, \tag{3.176}$$

and hence

$$A\mathbf{u} = \lambda(I - P)\mathbf{u} + AP\mathbf{u}. \tag{3.177}$$

Since $u_{||} = (I - P)\mathbf{u}$, being parallel to $\mathbf{a}$, it is an eigenvector of A with eigenvalue λ. Since $\mathbf{u}$ in (3.177) is arbitrary, it follows that

$$A = \lambda(I - P) + AP. \tag{3.178}$$

Since A is Hermitian with λ real, and $P^\dagger = P$, we have

$$\lambda(I - P) + AP = A = A^\dagger = \lambda(I - P) + PA^\dagger = \lambda(I - P) + PA. \tag{3.179}$$

We thus find that A and P commute, i.e.,

$$AP = PA. \tag{3.180}$$

It follows that in particular that

$$\mathrm{Im}PA = \mathrm{PA} = \left(u_{||}\right)^\perp. \tag{3.181}$$

Since I and A commute trivially, also $(I - P)$ and A commute: $[(I - P), A] = 0$ and the Hermitian matrix A operates completely independently on the one-dimensional subspace of vectors along an eigenvector $\mathbf{a}$ and on the subspace of vectors in the $n - 1$ dimensional hypersurface normal to $\mathbf{a}$. Equation (3.180) also shows that PA is Hermitian:

$$(PA)^\dagger = A^\dagger P^\dagger = AP = PA. \tag{3.182}$$

Therefore, we can repeat all the steps (3.171–3.180) for an eigenvector $\mathbf{a}'$ of $A_1 = AP$. Since $AP\mathbf{a}' = PA\mathbf{a}'$, this eigenvector is in the image space of P, and hence it is orthogonal to $\mathbf{a}$. By this orthogonality, P' associated with $\mathbf{a}'$ commutes with P. It follows that $A_1P' = APP'$ is Hermitian. Continuing in this fashion, we ultimately arrive at n mutually orthogonal eigenvectors $\mathbf{a}, \mathbf{a}', \cdots, \mathbf{a}^{''\cdots'}$.

Example 3.7. The 2×2 Hermitian matrix

$$A = \begin{pmatrix} 2 & 1 \\ 1 & 0 \end{pmatrix} \tag{3.183}$$

has eigenvalue-eigenvector pairs $\{(\lambda_1, \mathbf{a}_1), \ (\lambda_2, \mathbf{a}_2)\}$ with

$$\left\{1 \pm \sqrt{2}, \begin{pmatrix} 1 \pm \sqrt{2} \\ 1 \end{pmatrix}\right\}. \tag{3.184}$$

We wish to view the operation of A on a vector $\mathbf{u}$ as the sum of linearly independent operations associated with the directions $\mathbf{a}_1$ and $\mathbf{a}_2$. We first rewrite (3.185) in terms of the equivalent orthonormal pair

$$\hat{\mathbf{a}}_1 = \frac{\mathbf{a}_1}{\sqrt{\mathbf{a}_1^\dagger \mathbf{a}_1}}, \ \hat{\mathbf{a}}_2 = \frac{\mathbf{a}_2}{\sqrt{\mathbf{a}_2^\dagger \mathbf{a}_2}} \tag{3.185}$$

following (3.171). The $\{\hat{\mathbf{a}}_1, \hat{\mathbf{a}}_2\}$ form an *orthonormal basis* (a *complete* set of orthonormal vectors) for vectors $\mathbf{u}$ in our two-dimensional space. They satisfy the property

$$\hat{\mathbf{a}}_i^\dagger \hat{\mathbf{a}}_j = \delta_{ij} = \begin{cases} 1 \ (i = j) \\ 0 \ (i \neq j) \end{cases} \tag{3.186}$$

Here, δ_{ij} is the commonly used Kronecker delta symbol. For (3.183), we have

$$\hat{\mathbf{a}}_{1,2} = \frac{1}{2(2 \pm \sqrt{2})} \begin{pmatrix} 1 \pm \sqrt{2} \\ 1 \end{pmatrix}. \tag{3.187}$$

For an arbitrary vector, we can write

$$\mathbf{u} = \alpha \hat{\mathbf{a}}_1 + \beta \hat{\mathbf{a}}_2. \tag{3.188}$$

Multiplication by $\hat{\mathbf{a}}_{1,2}$ from the left obtains

$$\mathbf{a}_1^\dagger \mathbf{u} = \alpha \mathbf{a}_1^\dagger \hat{\mathbf{a}}_1 + \beta \hat{\mathbf{a}}_1^\dagger \mathbf{a}_2 = \alpha \tag{3.189}$$

$$\mathbf{a}_2^\dagger \mathbf{u} = \alpha \mathbf{a}_2^\dagger \hat{\mathbf{a}}_1 + \beta \hat{\mathbf{a}}_2^\dagger \mathbf{a}_2 = \beta. \tag{3.190}$$

Substitution of (3.190) into (3.188) gives the explicit expression

$$\mathbf{u} = \hat{\mathbf{a}}_1 \hat{\mathbf{a}}_1^\dagger \mathbf{u} + \hat{\mathbf{a}}_2 \hat{\mathbf{a}}_2^\dagger \mathbf{u}. \tag{3.191}$$

This represents the *Gram-Schmidt* orthogonal decomposition of $\mathbf{u}$ with respect to the eigenvectors of A. Accordingly, $A\mathbf{u}$ satisfies

$$A\mathbf{u} = \alpha A \hat{\mathbf{a}}_1 + \beta A \hat{\mathbf{a}}_2 = \alpha \lambda_1 \hat{\mathbf{a}}_1 + \beta \lambda_2 \hat{\mathbf{a}}_2 = \lambda_1 \hat{\mathbf{a}}_1 \hat{\mathbf{a}}_1^\dagger \mathbf{u} + \lambda_2 \hat{\mathbf{a}}_2 \hat{\mathbf{a}}_2^\dagger \mathbf{u}. \tag{3.192}$$

Since $\mathbf{u}$ is arbitrary, we conclude

$$A = \lambda_1 \hat{\mathbf{a}}_1 \hat{\mathbf{a}}_1^\dagger + \lambda_2 \hat{\mathbf{a}}_2 \hat{\mathbf{a}}_2^\dagger. \tag{3.193}$$

The same follows from $A = AI$, $I = \hat{\mathbf{a}}_1 \hat{\mathbf{a}}_1^\dagger + \hat{\mathbf{a}}_2 \hat{\mathbf{a}}_2^\dagger$.

Example 3.8. For Λ in (3.80) we have, according to (3.124, 3.125), a normalized pair of eigenvalues-eigenvectors given by

$$\left\{ e^{\pm\mu}, \frac{1}{\sqrt{2}} \begin{pmatrix} 1 \\ \pm 1 \end{pmatrix} \right\}. \tag{3.194}$$

Following (3.194), we consider

$$\begin{aligned} \lambda_1 \hat{\mathbf{a}}_1 \hat{\mathbf{a}}_1^\dagger &= \tfrac{e^\mu}{\sqrt{2}} (1 \;\; 1) \tfrac{1}{\sqrt{2}} \begin{pmatrix} 1 \\ 1 \end{pmatrix} = \tfrac{e^\mu}{\sqrt{2}} \begin{pmatrix} 1 & 1 \\ 1 & 1 \end{pmatrix}, \\ \lambda_2 \hat{\mathbf{a}}_2 \hat{\mathbf{a}}_2^\dagger &= \tfrac{e^{-\mu}}{\sqrt{2}} (1 \;\; {-1}) \tfrac{1}{\sqrt{2}} \begin{pmatrix} 1 \\ -1 \end{pmatrix} = \tfrac{e^{-\mu}}{\sqrt{2}} \begin{pmatrix} 1 & -1 \\ -1 & 1 \end{pmatrix}. \end{aligned} \tag{3.195}$$

Adding these expressions gives

$$\frac{e^\mu}{2} \begin{pmatrix} 1 & 1 \\ 1 & 1 \end{pmatrix} + \frac{e^{-\mu}}{2} \begin{pmatrix} 1 & -1 \\ -1 & 1 \end{pmatrix} = \frac{1}{2} \begin{pmatrix} e^\mu + e^{-\mu} & e^\mu - e^{-\mu} \\ e^\mu - e^{-\mu} & e^\mu - e^{-\mu} \end{pmatrix}, \tag{3.196}$$

i.e., we recover our definition of a Lorentz boost,

$$A = \begin{pmatrix} \cosh\mu & \sinh\mu \\ \cosh\mu & \sinh\mu \end{pmatrix}. \tag{3.197}$$

3.10 QR Factorization

In viewing an $n \times m$ matrix A as a *linear map* from the vector space $\mathbb{C}^m$ to $\mathbb{C}^n$, we frequently encounter the question if A is imaging $\mathbb{C}^m$ onto all of $\mathbb{C}^n$ or just a linear subspace of it. Similarly, A may map all nonzero vectors from $\mathbb{C}^m$ to nonzero vectors or map some linear subspace of $\mathbb{C}^m$ to the origin $\mathbf{0}$ in $\mathbb{C}^n$.

To streamline this discussion, we introduce the image of A, defined by the linear vector space

$$\mathrm{Im}\, A = \{\mathbf{v} \mid \mathbf{v} = A\mathbf{u}\,,\, \mathbf{u} \in \mathbb{C}^m\} \tag{3.198}$$

and the *kernel* of A, also known as the null space of A, defined by the linear vector space

$$\text{Ker}\, A = \{\mathbf{u} \mid A\mathbf{u} = \mathbf{0}, \mathbf{u} \in \mathbb{C}^m\}. \tag{3.199}$$

The image space is supported by the column vectors $\mathbf{a}_1, \mathbf{a}_2, \ldots, \mathbf{a}_m$ of A, e.g., for a 2×2 matrix

$$A = \begin{pmatrix} | & | \\ | & | \end{pmatrix} = (\mathbf{a}_1\ \mathbf{a}_2)\,, \tag{3.200}$$

The image space forms out of linear combinations

$$A\mathbf{u} = u_1\mathbf{a}_1 + u_2\mathbf{a}_2 + \cdots + u_m\mathbf{a}_m \tag{3.201}$$

by choice of $\mathbf{u}$ in $\mathbb{C}^m$,

$$\mathbf{u} = \begin{pmatrix} u_1 \\ u_2 \\ \cdot \\ u_m \end{pmatrix}. \tag{3.202}$$

The row space is supported by the row vectors $\mathbf{b}_1, \mathbf{b}_2, \cdots, \mathbf{b}_n$ of A, e.g., for a 2×2 matrix

$$A = \begin{pmatrix} \text{———} \\ \text{———} \end{pmatrix} = \begin{pmatrix} \mathbf{b}_1^T \\ \mathbf{b}_2^T \end{pmatrix} \tag{3.203}$$

that forms out of the linear combinations

$$\mathbf{r} = v_1\mathbf{b}_1 + v_2\mathbf{b}_2 + \cdots + v_n\mathbf{b}_n. \tag{3.204}$$

by choice of $\mathbf{v}$ in $\mathbb{C}^m$,

$$\mathbf{v} = \begin{pmatrix} v_1 \\ v_2 \\ \cdot \\ v_n \end{pmatrix}. \tag{3.205}$$

Following (3.199), the kernel of A consists of the vectors that are orthogonal to all of its row vectors $\mathbf{b}_i$ $(i = 1, 2, \ldots n)$, commonly written as

$$\text{Ker}\, A = \left(\text{Im}\, A^\dagger\right)^\perp, \tag{3.206}$$

where $\perp$ refers to orthogonality with respect to the inner product $\mathbf{a} \cdot \mathbf{b} = \mathbf{a}^\dagger\mathbf{b}$ for vectors in $\mathbb{C}^m$.

3.10.1 Examples of Image and Null Space

Let A be a nonsingular 2×2 matrix. We read off the columns vectors following (3.200), e.g.,

$$A = \begin{pmatrix} 1 & 2 \\ 2 & 0 \end{pmatrix}: \ \mathbf{a}_1 = \begin{pmatrix} 1 \\ 2 \end{pmatrix}, \ \mathbf{a}_2 = \begin{pmatrix} 2 \\ 0 \end{pmatrix}. \tag{3.207}$$

Hence, Im A is defined by all vectors obtained from linear combinations of the $\mathbf{a}_1$ and $\mathbf{a}_2$ in (3.207). We say

$$\text{Im } A = \text{span}\{\mathbf{a}_1, \mathbf{a}_2\} = \text{span}\left\{\begin{pmatrix} 1 \\ 2 \end{pmatrix}, \begin{pmatrix} 1 \\ 0 \end{pmatrix}\right\}. \tag{3.208}$$

Evidently, we have

$$\text{Im } A = \mathbb{R}^2 \tag{3.209}$$

(or $\mathbb{C}^2$) since the $\mathbf{a}_1$ and $\mathbf{a}_2$ in (3.207) point in different directions, whereby they are linearly independent. This may also be inferred from the fact that det $A \neq 0$.

Alternatively, consider the singular matrix

$$B = \begin{pmatrix} 1 & 2 \\ 2 & 4 \end{pmatrix}: \ \mathbf{a}_1 = \begin{pmatrix} 1 \\ 2 \end{pmatrix}, \ \mathbf{a}_2 = \begin{pmatrix} 2 \\ 4 \end{pmatrix}. \tag{3.210}$$

In this event, the second column satisfies

$$\mathbf{a}_2 = 2\mathbf{a}_1 \tag{3.211}$$

and the two columns are linearly dependent, as follows also from the fact that

$$\det B = 0, \tag{3.212}$$

Consequently, the image space is the one dimensional subspace given by

$$\text{Im } B = \text{span}\{\mathbf{a}_1, \mathbf{a}_2\} = \text{span}\left\{\begin{pmatrix} 1 \\ 2 \end{pmatrix}\right\}. \tag{3.213}$$

Proceeding with (3.207) above, we have, following (3.203), the row vectors

$$A = \begin{pmatrix} 1 & 2 \\ 2 & 0 \end{pmatrix}: \ \mathbf{b}_1 = \begin{pmatrix} 1 \\ 2 \end{pmatrix}, \ \mathbf{b}_2 = \begin{pmatrix} 2 \\ 0 \end{pmatrix}. \tag{3.214}$$

They happen to be the same as the column vectors since A is real-symmetric. Ker A is defined by vectors that are orthogonal to both $\mathbf{b}_1$ and $\mathbf{b}_2$. Since $\mathbf{b}_1$ and $\mathbf{b}_2$ are linearly independent, we have

$$\text{Ker}\, A = \mathbf{0}. \tag{3.215}$$

For the alternative (3.210), we have the row vectors

$$B = \begin{pmatrix} 1 & 2 \\ 2 & 4 \end{pmatrix}: \ \mathbf{b}_1 = \begin{pmatrix} 1 \\ 2 \end{pmatrix}, \ \mathbf{b}_2 = \begin{pmatrix} 2 \\ 4 \end{pmatrix}. \tag{3.216}$$

In this event, the second row satisfies $\mathbf{b}_2 = 2\mathbf{b}_1$, whereby they are linearly dependent. Consequently, the null space of A is the one dimensional subspace, given by the vectors orthogonal to $\mathbf{b}_1$, i.e.,

$$\text{Ker}\, B = \begin{pmatrix} 1 \\ 2 \end{pmatrix}^{\perp} = \text{span}\left\{\begin{pmatrix} -2 \\ 1 \end{pmatrix}\right\}. \tag{3.217}$$

The matrices (3.207) and (3.210) satisfy, respectively,

$$\begin{aligned} \dim \text{Im}\, A + \dim \text{Ker}\, A &= 2 + 0 = 2, \\ \dim \text{Im}\, B + \dim \text{Ker}\, B &= 1 + 1 = 2. \end{aligned} \tag{3.218}$$

3.10.2 *Dimensions of Image and Null Space*

In what follows, we will restrict our discussion to square matrices of size $n \times n$. In this event, the dimension $\dim \text{Im}\, A$ of (3.198) is n whenever A is of full rank, i.e., when $\det A \neq 0$. Complementary to this, we have that $\dim \text{Ker}\, A$ of (3.199) is 0 whenever A is of full rank, i.e., when $\det A \neq 0$. However, $\dim \text{Im}\, A < n$ and $\dim \text{Ker}\, A > 0$ when $\det A = 0$.

The matrices (3.207) and (3.210) exemplify a general relationship of $n \times n$ matrices, satisfying

$$\dim \text{Im}\, A + \dim \text{Ker}\, A = n \tag{3.219}$$

To derive this relationship, we begin by observing the invariance

$$\text{Im}\, A = \text{Im}\, A' \tag{3.220}$$

for $A' = \left(\mathbf{a}_1' \ \mathbf{a}_2' \ \dots \ \mathbf{a}_n'\right)$ obtained from $A = (\mathbf{a}_1 \ \mathbf{a}_2 \ \dots \ \mathbf{a}_n)$ by changing a column vector $\mathbf{a}_j$ by linear superposition with any of the other column vectors. Specifically, this may be by choice of $1 \leq j \leq n$ and a linear superposition

$$\mathbf{a}_i' = \mathbf{a}_i \ (i \neq j), \ \ \mathbf{a}_j' = \mathbf{a}_j - \sum_{i=1}^{j-1} \mu_i \mathbf{a}_i. \tag{3.221}$$

This transformation has a corresponding upper triangular transformation matrix U such that $A' = AU$. For instance, when $n = 3$ and $j = 2, 3$

$$U_2 = \begin{pmatrix} 1 & -\mu_1 & 0 \\ 0 & 1 & 0 \\ 0 & 0 & 1 \end{pmatrix}, \ \ U_3 = \begin{pmatrix} 1 & 0 & -\mu_1' \\ 0 & 1 & -\mu_2' \\ 0 & 0 & 1 \end{pmatrix}. \tag{3.222}$$

Since the AU_2 and AU_2U_3 form as superpositions of the column vectors of A, their image space remains Im A.

The *Gram-Schmidt orthogonalization* of $\mathbf{a}_j$ from A to mutually orthoginalized $\mathbf{a}_i$ $(1 \leq i \leq j-1)$ satisfies

$$\mathbf{a}_j' = \mathbf{a}_j - \sum_{i=1}^{j-1} \mu_i \mathbf{a}_i, \ \ \mu_i = \frac{\mathbf{a}_j^\dagger \mathbf{a}_i}{\mathbf{a}_i^\dagger \mathbf{a}_i}. \tag{3.223}$$

Here, we omit projections $\mu_i \mathbf{a}_i$ whenever $\mathbf{a}_i = \mathbf{0}$. Performing (3.223) for each $j = 2, 3, \ldots$ consecutively up to $j = n$ produces $A''^{\cdots\prime}$ with column vectors that are all orthogonal. For 2×2 matrix A, the Gram-Schmidt orthogonalization of its colum vectors obtains in one step

$$A' = AU_2 = \begin{pmatrix} \Big| & \Big| \end{pmatrix} \begin{pmatrix} 1 & -\mu_1 \\ 0 & 1 \end{pmatrix}, \tag{3.224}$$

If A has full rank, then so has A'. This may also be seen from the product rule

$$\det A' = \det A \det U = \det A, \tag{3.225}$$

since $\det U = 1$. The determinant of A' is nonzero iff the determinant of A is nonzero. For a 3×3 matrix, we apply (3.222). The product $U = U_2U_3$ is upper triangular,

$$U = U_2U_3 = \begin{pmatrix} 1 & -\mu_1 & -\mu_1' + \mu_1\mu_2' \\ 0 & 1 & -\mu_2' \\ 0 & 0 & 1 \end{pmatrix}. \tag{3.226}$$

The above is readily extended to $n \times n$ matrices

$$A' = AU = AU_2U_3 \cdots U_n \tag{3.227}$$

whose columns are mutually orthogonal, where U is upper triangular with unit determinant.

If $\det A = 0$, some of the columns of A' are zero, i.e.,

$$A' = \left(\mathbf{a}_1\ \mathbf{a}_2\ \ldots\ \mathbf{0}\ \ldots\ \mathbf{a}_j\ \ldots\ \mathbf{0}\ \ldots\ \mathbf{a}_n\right). \tag{3.228}$$

The null vectors of A' are of the form $\mathbf{u}' = (0\ \cdots\ 1\ \cdots 0)^T$, where 1 appears at a position j where $\mathbf{a}_j = \mathbf{0}$. Since U is invertible, $A = A'U^{-1}$, and hence $\mathbf{u} = U\mathbf{u}'$ is a null vector of A. Our theorem (3.219) now readily follows: the number of non-zero columns in A' define the dimension of the image space of A and the number of zero columns of A' define the dimension of the null space of A.

Example 3.9. Consider the non-singular matrix

$$A = \begin{pmatrix} 1 & 2 & 1 \\ 2 & 1 & 0 \\ 0 & 2 & 3 \end{pmatrix}. \tag{3.229}$$

The first and second steps in (3.223) produce, respectively,

$$A' = AU_2 = \begin{pmatrix} 1 & 6/5 & 1 \\ 2 & -3/5 & 0 \\ 0 & 2 & 3 \end{pmatrix}, \quad U_2 = \begin{pmatrix} 1 & -4/5 & 0 \\ 0 & 1 & 0 \\ 0 & 0 & 1 \end{pmatrix}, \tag{3.230}$$

$$A'' = A'U_3 = \begin{pmatrix} 1 & 6/5 & -\frac{20}{29} \\ 2 & -3/5 & \frac{10}{29} \\ 0 & 2 & \frac{15}{29} \end{pmatrix}, \quad U_3 = \begin{pmatrix} 1 & 0 & -1/5 \\ 0 & 1 & -\frac{36}{29} \\ 0 & 0 & 1 \end{pmatrix}. \tag{3.231}$$

It follows that

$$\begin{pmatrix} 1 & 6/5 & -\frac{20}{29} \\ 2 & -3/5 & \frac{10}{29} \\ 0 & 2 & \frac{15}{29} \end{pmatrix} = \begin{pmatrix} 1 & 2 & 1 \\ 2 & 1 & 0 \\ 0 & 2 & 3 \end{pmatrix} \begin{pmatrix} 1 & -4/5 & \frac{23}{29} \\ 0 & 1 & -\frac{36}{29} \\ 0 & 0 & 1 \end{pmatrix}, \tag{3.232}$$

where the second matrix on the right hand side is $U = U_2U_3$. Similarly, consider the singular matrix

$$B = \begin{pmatrix} 1 & 2 & 1 \\ 2 & 4 & 0 \\ 0 & 0 & 3 \end{pmatrix}. \tag{3.233}$$

The first and second step in (3.223) produce, respectively,

$$B' = BU_2 \begin{pmatrix} 1\,0\,1 \\ 2\,0\,0 \\ 0\,0\,3 \end{pmatrix}, \; U_2 = \begin{pmatrix} 1 & -2 & 0 \\ 0 & 1 & 0 \\ 0 & 0 & 1 \end{pmatrix}, \tag{3.234}$$

$$B'' = B'U_3 = \begin{pmatrix} 1 & 0 & 4/5 \\ 2 & 0 & -2/5 \\ 0 & 0 & 3 \end{pmatrix}, \; U_3 = \begin{pmatrix} 1 & 0 & -1/5 \\ 0 & 1 & 0 \\ 0 & 0 & 1 \end{pmatrix}. \tag{3.235}$$

It follows that

$$\begin{pmatrix} 1 & 0 & 4/5 \\ 2 & 0 & -2/5 \\ 0 & 0 & 3 \end{pmatrix} = \begin{pmatrix} 1 & 2 & 1 \\ 2 & 4 & 0 \\ 0 & 0 & 3 \end{pmatrix} \begin{pmatrix} 1 & -2 & -1/5 \\ 0 & 1 & 0 \\ 0 & 0 & 1 \end{pmatrix}, \tag{3.236}$$

where the second matrix on the right hand side is $U = U_1 U_2$.

Comparing (3.232–3.236), we see that A in (3.229) has full rank with the trivial null space $\text{Ker}\, A = \mathbf{0}$, whereas B in (3.233) is of rank 2 with the nontrivial null space given by the second column of $U = U_2 U_3$, i.e.,

$$\text{Ker}\, B = \text{span}\, U \begin{pmatrix} 0 \\ 1 \\ 0 \end{pmatrix} = \text{span} \begin{pmatrix} -2 \\ 1 \\ 0 \end{pmatrix}. \tag{3.237}$$

3.10.3 *QR* Factorization by Gram-Schmidt

The above is more commonly used to derive the QR factorization of a matrix upon including normalization in each step of the Gram-Schmidt procedure,

$$\mathbf{a}'_j \to \hat{\mathbf{a}}_j = \frac{\mathbf{a}'_j}{\sqrt{(\mathbf{a}')^\dagger \mathbf{a}'}} \tag{3.238}$$

if $\mathbf{a}'_j \neq \mathbf{0}$ (otherwise, we skip this step). The result $A = QR$ has column vectors of Q forming an orthonormal bases for Im A and R upper triangular. If A is square and invertible, then Q is unitary, $Q^\dagger Q = I$, whereby $R = Q^\dagger A$.

Example 3.10. Consider the QR factorizations of the non-singular 2×2 matrix. Let $A' = AU_2$ be the outcome of the Gram-Schmidt procedure and D_2 denote the diagonal matrix D_2 containing the norms its column vectors. Then $A' = QD_2$ defines

$$\begin{pmatrix} 1 & 2 \\ 2 & 1 \end{pmatrix} = \begin{pmatrix} \frac{1}{\sqrt{5}} & \frac{2}{\sqrt{5}} \\ \frac{1}{\sqrt{5}} & -\frac{1}{\sqrt{5}} \end{pmatrix} \begin{pmatrix} \sqrt{5} & \frac{4}{\sqrt{5}} \\ 0 & \frac{3}{\sqrt{5}} \end{pmatrix} \equiv QR. \tag{3.239}$$

For a singular matrix, we similarly obtain

$$\begin{pmatrix} 1 & 2 \\ 2 & 4 \end{pmatrix} = \begin{pmatrix} \frac{1}{\sqrt{5}} & \frac{2}{\sqrt{5}} \\ \frac{2}{\sqrt{5}} & -\frac{1}{\sqrt{5}} \end{pmatrix} \begin{pmatrix} \sqrt{5} & 2\sqrt{5} \\ 0 & 0 \end{pmatrix}. \tag{3.240}$$

For the A and B in (3.229) and (3.233), the QR factorizations are

$$A = \begin{pmatrix} 1/5\,\sqrt{5} & \frac{6}{145}\sqrt{145} & -\frac{4}{29}\sqrt{29} \\ 2/5\,\sqrt{5} & -\frac{3}{145}\sqrt{145} & \frac{2}{29}\sqrt{29} \\ 0 & \frac{2}{29}\sqrt{145} & \frac{3}{29}\sqrt{29} \end{pmatrix} \begin{pmatrix} \sqrt{5} & 4/\sqrt{5} & 1/\sqrt{5} \\ 0 & 1/\sqrt{145} & \frac{36}{145}\sqrt{145} \\ 0 & 0 & \frac{5}{29}\sqrt{29} \end{pmatrix}, \tag{3.241}$$

$$B = \begin{pmatrix} 1/\sqrt{5} & 2/\sqrt{5} & 0 \\ 2/\sqrt{5} & -1/\sqrt{5} & 0 \\ 0 & 0 & 1 \end{pmatrix} \begin{pmatrix} \sqrt{5} & 4/\sqrt{5} & 1/\sqrt{5} \\ 0 & 3/\sqrt{5} & 2/\sqrt{5} \\ 0 & 0 & 3 \end{pmatrix}. \tag{3.242}$$

Table 3.1 summarizes this discussion.

3.11 Exercises

3.1. Let

$$\mathbf{a} = \begin{pmatrix} 1 \\ 2 \\ 0 \end{pmatrix}, \ \mathbf{b} = \begin{pmatrix} 0 \\ 1 \\ 2 \end{pmatrix}, \ \mathbf{c} = \begin{pmatrix} 2 \\ 0 \\ 1 \end{pmatrix}. \tag{3.243}$$

Calculate

Table 3.1 Matrices and some symmetry properties

1. Angular momentum is a vector of dimension mass times area per unit time. In flat spacetime, angular momentum is proportional to angular velocity relative to the distant stars. Angular momentum is a conserved quantity. Torque is the time rate of change of angular momentum.

2. An $n \times m$ matrix A of n row and m column vectors may be real-symmetric ($A^T = A$), skew-symmetric ($A^T = -A$), Hermitian ($A^\dagger = A$) or unitary ($A^\dagger A = I$). A real-symmetric or Hermitian matrix has real eigenvalues and eigenvectors with different eigenvalues are orthogonal.

3. The product of eigenvalues is the determinant of a matrix A. If A is unitary ($A^\dagger A = I$) or Hermitian ($A^\dagger = A$), the eigenvalues are on the unit circle, respectively, the real line. The Householder matrix (3.139) exemplifies a unitary and real-symmetric matrix with eigenvalues ± 1. An $n \times n$ Hermitian matrix A has a complete set of orthonormal eigenvectors, i.e., an orthonormal basis of $\mathbb{R}^n$.

4. Every vector $\mathbf{u}$ has a unique Gram-Schmidt decomposition in terms of a sum of vectors along orthonormal eigenvectors $\hat{\mathbf{a}}$ of A. A operates on $\mathbf{u}$ by multiplication along $\hat{\mathbf{a}}$ by its real eigenvalues.

5. The Minkowski metric on four-vectors is invariant under Lorentz transformations, preserving length of and inner products between four-vectors. They are realized by elements of $SL(2, \mathbb{C})$ acting on spinors, given by rows of two complex numbers. Spinors are "square roots" of null-vectors, that change sign following rotation over 2π in Minkowski space.

6. The image space of an $n \times m$ matrix A is spanned by its m column vectors $\mathbf{a}_i$, $\mathrm{Im}\, A = \mathrm{span}\{\mathbf{a}_i\}$, and the null space of A is orthogonal to the span of its n row vectors $\mathbf{b}_i$, $\mathrm{Ker}\, A = (\mathrm{span}\{\mathbf{b}_i\})^\perp$. Square matrices of $n \times n$ satisfy (3.219), $\dim \mathrm{Im}\, A + \dim \mathrm{Ker}\, A = n$. If $\det A = 0$, image space has rank less than n and the null space of A has positive rank: $\dim \mathrm{Im}\, A \leq n - 1, \quad \dim \mathrm{Ker}\, A \geq 1$.

7. Image and null space obtain by Gram-Schmidt orthogonalization on column vectors of a matrix A, producing $A' = AU$ or $A = QR$ by upper triangular matrices U and R satisfying $\mathrm{Im}\, A' = \mathrm{Im}\, A = \mathrm{Im}\, Q$. The null space of A is given by the orthogonal complement to the span of A's columns in U, by zero-columns $\mathbf{a}'_i = \mathbf{0}$ in A'.

$$(i): \mathbf{a} \times \mathbf{b} \cdot \mathbf{c}, \ \ \mathbf{a} \cdot \mathbf{b} \times \mathbf{c}; \ \ (ii): \mathbf{a} \times (\mathbf{b} \times \mathbf{c}), \ \ (\mathbf{a} \times \mathbf{b}) \times \mathbf{c}. \tag{3.244}$$

Compare your answers to (i) and explain.

3.2. Consider a Cartesian coordinate system (x, y, z) and rotation of a vector $\mathbf{r}$ about the z-axis with angular velocity $\omega = \Omega i_z$, where $\mathbf{i}_z$ denotes the unit vector along the z-axis. The velocity of $\mathbf{r}$ satisfies $\mathbf{v} = \omega \times \mathbf{r}$, where $\times$ denotes the outer product.
(i) If $\mathbf{r} = 2\mathbf{i}_x + 3\mathbf{i}_z$, calculate the velocity $\mathbf{v}$.
(ii) Show that $|\mathbf{v}| = \Omega\sigma$, where σ denotes the distance to the axis of rotation.

3.3. Derive the equivalent expression for the dip angle (3.245), given by

$$\sin\theta = \frac{M\sigma^2\omega_p}{J} = \frac{M^2 g\sigma^3}{J^2} = \left(\frac{g}{\sigma}\right)\left(\frac{\sigma}{b}\right)^4 \frac{1}{\omega^2}. \tag{3.245}$$

3.4. For the each of the following transformations in the two-dimensional plane, state which are projections, reflections and rotations:

$$A_1 = \begin{pmatrix} 1 & 0 \\ 0 & 0 \end{pmatrix}, \ A_2 = \begin{pmatrix} 1 & 0 \\ 0 & 1 \end{pmatrix}, \ A_3 = \begin{pmatrix} 1 & 0 \\ 0 & -1 \end{pmatrix}, \ A_4 = \begin{pmatrix} 0 & 1 \\ 1 & 0 \end{pmatrix},$$
$$A_5 = \frac{1}{\sqrt{2}}\begin{pmatrix} 1 & -1 \\ 1 & 1 \end{pmatrix}. \tag{3.246}$$

3.5. Show that complex numbers $z = x + iy$ can be written in terms of the matrices

$$A(z) = \begin{pmatrix} x & -y \\ y & x \end{pmatrix} \tag{3.247}$$

satisfying $A(z)A(w) = A(zw)$ by the rules of matrix multiplication. In particular, show that for $z = i$, (3.247) satisfies $I + A^2 = 0$, where I denotes the identify matrix $(z = 1)$.

3.6. Consider the matrix

$$A = \begin{pmatrix} 1 & 2 \\ 2 & 0 \end{pmatrix}. \tag{3.248}$$

Compute the, determinant, the eigenvalues and eigenvectors.

3.7. Permutation of the x- and y-coordinates is described by

$$\mathbf{z} = x\mathbf{i}_x + y\mathbf{i}_y \rightarrow \mathbf{w} = y\mathbf{i}_x + z\mathbf{i}_y, \ \ w = y + ix. \tag{3.249}$$

Show that $w(z)$ is not analytic in z, i.e., the Cauchy-Riemann relations are not satisfied. Derive the equivalent 2×2 matrix equation for $x' = y$ and $y' = x$.

3.8. Consider the matrix

$$A = \begin{pmatrix} 1 & 2 \\ 2 & 1 \end{pmatrix}. \tag{3.250}$$

Obtain the eigenvalues λ_i and eigenvectors $\mathbf{a}_i$ $(i = 1, 2)$ and decompose A in the form

$$A = \lambda_1 A_1 + \lambda_2 A_2, \quad A_i = \hat{\mathbf{a}}_i \hat{\mathbf{a}}_i^\dagger, \tag{3.251}$$

where hat refers to normalization to unit norm.

3.9. Show the orthogonality (3.119).

3.10. If A is both unitary and Hermitian, show that $A = A^{-1}$.

3.11. Consider the Householder matrix (3.139). Show that H is Hermitian ($H^\dagger = H$) and unitary ($H^\dagger H = I$), whence it is *involuntary* (H is its own inverse): $H^2 = I$. In two dimensions, determine its eigenvalue-eigenvector pairs for a general direction $\mathbf{u}$ and interpret the result geometrically. What happens in three dimensions to the multiplicity of the eigenvalues?

3.12. Let A be Hermitian, i.e., $A^\dagger = A$. Show that A is diagonalizable according to

$$A = U \Lambda U^\dagger, \tag{3.252}$$

where U is the unitary matrix satisfying $U^\dagger U = I$. Compute U for A in (3.250). [*Hint*: Compose U from the eigenvectors of A.]

3.13. Consider the matrix

$$A = \begin{pmatrix} 1 & 2 \\ 2 & 4 + a \end{pmatrix}. \tag{3.253}$$

Compute the determinant and determine the *condition number* of A, defined by the ratio of the maximal to the minimal square root of the eigenvalues of $A^\dagger A$. [*Hint*: use (3.252).] What happens when a approaches zero? Compute the solution to the system of equations

$$A\mathbf{u} = \mathbf{v}. \tag{3.254}$$

Show that the solution is regular, respectively, ill-behaved as a approaches zero when

$$\mathbf{v} = \begin{pmatrix} 1 \\ 2 \end{pmatrix}, \begin{pmatrix} -2 \\ 1 \end{pmatrix}. \tag{3.255}$$

What is the condition number of A^2. How does it generalize to A^n $(n \geq 3)$?

3.14. Show that $U(1)$ can be identified with the tangents of complex numbers on the unit circle S^1. [*Hint*: Express elements of S^1 by $e^{i\theta}$ and generalize the Taylor expansion θ,

$$e^{i\theta} = 1 + i\theta + O\left(\theta^2\right), \tag{3.256}$$

about the identity $\theta = 0$ to arbitrary θ.][13]

3.15. Show that $U(1)$ is *abelian*:

$$AB - BA = 0 \quad (A, B \in U(1)). \tag{3.257}$$

3.16. Show that the elements of $U(2)$ are of the form

$$A = e^{i\theta} \begin{pmatrix} z & -\bar{w} \\ w & \bar{z} \end{pmatrix}, \quad z\bar{z} + w\bar{w} = 1 \tag{3.258}$$

and that they are in general not Hermitian.

3.17. Following (3.258), specialize to det $A = 1$.[14] Give a general representation of $SU(2)$ in terms of *traceless* 2×2 matrices (the sum of diagonal elements being zero). Determine the number of degrees of freedom in view of the conditions $A^\dagger A = I$ and det $A = 1$. Show that these traceless matrices are Hermitian and derive their eigenvalues.

3.18. Illustrate a double cover of S^1 by way of a curve on a two-torus.

3.19. From the definition of the inner product of two arbitrary spinors o^A and ι^A and the definition of lowering indices, show that

$$o^A \iota_A = -o_A \iota^A. \tag{3.259}$$

[13] S^1 is illustrative of a one-dimensional manifold which is compact and simply connected. It has nontrivial topology, since the winding number of a loop in S^1 can take any value in $\mathbb{Z}$. By homotopy, the topology of S^1 is the same as that of the punctured disk $0 < |z| \leq 1$.

[14] As a result, we say $SU(2) \subset U(2) \cong SU(2) \times U(1)$.

3.20. In (3.165), obtain ι^A from a rotation of o^A and evaluate their inner product as a function of the rotation angle. Next, consider a spinor basis

$$o^A = \frac{1}{\sqrt{2}}\left(1\ i\right),\quad \iota^A = \frac{1}{\sqrt{2}}\left(i\ 1\right). \tag{3.260}$$

Show that $o^A \iota_A = 1$ in (3.166). Rank-one matrices of the form $o^A \bar{\iota}^{A'}$ may be expanded as[15]

$$o^A \bar{\iota}^{A'} = \frac{1}{2}\left[\begin{pmatrix} -i \\ 1 \end{pmatrix}\left(1\ i\right)\right]^T = \begin{pmatrix} -i & 1 \\ 1 & i \end{pmatrix}^T = \begin{pmatrix} -i & 1 \\ 1 & i \end{pmatrix}. \tag{3.261}$$

where T denotes the ordinary matrix transpose. Express the Pauli spin matrices in this new basis similar to (3.170).

3.21. Occasionally, we allow coordinates to become complex. For reference, recall the line-element

$$ds^2 = dx^2 + dy^2 = dr^2 + r^2 d\theta^2 \tag{3.262}$$

of the Euclidean plane, expressed in Cartesian and, respectively, polar coordinates. The Euclidean is flat, like an ordinary sheet of paper. Consider[16]

$$ds^2 = -x^2 dt^2 + dx^2. \tag{3.263}$$

Show that (3.263) again is the line-element of a flat two-surface using analytic continuation in t.[17]

3.22. Consider the matrix

$$A = \begin{pmatrix} 1 & 2 \\ 2 & 4+a \end{pmatrix}. \tag{3.264}$$

(i) Obtain the image space and the null space of the matrix for all a. [*Hint*: Distinguish between $a = 0$ and $a \neq 0$.];
(ii) Apply Gram-Schidt orthogonalization to obtain $A' = AU$, where the column factors of A' are orthogonal and U is upper triangular;
(iii) Obtain the QR factorization of A.

---

[15] The symbols $o^A \bar{\iota}^{A'}$ and $\bar{\iota}^{A'} o^A$ are the same, i.e., there is no ordering between unprimed and primed indices. Only upon expansion into a matrix, a choice of ordering is made.

[16] A so-called Rindler space.

[17] It can be shown that flatness is preserved under analytic continuation, whereby the Lorentz metric $ds^2 = -dt^2 + dx^2 = (idt)^2 + dx^2$ is trivially flat.

3.23. Following (3.239), obtain the QR factorizations of the 2×2 rotation matrix $R(\varphi)$ and the Lorentz boost $\Lambda(\mu)$ of Example 3.8.

3.24. Obtain the QR factorizations of general 2×2 matrices that are (a) Hermitian or (b) unitary.

3.25. Let $i = 0, 1, 2, 3$ correspond to (t, x, y, z). For the Pauli spin matrices (3.143), show or calculate
(i) The σ_i are involutory: $\sigma_1^2 = \sigma_2^2 = \sigma_3^2 = -i\sigma_1\sigma_2\sigma_3 = I$; det $\sigma_i = -1$ for $i = 1, 2, 3$ and they are trace-free, $\mathrm{Tr}(\sigma_i) = 0$.
(ii) The σ_i satisfy $\sigma_i\sigma_j + \sigma_j\sigma_i = 2\delta_{ij}I$ $(i, j = 1, 2, 3)$.
(iii) The σ_a $(a = 0, 1, 2, 3)$ form a basis of the 2×2 Hermitian matrices.
(iv) The eigenvalues and eigenvectors of the σ_i $(i = 1, 2, 3)$.
(v) The commutator $[\sigma_i, \sigma_j]$ for all $i, j = 1, 2, 3$.

References

1. Everitt, C.W.F., et al, 2011, Phys. Rev. Lett., 106, 221101, http://www.einstein.stanford.edu.
2. Ciufolini, I., & Pavlis, E.C., 2004, Nature, 431, 958.
3. Kerr, R.P., 1963, Phys. Rev. Lett., 11, 237.
4. van Putten, M.H.P.M., 2005, Nuov. Cim. B, 28, 597; van Putten, M.H.P.M., & Gupta, A.C., 2009, Mon. Not. R. Astron. Soc., 394, 2238.
5. Papapetrou A., 1951, Proc. R. Soc., 209, 248.
6. Feynman, R.P., 1963, *Lectures on Physics*, Vol. I (Addison-Wesley Publishing Co.), Ch. 20.
7. van Putten, M.H.P.M., 2017, NewA, 54, 115, arXiv:1609.07474.

Chapter 4
Linear Partial Differential Equations

Various physical processes and phenomena are described by partial differential equations (PDEs). Their principle part is often linear, which governs the behavior of small amplitude perturbations. The theory of *linear* PDEs is rather well developed, both in formulation and computation. Its classification falls into three groups: *hyperbolic*, *parabolic* and *elliptic* equations.

Illustrative is dynamics of accretion flows. Internal viscosity, possibly augmented by outflows, mediates angular momentum transport outwards, allowing mass flow inwards. The former is diffusive described by a parabolic equation. Instabilities in the disk tend to produce wave motion, described by a hyperbolic equation. Wave motion is of great interest to quasi-periodic oscillations and their propagation may transport angular momentum outwards in accord with the Rayleigh criterion of rotating fluids. The steady-state of parabolic and wave equations satisfies an elliptic equation, that may evolve on a secular time scale of evolution of accretion rate or the central mass.

Below, we review the mathematical structure of these three types of equations and some of the solution methods.

4.1 Hyperbolic Equations

Hyperbolic equations describe wave motion expressed in terms of an amplitude $u = u(t, x, y, z)$ as a function time and space, e.g., the three space dimensions in Cartesian coordinates (x, y, z). Wave motion may derive as a macroscopic feature of systems with a large number of degrees of freedom,[1] like sound waves in air or surface waves in water, or it may be fundamental, as in the theory of electromagnetic waves.

[1]Commonly referred to as *emergent*.

M.H.P.M. van Putten, *Introduction to Methods of Approximation in Physics and Astronomy*, Undergraduate Lecture Notes in Physics,
DOI 10.1007/978-981-10-2932-5_4

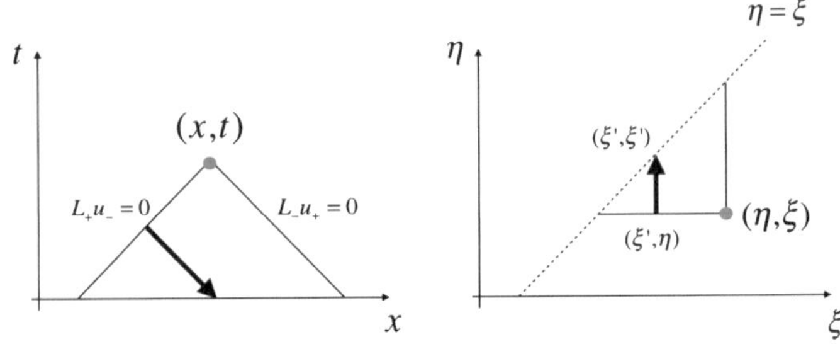

Fig. 4.1 (*Left*) Initial data on amplitude propagate along the characteristics $dx/dt = \pm c$. The complete past domain of dependence D^- of (x, t) is the interval $[x - t, x + t]$, that includes initial data on velocities in (4.7). (*Right*) Domain of integration in deriving $u(x, t)$ from $w(\xi, \eta)$ in the Duhamel integral (4.27)

Small amplitude variations propagate along a *Monge cone*[2] at each point (t, x, y, z) in time and space. For small times, the Monge cone [1] is locally generated by straight lines along the directions of wave motion, generalizing the notion of a light cone in the theory of electromagnetic waves.

By definition, the principle part—ignoring terms lower order in differentiation—of a hyperbolic PDE is of the form

$$u_{tt} = c^2 \Delta u, \tag{4.1}$$

where c denotes the velocity of wave propagation and Δ denotes the Laplace operator. In Cartesian coordinates for flat space, we have

$$\Delta u = u_{xx} + u_{yy} + y_{zz}. \tag{4.2}$$

Here, and in what follows, we use a common notation $u_i = \partial u/\partial x^i$ for partial derivatives and $x^i = (x, y, z)$. In one dimension, (4.1) reduces to

$$u_{tt} = c^2 u_{xx} \tag{4.3}$$

for $u = u(t, x)$.

The one-dimensional wave equation (4.3) can be written as

$$L_+ L_- u = 0, \quad L_\pm = \partial_t \pm \partial_x \tag{4.4}$$

where $L_\pm$ refer to the *left* and *right* movers, defined by the *simple waves* of the form (Fig. 4.1)

$$L_\pm u_\mp = 0: \quad u_+ = \Phi_+(x + t), \quad u_- = \Phi_-(x - t). \tag{4.5}$$

[2]Gaspard Monge (1746–1818).

The general Initial Value Problem (IVP) for (4.4) is set by initial data

$$u(0,x) = f(x), \quad u_t(0,x) = g(x). \tag{4.6}$$

Together, (4.4) and (4.6) completely define the solution for all time. The explicit analytic solution of d'Alembert is instructive,[3] given by

$$u(t,x) = \frac{1}{2}\left[f(x+t) + f(x-t)\right] + \frac{1}{2}\int_{x-t}^{x+t} g(s)ds, \tag{4.7}$$

where we normalize to $c = 1$ for ease of notation. This solution brings about the *past domain of dependence* D^-, given by the triangle with base $[x-t, x+t]$ on the $x-$axis and edges given by the *characteristics* with directions

$$\frac{dx}{dt} = \pm 1 \tag{4.8}$$

passing through the vertex at (x,t). The propagation of initial data $f(x)$ on the wave amplitude along the edges (4.8) and it invokes the initial data $g(x)$ restricted to the finite base $[x-t, x+t]$ of D^-. These are causality constraints that are defining properties of hyperbolic equations.

The d'Alambert formula (4.7) can be verified by direct substitution into (4.6). It can also be derived by first aligning the coordinates with the directions (4.8) by a rotation (Fig. 4.1),

$$\xi = x + t, \quad \eta = x - t. \tag{4.9}$$

Up to a factor $1/\sqrt{2}$, (4.9) will be recognized as a coordinate rotation over 45°. In these coordinates, $w(\xi,\eta) = u(t,x)$ and

$$u_{tt} - u_{xx} = -4w_{\xi\eta}. \tag{4.10}$$

Integration of $w_{\xi\eta} = 0$ with respect to η obtains the general solution $w_\xi = F'(\xi)$ for some function $F(\xi)$. Integration once more, now with respect to ξ, gives

$$w(\xi,\eta) = F(\xi) + G(\eta) : \; u(t,x) = F(x+t) + G(x-t). \tag{4.11}$$

Imposing the initial data (4.6), we have

$$F(x) + G(x) = f(x), \quad F'(x) - G'(x) = g(x), \tag{4.12}$$

whereby $F'(x) = \frac{1}{2}\left(f'(x) + g(x)\right)$ and $G'(x) = \frac{1}{2}\left(f'(x) - g(x)\right)$. Integration gives

---

[3] Jean-Baptiste le Rond d'Alembert (1717–1783).

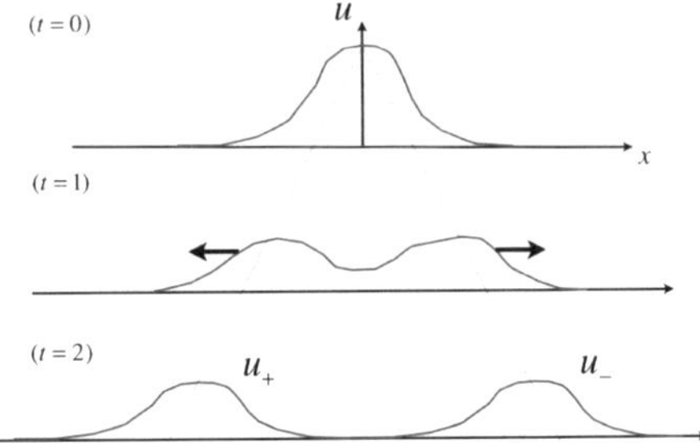

Fig. 4.2 Formation of left and right traveling waves from an initial hump with zero initial velocity by linear superposition in (4.15)

$$F(x) = \frac{1}{2}f'(x) + \int^x g(s)ds, \quad G(x) = \frac{1}{2}f(x) - \int^x g(s)ds, \tag{4.13}$$

the sum of which is (4.7).

Example 4.1. To exemplify, consider the data

$$u(0, x) = e^{-x^2}, \quad u_t(0, x) = 0 \tag{4.14}$$

for an initial value problem on $\mathbb{R}$. By (4.7), the solution (Fig. 4.2)

$$u(t, x) = \frac{1}{2}\left[e^{-(x+t)^2} + e^{-(x-t)^2}\right] \tag{4.15}$$

is explicitly the sum of a left and right moving simple wave (Fig. 4.2); the data (4.14) represent the instantaneous superposition of these waves, reinforcing each other in amplitude with cancellation of velocities at $t = 0$.

The result (4.15) illustrates a wave decomposition $u = u_- + u_+$ into left and right moving waves with a commensurate split of the total energy in the initial data. For the canonical wave equation (4.3) with $c = 1$, consider an energy integral of kinetic and potential energy

$$H(t) = \frac{1}{2}\int_{\alpha(t)}^{\beta(t)} \left[u_t^2 + u_x^2\right] dx \tag{4.16}$$

with moving boundaries $\alpha(t) = a + t$ and $\beta = b - t$, defined by the characteristics $dx/dt = \pm$ associated with $L_\mp$ (Fig. 4.3). The moving boundaries have the property of zero influx: $L_\mp u_\pm = 0$. As a result, wave energy can leave but not enter the domain of integration $[\alpha(t), \beta(t)]$. By this construction,

$$\frac{dH(t)}{dt} \leq 0, \tag{4.17}$$

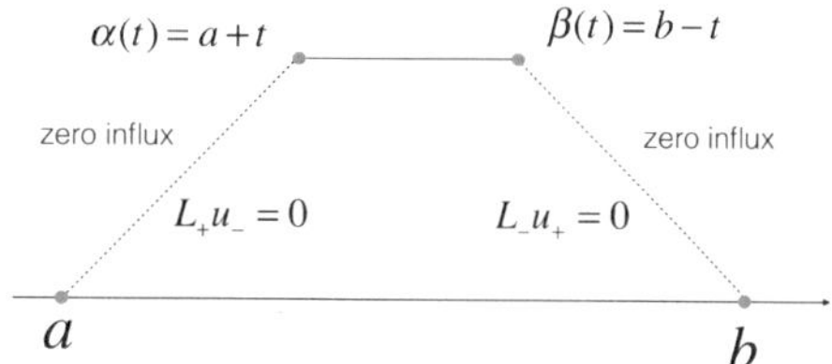

Fig. 4.3 The energy integral $H(t)$ in (4.16) is non-decreasing on a domain with moving boundaries with zero influx defined by the characteristics of $L_{\pm}$

which may also be seen by differentiation of (4.16):

$$\frac{dH(t)}{dt} = \int_{\alpha(t)}^{\beta(t)} [u_t u_{tt} + u_x u_{xt}]\, dx - \frac{1}{2}\left(u_t^2 + u_x^2\right)_\alpha - \frac{1}{2}\left(u_t^2 + u_x^2\right)_\beta \quad (4.18)$$

where the subscripts refer to (α, t) and, respectively, (β, t), and

$$\int_{\alpha(t)}^{\beta(t)} [u_t u_{tt} + u_x u_{xt}]\, dx = \int_{\alpha(t)}^{\beta(t)} [u_t (u_{tt} - u_{xx})]\, dx - (u_t u_x)_\alpha + (u_t u_x)_\beta\,. \quad (4.19)$$

With $u(t, x)$ satisfying (4.3), we thus arrive at

$$\frac{dH(t)}{dt} = -\frac{1}{2}(u_t + u_x)_\alpha^2 - \frac{1}{2}(u_t - u_x)_\beta^2 \le 0, \quad (4.20)$$

which establishes (4.17).

The split of initial data into two waves in (4.2) hereby carries energies

$$H = H_L + H_R, \;\; H_L = H_R = \frac{1}{2}H \quad (4.21)$$

in left and right moving waves, each with kinetic and potential energies equal to $(1/4)H$ of the total energy defined by the initial data.

4.1.1 Inhomogeneous Wave Equation (Duhamel)

The above can be extended to inhomogeneous wave equations, that include an excitation $h = h(x, t)$,

$$u_{tt} - c^2 u_{xx} = h(t, x). \quad (4.22)$$

Since (4.22) is linear, we focus on the role of $h(t, x)$ by considering an IVP with homogeneous initial data, i.e.,

$$u(0, x) = 0, \quad u_t(0, x) = 0. \tag{4.23}$$

Following (4.9) and putting $c = 1$, we have

$$w_{\xi\eta} = -\frac{1}{4} h(\xi, \eta), \tag{4.24}$$

were the transformation to $h(\xi, \eta)$ is understood. By (4.23), $u_x(0, x) = u_t(0, x) = 0$ at $t = 0$, whereby the initial data for $w(\xi, \eta)$ on $\xi = \eta$ (with $t > 0$ in $\xi > \eta$ and $t < 0$ in $\xi < \eta$) are

$$w(\xi, \xi) = 0, \quad w_\xi(\xi, \xi) = w_\eta(\xi, \xi) = 0. \tag{4.25}$$

Integration with respect to η obtains

$$w_\xi = -\frac{1}{4} \int_{A(\xi)}^{\eta} h(\xi, \eta') d\eta' = \frac{1}{4} \int_{\eta}^{\xi} h(\xi, \eta') d\eta'. \tag{4.26}$$

Here, $A = A(\xi)$ serves as an integration constant. By (4.25), $w_\xi = 0$ at $\xi = \eta$, and hence $A(\xi) = \xi$. Integration with respect to ξ subject to $w(\xi, \xi) = 0$, we have, similarly,

$$w = \frac{1}{4} \int_{B(\eta)}^{\xi} \int_{\eta}^{\xi} h(\xi, \eta') d\eta' d\xi' = \frac{1}{4} \int_{\eta}^{\xi} \int_{\eta}^{\xi'} h(\xi', \eta') d\eta' d\xi'. \tag{4.27}$$

The domain covered in the double integral is the region with the triangle formed by η, ξ and the interval $[\eta, \xi]$ on the real axis at $t = 0$, i.e., the full past domain of dependence D^-. Following in inverse of our coordinate transformation (4.9) and noting the Jacobian

$$J = \frac{\partial(\eta, \xi)}{\partial(x, t)} = \begin{vmatrix} 1 & 1 \\ 1 & -1 \end{vmatrix} = 2, \tag{4.28}$$

we arrive at

$$u_P(t, x) = \frac{1}{2} \int_0^t \int_{x-(t-\tau)}^{x+(t-\tau)} h(\tau, s) ds d\tau. \tag{4.29}$$

This Duhamel integral[4] is the sum of the response to inhomogeneous initial data and the excitation $h(t, x)$, i.e.,

$$u(t, x) = u_H(t, x) + u_P(t, x) \tag{4.30}$$

where $u_H(t, x)$ is given by (4.7).

[4] Jean-Marie Constant Duhamel (1797–1872).

4.2 Diffusion Equation

Diffusive transport phenomena are described by parabolic equations. In fluids, diffusion is commonly encountered in the flux of heat, momentum or vorticity, e.g., in the formation and dissipation of swirling flows past an obstacle. As such, they typically derive from some underlying microscopy theory by a process known as random walks.

In brief, a random walk of a particle arises from a series of random steps $\mathbf{\Delta}_i = \mathbf{x}_i - \mathbf{x}_{i-1}$,

$$\mathbf{x}_N = \mathbf{x}_0 + (\mathbf{x}_1 - \mathbf{x}_0) + \cdots + (\mathbf{x}_N - \mathbf{x}_{N-1}) = \mathbf{x}_0 + \sum_{i=1}^{N} \mathbf{\Delta}_i, \tag{4.31}$$

where $\mathbf{x} = x\mathbf{i} + y\mathbf{j} + z\mathbf{k}$ denotes the position vector in a Cartesian coordinate system (x, y, z) with orthonormal basis $\{\mathbf{i}, \mathbf{j}, \mathbf{k}\}$. In the approximation of an isotropic, stationary and uncorrelated ("no memory") process, we have the expectation values

$$\langle \mathbf{x}_N \rangle = \mathbf{x}_0, \quad \langle \mathbf{\Delta}_i \rangle = \mathbf{0}, \quad \left\langle \mathbf{\Delta}_i \cdot \mathbf{\Delta}_j \right\rangle = \begin{cases} \sigma^2 \ (i = j) \\ 0 \ \ (i \neq j) \end{cases}, \tag{4.32}$$

where the variance σ^2 is a constant. If so, we have

$$\left\langle (\mathbf{x}_N - \mathbf{x}_0)^2 \right\rangle = 2 \left\langle \sum_{1 \le i \neq j \le N} \mathbf{\Delta}_i \cdot \mathbf{\Delta}_j \right\rangle + \left\langle \sum_{i=1}^{N} \mathbf{\Delta}_i^2 \right\rangle = N\sigma^2. \tag{4.33}$$

With

$$\left\langle (\mathbf{x}_N - \mathbf{x}_0)^2 \right\rangle = \left\langle \mathbf{x}_N^2 \right\rangle - 2 \left\langle \mathbf{x_N} \cdot \mathbf{x_0} \right\rangle + \mathbf{x}_0^2 = \left\langle \mathbf{x}_N^2 \right\rangle - \mathbf{x}_0^2, \tag{4.34}$$

we arrive at

$$R_N = \sqrt{\left\langle (\mathbf{x}_N - \mathbf{x}_0)^2 \right\rangle} = \sigma \sqrt{N} \tag{4.35}$$

By the rather minimal assumptions (4.32), this key result on random walks finds broad applications to a wide variety of transport processes.

Applied to fluids, we go one step further by identifying σ with the mean free path length at finite temperature T. Let $c_s = \sqrt{k_B T/m}$ denote the isothermal sound speed, giving the mean thermal velocity of the molecules of mass m in each direction, where k_B denotes the Boltzmann constant.[5] With $N = t/\tau$, where τ denotes the mean time between collisions, $\sigma = c_s \tau$, we write (4.35) as

[5]The mean kinetic energy satisfies $\frac{1}{2}m\left\langle \mathbf{v}^2 \right\rangle = \frac{1}{2}\left\langle v_x^2 + v_y^2 + v_z^2 \right\rangle = \frac{3}{2}k_B T$, where $\left\langle v_x^2 \right\rangle = \left\langle v_y^2 \right\rangle = \left\langle v_z^2 \right\rangle = k_B T/m$ for an isotropic thermal distribution at temperature T.

$$R = \sqrt{\sigma^2 N} = \sqrt{\sigma c_s t} \equiv \sqrt{Dt} \tag{4.36}$$

with diffusion constant

$$D = \sigma c_s \tag{4.37}$$

of dimension $\text{cm}^2\,\text{s}^{-1}$. The relation between mean free path length and the particle number density n is defined by the particle collisional cross-section σ_c, according to which

$$\lambda n \sigma_c = 1 \tag{4.38}$$

with $\sigma = \sqrt{3}\lambda$.[6]

For a dilute solution at a concentration c in a solvent at temperature T, the solution contributes a partial pressure p to the total pressure in the fluid accord with van 't Hoff's equation[7] for partial pressures,

$$p = c k_B T, \tag{4.39}$$

where we recall the dimension cm^{-3} of c. According to van 't Hoff's equation, solutions satisfy the ideal gas law, whether the solvent is a gas or a liquid, whenever concentrations are small. Across a (real or virtual) surface, a concentration difference $[c] = (c)^+ - (c)^-$ carries an associated pressure difference

$$[p] = [c] k_B T. \tag{4.40}$$

The total pressure of solution and solvent being constant across the surface, (4.40) implies an *osmotic pressure jump* in the solvent of $-[c]k_B T$. For instance, a semi-permeable membrane of a chicken egg is permeable to water but a poor conductor to all else. Exposed to an environment of pure water (*hypotonic solution*), it accumulates more water in an attempt to lower concentrations of various solutions inside. In contrast, it dispels of water to an environment of low water concentration (*hypertonic solution*), e.g., syrup. These osmotic pressures need not be small, and they are key to the functioning of our kidneys.

In case of a non-uniform distribution described by a concentration $c = c(t, x, y, z)$, concentration gradients ∇c tend to smooth out by random thermal motions of the solution itself, in accord with (4.39),

$$\nabla p = k_B T \nabla c. \tag{4.41}$$

The resulting transport of N_c particles of the solution in the direction of $-\nabla c$ through a surface element $\delta \mathbf{A} = \mathbf{n} \delta A$ per unit time, is described by Fick's law [3]

---

[6] λ reduces by $\sqrt{2}$ in a thermal distribution, see [2].

[7] Jacobus H. van 't Hoff (1852–1911).

$$\frac{d\delta N_c}{dt} = -D\nabla c \cdot \delta \mathbf{A}, \tag{4.42}$$

This linear transport equation serves as a defining relationship for the diffusion constant D in the continuum limit, as opposed to (4.37) describing our random walk model. Taking a further divergence of (4.42) assuming D is constant throughout the solvent, we arrive at the convective-diffusion equation[8]

$$\partial_t c + \partial_x (v_d c) = D\Delta c, \tag{4.43}$$

where we include the presence of a drift velocity in the solution. Drift velocities occur, for instance, when ionized solutions are exposed to electric fields.

To identify D in (4.43) with the underlying microphysics of diffusion at finite temperature, we next turn to the *mobility* μ of the solution in the solvent, describing a proportional relation of drift with force per particle,

$$v_d = \mu f, \tag{4.44}$$

here assumed to be along the x-axis. As pressure P is force per unit surface area, $\partial_x P$ is force per unit volume. An external force f on the solution, therefore, has an associated gradient in partial pressure

$$\partial_x p = cf. \tag{4.45}$$

By van 't Hoff's equation (4.39), $\partial_x p = k_B T \partial_x c$, we therefore have $k_B T \partial_x c = cf$. Integrating the time-independent limit $\partial_t = 0$ of (4.43) once, we also have

$$v_d c = D c_x. \tag{4.46}$$

With (4.44), we now arrive at the *Einstein relationship*[9]

$$D = \mu k_B T. \tag{4.47}$$

Resorting to random motions with no drift, the derivation of Einstein's relation can be sketched as follows. Particle positions evolve by random kicks with typical velocities set by the temperature, followed by viscous decay in accord with Stokes' flows at low Reynolds numbers, i.e.,

$$m\dot{v} = -fv: \;\; v(t) = v_0 e^{-t/\tau_d}, \;\; \tau_d = \frac{m}{f} \tag{4.48}$$

over time intervals $0 \le t \le \tau$ between subsequent next collisions, where m denotes the mass of the particles as before and the initial velocities v_0^2 are set by solvent's

[8] Fick's second law with convection.

[9] For a related discussion, see [4].

temperature. For spherical particles of radius R much larger than the solvent's particles, Stokes' relation $f = 6\pi\eta R$ holds with dynamical viscosity η. Even though viscous decay slows particles down, their mean velocity squared must correspond to the solvent's temperature,

$$\frac{1}{\tau}\int_0^\tau \frac{1}{2}\left\langle \mathbf{v}(t)^2\right\rangle dt = \frac{3k_BT}{2m}. \tag{4.49}$$

A common temperature of solvent and particles implies that time scales of viscous decay, respectively, the rate of kicks are similar: $\tau \simeq \tau_d$. The variance in displacement over times $t \gg \tau$, $t = N\tau$, hereby satisfies the scaling

$$\left\langle (\mathbf{x}-\mathbf{x}_0)^2\right\rangle \propto (vt)^2N^{-1} \propto \frac{k_BT}{f}t. \tag{4.50}$$

According to (4.36) and (4.37), we therefore have

$$D \propto \frac{k_BT}{f} = \mu k_BT \tag{4.51}$$

with $\mu \equiv 1/f = \tau_d/m$.

With no drift, (4.43) reduces to the *homogeneous heat equation*

$$\partial_t u = Du_{xx}, \tag{4.52}$$

here reduced to $c = u(t,x)$. Since (4.52) is first order in time, its IVP is defined by initial data for $u(t,x)$

$$u(0,x) = f(x), \tag{4.53}$$

such as an initial distribution in temperature or concentration of a species in a solvent. Note that the diffusion equation is conservative in that it preserves the integral $\mathcal{A} = \int_{-\infty}^{\infty} u(x,t)dx$ in time by

$$\frac{d}{dt}\mathcal{A} = \int_{-\infty}^{\infty} u_t(x,t)dx = D\int_{-\infty}^{\infty} u_{xx}(x,t)dx \tag{4.54}$$

and so

$$\frac{d}{dt}\mathcal{A} = \lim_{L\to\infty}\left(u_x(-L,t) - u_x(L,t)\right) = 0. \tag{4.55}$$

Since (4.53) represents the continuum limit of a random walk, we expect the existence of solutions that describe the propagation of diffusion fronts as a function of the similarity variable

$$\xi = \frac{x^2}{Dt}. \tag{4.56}$$

Example 4.2. The *fundamental solution*

$$G(t,x) = A(t)e^{-\frac{1}{4}\xi} = \frac{1}{2\sqrt{\pi Dt}}e^{-\frac{x^2}{4Dt}}, \tag{4.57}$$

is a function of (4.56), where the time dependent coefficient $A(t)$ is included to preserve the total amount of particles in the solvent: for each $t > 0$, $G(t,x)$ is a Gaussian in x with standard deviation $\sqrt{2Dt}$ such that

$$\int_{-\infty}^{\infty} G(t,x)\,dx = 1. \tag{4.58}$$

In the limit as t approaches zero, $G(t,x)$ hereby becomes the Dirac delta function $\delta(x)$, since

$$f(x) = \lim_{t\to 0^+}\int_{-\infty}^{\infty} G(t,x-s)f(s)ds \tag{4.59}$$

for any continuous and integrable function $f(x)$ on $\mathbb{R}$. For $t > 0$, $G(t,x)$ satisfies the homogeneous heat equation, as follows from the explicit evaluation

$$G_t(t,x) = \left[-\frac{1}{2t^{\frac{3}{2}}} + \frac{x^2}{4Dt^2}\right]\frac{1}{2\sqrt{\pi D}}e^{-\frac{x^2}{4Dt}}, \tag{4.60}$$

$$G_x(t,x) = \frac{1}{2\sqrt{\pi Dt}} \times \frac{x}{2Dt}e^{-\frac{x^2}{4Dt}}, \tag{4.61}$$

$$G_{xx}(t,x) = \left[-\frac{1}{2t^{\frac{3}{2}}} + \left(\frac{x}{2Dt}\right)^2\right]\frac{1}{2\sqrt{\pi D}}e^{-\frac{x^2}{4Dt}}. \tag{4.62}$$

The change to a smooth Gaussian from an initially non-smooth distribution function in $G(t,x)$ is perhaps the most characteristic behavior described by the heat equation. $G(t,x)$ allows us to express the general solution as a convolution

$$u(t,x) = \int_{-\infty}^{\infty} G(t,x-s)f(s)ds, \tag{4.63}$$

solving the IVP (4.52) and (4.53).

Alternatively, solutions on the real line or a finite interval are readily obtained using the Fourier transform, to be discussed in detail later. Following

$$\tilde{u}(t,k) = \int_{-\infty}^{\infty} u(t,x)e^{-ikx}dx, \tag{4.64}$$

we have

$$\tilde{u}_t(t,k) = -k^2 D\tilde{u}(t,k): \ \ \tilde{u}(t,k) = \tilde{u}(0,k)e^{-Dk^2t} = \tilde{f}(k)e^{-Dk^2t} \tag{4.65}$$

by the Fourier transform of (4.53). The solution to our IVP, therefore, is

$$u(t,x) = \frac{1}{2\pi}\int_{-\infty}^{\infty} \tilde{f}(k)e^{-Dk^2t+ikx}dk. \tag{4.66}$$

It brings about a typical behavior in time-dependent solutions of the heat equations, wherein relatively high frequency variations (large k) are damped rather strongly, following suppression by e^{-Dk^2t}. After some time, only the far infrared of the initial data survives in a smooth solution with only long wave length variations.

Figure 4.4 illustrates the smoothing behavior by heat flux arising from temperature discontinuities, here by initial data of the form $u(x,0) = K \ (-a < x < a)$ and $u(x,0) = 0$ elsewhere. With period L, the

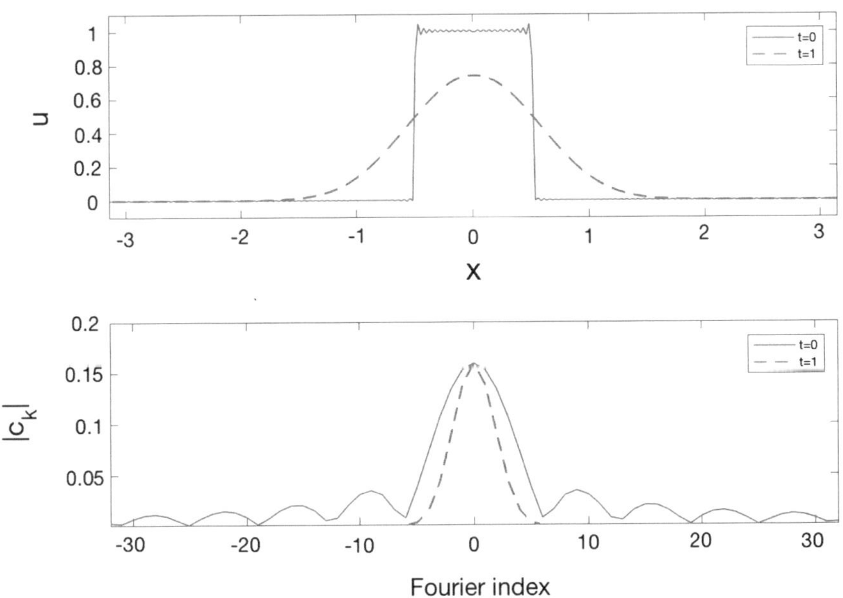

Fig. 4.4 Smoothing of an initially sharp transition ($t = 0$), here shown in block type function represented by a finite Fourier series ($N = 256$). Fourier coefficients with positive Fourier index $k > 0$ are suppressed in time by diffusion, while c_0 remains constant. At late times, only c_0 survives

$$u(x,t) = \sum_{n=0}^{\infty} C_n \cos(k_n x) e^{-Dk_n^2 t} \tag{4.67}$$

defined by the Fourier coefficients

$$C_n = K \int_{-a}^{a} e^{-ik_n x} dx = \mathcal{A} \frac{\sin k_n a}{k_n a} \tag{4.68}$$

where $\mathcal{A} = 2Ka/L$ and $k_n = nk_0$ with fundamental period $k_0 = 2\pi/L$. At $t > 0$, the sharp transitions become smooth (Fig. 4.4) and, at late times,

$$u(t,x) \simeq \mathcal{A}. \tag{4.69}$$

4.2.1 Photon Diffusion in the Sun

Solar radiation is made up of relic photons from their time of creation in the process of nuclear fusion in the core of the sun. Their escape is an illustrative example of random walks by elastic scattering with protons,

$$\gamma + p \rightarrow \gamma + p \tag{4.70}$$

with Thomson cross section $\sigma_T = 6.65 \times 10^{-25}$ cm^2, that defines a migration time as a function of various mean physical properties of the sun.

The sun is a virialized self-gravitating object with total thermal energy E_{th} mostly in protons of mass $m_p = 1.67 \times 10^{-24}$ g, satisfies[10]

$$E_{th} = -\frac{1}{2}U = \frac{GM^2}{4R_\odot}, \tag{4.71}$$

where $M = 2 \times 10^{33}$ g and $R_\odot = 7 \times 10^{10}$ cm denote the mass and radius of the Sun. According to the ideal gas law, $E_{th} = PV_\odot$, $V_\odot = 4\pi R_\odot^3/3$, giving the mean pressure

$$P = \frac{GM^2}{2R_\odot V_\odot} = nk_B T, \quad n = \frac{\rho}{m_p}, \quad \rho = \frac{M}{V_\odot}, \tag{4.72}$$

where $k_B = 1.38 \times 10^{-16}$ erg K^{-1} denotes the Boltzmann constant, n is the number density and ρ is the mass density of the protons. Accordingly, the mean free path length of photons inside satisfies

[10]Here, the factor of one-half can be seen to derive from integration of the binding energy $dU = -(GM/R)dM$ in case of a constant radius R.

$$\lambda = \frac{1}{n\sigma_T} \simeq 2\,\text{cm} \tag{4.73}$$

The total number of scatterings N involved in the migration of photons to the surface of the Sun satisfies $R_\odot = \sqrt{N}\lambda$. Given that photons propagate with the velocity of light, the diffusion time satisfies[11]

$$t_d = \frac{N\lambda}{c} = \frac{R_\odot^2}{\lambda c} \simeq 10^{11}\,\text{s} \simeq 3\,\text{kyr}. \tag{4.74}$$

This result may be compared with the *Kelvin-Helmholtz* timescale

$$t_{KH} = \frac{E_{th}}{L_\odot} = \frac{GM^2}{4R_\odot L_\odot} \simeq 8\,\text{Myr}, \tag{4.75}$$

where $L_\odot \simeq 3.38 \times 10^{33}$ erg s^{-1} is the solar luminosity. It follows that

$$\frac{t_d}{t_{KH}} \simeq 10^{-3}. \tag{4.76}$$

Photon diffusion by random walks is sufficiently fast to keep the Sun in thermodynamic equilibrium, between heat produced in the center and heat radiated off from its surface.

Finally, the mean temperature of the sun satisfies

$$k_B T = \frac{Pm_p}{\rho} = \frac{GMm_p}{4R_\odot} \simeq 8 \times 10^{-10}\,\text{erg}: \quad T = 5 \times 10^6\,\text{K}, \tag{4.77}$$

giving a sound crossing time scale $t_s = R_\odot/c_s$ of about 1 h, $c_s = \sqrt{k_B T/m_p} \simeq 2 \times 10^7$ cm s^{-1}. Summarizing, we have the following time scales:

$$t_s \ll t_d \ll t_{KH} \ll T_{MS}, \tag{4.78}$$

where we include the sound crossing time scale t_s an the main sequence lifetime T_{MS} of the sun.

4.3 Elliptic Equations

Elliptic equations arise naturally as the time-independent limit of the wave equation and the diffusion equation,

---

[11] A more refined analysis takes into account a density in the core that is about 58 times the mean density, showing a significant increase in the diffusion time; see [5].

$$\Delta u = 0. \tag{4.79}$$

In two dimensions, (4.79) is satisfied by the real and imaginary parts of complex analytic functions. What is the appropriate formulation for (4.79)?

Hadamard pointed out that (4.79) should not be given an initial value problem because of *ill-posedness* [6].

Consider, tentatively, the initial boundary value problem (IBVP) for (4.79) on a rectangle of height L and width π two dimensions,

$$u_{yy} = -u_{xx} \text{ on } 0 \leq y \leq L, \ -0 < x < \pi \tag{4.80}$$

subject to the homogeneous boundary values $u(0, y) = u(\pi, y) = 0$. Non-zero solutions are

$$u_n(x, y) = \frac{1}{n^2} \sinh ny \sin nx \tag{4.81}$$

for any integers n, satisfying

$$u_n(x, 0) = 0, \quad \lim_{n\to\infty} \partial_y u_n(x, 0) = 0. \tag{4.82}$$

Despite this regular convergence of boundary conditions to zero on $y = 0$, the solution

$$u_n(x, L) \tag{4.83}$$

at $y = L$ diverges in the same limit as n approaches infinity. Thus, arbitrarily small initial data give rise to large data at $y = L$ in the limit as n becomes large. Associating the former with small wave length perturbations, the IBVP for the elliptic equation is unduly sensitive to the initial data, i.e., the initial-boundary value problem is *ill-posed* in the sense of Hadamard.

A suitable formulation for (4.79) is by prescribing, instead, data to the full boundary of the domain of interest. Solutions in the domain of interest then have regular dependence on data.

Continuing with the Hadamard example above, we may prescribe

$$u(x, 0) = 0, \quad u(0, y) = u(\pi, y), \quad u(x, L) = f(x). \tag{4.84}$$

The solution

$$u(x, y) = \sum_{n\in\mathbb{Z}} a_n \frac{\sinh ny}{\sinh nL} \sin nx, \quad a_n = \frac{2}{\pi} \int_0^\pi f(x) \sin nx dx \tag{4.85}$$

now shows explicitly regular dependence on data (4.84) by linear superposition.

Example 4.3. Of considerable practical and theoretical interest is the problem of elliptic equations on the unit disk in two dimensions, $D: x^2 + y^2 \leq 1$. In polar coordinates (r, φ), (4.79) becomes

$$u_{rr} + \frac{1}{r}u_r + \frac{1}{r^2}u_{\varphi\varphi} = 0. \tag{4.86}$$

By separation of variables, $u = r^n e^{im\phi}$, we find $n(n-1) + n - m^2 = 0$, i.e., $m = \pm n$. Focusing on non-singular solutions within $r \leq R$ for some R, we have the general solution

$$u = \sum_{n\epsilon\mathbb{Z}}^{\infty} c_n r^{|n|} e^{in\varphi} = c_0 + 2\sum_{n=1}^{\infty} r^n \left(a_n \cos n\varphi - b_n \sin n\varphi\right) \tag{4.87}$$

in terms of Fourier coefficients $c_n = a_n + ib_n$ satisfying $c_{-n} = \bar{c}_n$ assuming real data. Prescribing data on $r = 1$,

$$u(1, \varphi) = f(\varphi), \tag{4.88}$$

whereby

$$c_n = \frac{1}{2\pi}\int_0^{2\pi} f(\varphi)e^{-in\phi}d\varphi, \tag{4.89}$$

The above is closely related to the theory of complex functions. If $g(z)$ is a function of a complex variable $z = x + iy$ which is analytic at $z = z_0$, then its real and imaginary parts $u = u(x, y)$, respectively $v(x, y)$ satisfy the Cauchy-Riemann relations:

$$g(z) = u(x, y) + iv(x, y): \ \ u_x = v_y, \ u_y = -v_x \text{ at } (x, y) = (x_0, y_0). \tag{4.90}$$

If $g(z)$ is analytic in an open neighborhood of z_0, then differentiation of the Cauchy-Riemann relations show that both u and v are harmonic (cf. Chap. 2)

$$u_{xx} + u_{yy} = 0, \ \ v_{xx} + v_{yy} = 0. \tag{4.91}$$

Without loss of generality, we may consider $z_0 = 0$. Analyticity of $g(z)$ in a neighborhood of the origin is equivalent to having a Taylor series expansion (a point of view emphasized by Weierstrass),

$$g(z) = \sum_{n=0}^{\infty} d_n z^n, \tag{4.92}$$

where we may assume the radius of convergence to be greater or equal to 1, possibly after a rescaling of z. Using $z = re^{in\phi}$ we have at $|z| = 1$

$$g\left(e^{i\varphi}\right) = \sum_{n=0}^{\infty} c_n e^{in\varphi} \tag{4.93}$$

i.e.,

$$\begin{aligned} u(1,\varphi) &= \textstyle\sum_{n=0}^{\infty} \alpha_n \cos n\varphi - \beta_n \sin n\varphi, \\ v(1,\varphi) &= \textstyle\sum_{n=0}^{\infty} \alpha_n \cos n\varphi + \beta_n \sin n\varphi. \end{aligned} \tag{4.94}$$

where $d_n = \alpha_n + i\beta_n$. Thus, (4.87) and (4.94) are the same with $d_n = 2c_n$.

As a result, solutions within the disk ($r < 1$) represent weighted averages of the data on the boundary ($r = 1$) conform Cauchy's integral formula. The solution inside is hereby analytic, even if the data on the boundary are merely integrable. The solution inside is therefore smooth in the sense of being infinitely differentiable. This smoothing by averaging should be contrasted with that in diffusion, by suppressing higher order Fourier coefficients progressively in time.

To conclude, given a solution $u(r, \varphi)$ to the elliptic problem (4.79) on the unit disk in two dimensions subject to $f(x)$ on the boundary, there is a corresponding complex function $g(z)$ that is analytic inside the unit disk. Solutions $u(r, \varphi)$ are entirely determined by the Fourier coefficients of $u(1, \varphi) = f(\varphi)$. They are, within a factor of two, the Taylor coefficients of $g(z)$ when expanded about the origin.

4.4 Characteristics of Hyperbolic Systems

The wave equation (4.3) can be expressed as a first-order system of equations in terms of a state variable given by all first derivatives. Letting $v = u_x$ and $w = u_t$,

$$\begin{pmatrix} v \\ w \end{pmatrix}_t = \begin{pmatrix} 0 & 1 \\ 1 & 0 \end{pmatrix} \begin{pmatrix} v \\ w \end{pmatrix}_x, \tag{4.95}$$

where we put the velocity of light c equal to 1. This first-order system describes the propagation of the normal $\nabla u = u_x\mathbf{i} + u_t\mathbf{j}$, e.g., wave fronts of constant total phase $u(x,t) = \phi(x,t)$, $d\phi = kdx + \omega dt$ with $k = \phi_x$ and $\omega = \phi_t$ and dispersion relation $\omega = ck$. Written abstractly as

$$A^a \partial_a \mathbf{U} = \mathbf{0}, \quad \mathbf{U} = u_x\mathbf{i} + u_t\mathbf{j}, \tag{4.96}$$

all A^a are real symmetric. Given that (4.96) satisfies (4.3), (4.96) is commonly referred to as symmetric hyperbolic.

Systems of the form (4.96) facilitate investigations of various properties. Consider for instance simple waves $U = U(\phi)$, where ϕ is a scalar as a function of coordinates (t, x, y, z). When U' is non-zero, (4.96) implies

$$A^a \nu_a \mathbf{U}' = 0 \tag{4.97}$$

with $\mathbf{U}' = d\mathbf{U}/d\phi$, $\nu_a = \partial_a \phi$. This linear equation on the normal ν_a of wave fronts of constant phase defines characteristic directions of the system, i.e., cones of (4.3) satisfying system satisfies

$$\det A^a \partial_a \nu. \tag{4.98}$$

We say (4.95) satisfies causality whenever all wave front velocities defined by (4.98) are less than or equal to the velocity of light, i.e., ν_a is space-like or null.

It should be mentioned that covariant formulations of Maxwell's equations and general relativity include gauge freedom in their fields. Consequently, A^t will be singular in any direct formulation. Even so, A^t can be given suitable rank-one updates by invoking constraints or a choice of gauge, so that (4.96) becomes regular allowing for direct numerical implementation or analytic studies, e.g., studies of stability and well-posedness based on (4.98) [7].

A further connection with the hyperbolic structure of Minkowski spacetime considers first-order systems with Hermitian matrices following Sect. 3.6.

4.5 Weyl Equation

Consider a spinor $\kappa^A \in \mathbb{C}^2$ (a row vector κ^A, $A = 1, 2$) and its extension to a spinor field over Minkowski space-time. Recall that κ^A represents a null-vector in four-dimensional space-time of a four-momentum of a massless particle. A general four-covariant first-order system for κ is

$$G^{ab} \sigma_a^{AA'} \partial_b \bar{\kappa}_{A'} = 0^A, \tag{4.99}$$

where G^{ab} is a symmetric real-valued tensor over Minkowki spacetime and the $\sigma_a^{AA'}$ are the Pauli spin matrices (3.170).

Focusing on plane waves propagating along each of the three spatial coordinate axis, the simplest choice of

$$G^{ab} = \begin{pmatrix} 1 & 0 & 0 & 0 \\ 0 & -\lambda_x & 0 & 0 \\ 0 & 0 & -\lambda_y & 0 \\ 0 & 0 & 0 & -\lambda_z \end{pmatrix} \tag{4.100}$$

in (4.99) can be seen by inspection. Write

$$\bar{\kappa}_{A'} = \begin{pmatrix} \xi \\ \eta \end{pmatrix}. \tag{4.101}$$

A simple waves along the x^i-axis ($x^1 = x, x^2 = y, x^3 = z$) satisfies

$$\begin{pmatrix} \xi \\ \eta \end{pmatrix}_{,t} = \lambda_i \sigma_i \begin{pmatrix} \xi \\ \eta \end{pmatrix}_{,i} \quad (i = 1, 2, 3), \tag{4.102}$$

where the subscript $,t$ and $,i$ refers to differentiation with respect to time and, respectively, x^i. These three $1+1$ systems are hyperbolic, provided that they each reduce to a second order wave equation (4.3) or a pair of first-order equations (4.5), all with velocity equal to 1. This will hold provided that

$$\lambda_x = 1, \quad \lambda_y = 1, \quad \lambda_z = 1, \tag{4.103}$$

each up to a sign. The resulting *Weyl equation*

$$\sigma^{cAA'} \partial_c \bar{\kappa}_{A'} = 0^A \tag{4.104}$$

hereby satisfies causality, in having characteristics coincident with the null-cone of Minkowski space-time.

We can form a second order equation upon multiplying (4.104) with $\sigma^{aBB'}\partial_a$,

$$\left(\sigma^{aBB'} \partial_a \sigma^{cAA'}\right) \epsilon_{AB} \partial_c \bar{\kappa}_{A'} = 0^{B'}. \tag{4.105}$$

Expanding in $1+1$, we have

$$(\partial_t - \sigma_x \partial_x) \begin{pmatrix} 0 & -1 \\ 1 & 0 \end{pmatrix} (\partial_t - \sigma_x \partial_x) \bar{\kappa}_{A'} = \begin{pmatrix} 0 \\ 0 \end{pmatrix}, \tag{4.106}$$

where we made use of the fact that σ_0 and σ_x are symmetric. Expansion gives, in the notation of (4.4) and (4.5),

$$\begin{pmatrix} 0 & -L_- L_+ \\ L_- L_+ & 0 \end{pmatrix} \bar{\kappa}_{A'} = \begin{pmatrix} 0 \\ 0 \end{pmatrix}, \tag{4.107}$$

Table 4.1 Linear partial differential equations

1. Hyperbolic equations describe wave motion. They can be second- or first-order symmetric systems of equations.

2. A hyperbolic equation satisfies causality if all wave velocities are less than or equal to the velocity of light, as defined by the Minkowski metric. Hyperbolicity ensures well-posedness of the solution in initial value problem, defined by Cauchy data on an initial time slice.

3. Parabolic equations describe diffusive transport phenomena. They emerge as effective descriptions of underlying microscopic physics, e.g., viscosity at finite temperature. They are first order in time and give well-posed initial value problems.

4. Elliptic problems describe time-independent limits of hyperbolic and parabolic problems. In two dimesions, they are also the real and imaginary part of complex analytic functions. They give rise to boundary value problems in close connection with complex function theory.

5. Minkowski space-time has a Lorentz invariant hyperbolic structure. Velocity four-vectors in Minkowski space-time can be represented by 2×2 Hermitian matrices, whose unimodular transformations give a double cover of the Lorentz transformations. In particular, SU(2) gives a double cover of the proper rotations SO(3) on the celestial sphere S^2.

6. Spinors in $\mathbb{C}^2$ represent square roots of null-vectors in Minkowski space-time. The simplest linear Hermitian hyperbolic system of equations is Weyl's equation of a massless spinor. It describes a particle of spin $\frac{1}{2}$, providing a starting point for the wave functions of neutrinos and electrons.

showing that $\bar{\kappa}_{A'}$ satisfies the canonical wave equation (4.3). To obtain the general result in 3+1, let

$$\Sigma^{AA'} = \sigma_c^{AA'}\partial^c \tag{4.108}$$

denote our embedding (3.142) of $\partial^c = \eta^{ac}\partial_a$. By (3.162), it follows that

$$\epsilon_{AB}\sigma^{aBB'}\partial_a\sigma^{cAA'}\partial_c = \epsilon_{AB}\Sigma^{BB'}\Sigma^{AA'} = \frac{1}{2}\epsilon^{A'B'}\Box, \tag{4.109}$$

where $\Box = \det\Sigma = \partial_t\partial_t - \partial_x\partial_x - \partial_y\partial_y - \partial_z\partial_z$ denotes the d'Alembert wave operator. Consequently, (4.105) reduces to

$$\Box\bar{\kappa}^{B'} = 0^{B'}. \tag{4.110}$$

The Weyl equation describes the wave function of a massless Fermion, a particle with angular momentum $\pm\frac{1}{2}\hbar$ according to the eigenvalues ± 1 of the Pauli spin matrices. Because it described by Hermitian operator, the conjugate of a spinor satisfies the same. The Weyl equation hereby introduces two particles, the spinor and its conjugate being distinct.

It is not known whether such massless spin $\frac{1}{2}$ particles exist. The closest to (4.99) is the equation for neutrinos, that, however, are experimentally known to have a small mass responsible for mixing between different neutrino species. Dirac extended (4.99) to two massive spinors. The Dirac equation successfully describes electrons and positrons, both of spin $\frac{1}{2}$, that defines the basis of quantum electromagnetics. A further extension of the above allows for spin 1 particles, described by spinors with 2 indices, representing photons. The extension to gravitation is very much a research topic, focused on novel descriptions of null-geodesics.

Table 4.1 summarizes this discussion.

4.6 Exercises

4.1. Solve the IBVP for the one-dimensional homogeneous wave equation

$$u_{tt} = c^2 u_{xx} \tag{4.111}$$

subject to

$$u(0, x) = 0, \; u_t(0, x) = \sin x, \; u(t, 0) = u(t, \pi) = 0. \tag{4.112}$$

4.2. Because of their rotation with angular velocity ω, planets around the Sun are exposed to a time-harmonic variation in radiation flux. The soil in, e.g., the Earth and Mars, is hereby heated, further leading to heating of the atmosphere. The average temperature of the surface of the soil is hereby higher that the average weather temperature. Solve for the temperature $u = u(t, z)$ in the soil varies as a function of depth, described by the heat equation

$$u_t = Du_{zz}, \quad (-\infty < z \leq 0) \tag{4.113}$$

subject to the *Neumann boundary condition*

$$u_z(t,0) = A\sin\omega t. \tag{4.114}$$

4.3. If only by molecular diffusion, how long does it take for perfume to reach a friend at a distance of 1 m? In reality, perfume is smelled rather swiftly. What mechanism accounts for significantly shorter diffusion times?

4.4. In (4.78), show the first and last inequalities for t_s an T_{MS}.

4.5. Consider two bars of equal material and size at temperatures T_i $(i = 1, 2)$. Brought into contact head-to-head, heat will diffuse from the warm to cold bar. Based on Fourier analysis of the initial value problem for the one-dimensional heat equation, show that at late times the temperature in the two bars becomes a constant. What is their common temperature at large time?

4.6. Consider the inhomogeneous wave equation (4.22) with homogeneous initial data. Calculate (4.29) for $h(t,x) = e^{-t}e^{-x^2}$.

4.7. Obtain the solution to the Cauchy problem for (4.3) with $f(x) = 0$ and $g(x) = e^{-x^2}$ and interpret the result.

4.8. In spherical symmetry, the scalar wave equation for a potential u in spherical coordinates (r, θ, φ) is

$$u_{tt} = \Delta u, \quad \Delta u = \frac{1}{r^2}\left(r^2 u_r\right)_r, \tag{4.115}$$

where subscript r refers to differentiation. For a time-harmonic solution $u(t,r) = \tilde{u}e^{i\omega t}$, derive the fundamental solution to the *Helmholtz equation*,

$$\omega^2\tilde{u} + \Delta\tilde{u} = 0 \tag{4.116}$$

satisfying the Sommerfeld outgoing radiation condition at infinity,

$$\lim_{r\to\infty}\left(\tilde{u}_r - ik\tilde{u}\right) = 0. \tag{4.117}$$

4.9.[12] The Cauchy problem for (4.115) poses the propagation of initial amplitude $u_0(r) = u(0,r)$ and velocity $u_1(r) = u_t(0,r)$ at $t = 0$ to $t > 0$. If $u_0(r)$ and $u_1(r)$ are analytic defined on $r \in \mathbb{R}$, we may consider an analytic extension to the line $z = r + is$ $(-\infty < r < \infty)$ for some $s > 0$. The conformal map $w = 1/z$ maps $z = r + is$ onto a circle C. Formulate the equivalent Cauchy problem for $u = u(w)$ on C. How is the solution $u(r)$ on $r \geq 0$ recovered from the solution on C?

4.10. The one-dimensional inviscid Burgers' equation for a velocity field $u = u(t,x)$

$$u_t + uu_x = 0 \tag{4.118}$$

[12]See, e.g., [8].

is a first-order *nonlinear* hyperbolic equation on $-\infty < x < \infty$. Cauchy problems propagate initial velocities $u_0(x) = u(0, x)$ at $t = 0$ to $t > 0$. Show that the data $u_0(x)$ are preserved along the characteristics with slope $dx/dt = u_0(x)$.[13] [*Hint*: consider $u(X(t, \xi), t)$ with $X(0, \xi) = u_0(\xi)$ at $t = 0$ and its (total) derivative.] When $u_0'(x) < 0$ at some x, derive the associated crossing time of characteristics

$$t_c(x) = -\left[u_0'(x)\right]^{-1} \tag{4.119}$$

and interpret the result in terms of traffic flow.

4.11. For the boundary value problem (4.86) in $r \leq R$, show that a non-constant u attains its extrema on $r = R$. [*Hint:* Consider u as the real part of an analytic function $f(z)$ and use Mean Value Theorem in $r < R$.]

4.12. Consider Green's identity for the Laplace's operator

$$\int_{\Omega} \Delta u \, d\Omega = \int_{\partial\Omega} u_n dS, \tag{4.120}$$

where Ω is a domain with boundary $\partial\Omega$ with outgoing unit normal $\mathbf{n}$, and $u_n = \mathbf{n} \cdot \nabla u$. Illustrate this for $u = r^2$ in n dimensions.

4.13. Estimate the adiabatic sound speed $c_s = \sqrt{\gamma k_B T/m_p}$ for the Sun, using the adiabatic index $\gamma = 5/3$ of fully ionized hydrogen at the mean temperature and compute the sound speed crossing time $t_s = R_\odot/c_s$.

References

1. Garabedian, P.R., *Partial Differential Equations*, Chelsia Publishing, New York (1986).
2. S. Chapman and T.G. Cowling, *The Mathematical theory of non-uniform gases* (Cambridge University Press, 1990).
3. Fick, A.E., 1855, Über Diffusion, *Annal. Physik und Chemie*, 94, 59.
4. Pekir, G., 2003, *Stoch. Models*, 19, 383.
5. Mitalas, R., & Sills, K.R., 1992, ApJ, 401, 759.
6. Hadamard, J.S., 1923, *Lectures on Cauchy's problem in linear differential equations* (Yale University Press, 1923; Reprint Dover 2003).
7. van Putten, M.H.P.M., 1994, Phys. Rev. D, 50, 6640.
8. van Putten, M.H.P.M., 2006, PNAS, 103, 519.

[13]Conserved quantities along characteristic directions are *Riemann invariants, that allow analytic solutions to, e.g., the nonlinear equations of compressible gas dynamics.*

Part II
Methods of Approximation

Chapter 5
Projections and Minimal Distances

In describing real-world phenomena, we often seek a local or leading order approximation, e.g., by an orthogonal projection onto a finite dimensional linear subspace of vectors or functions. For instance, a "best-fit" linear approximation to scattered data defined by minimal residual errors is a common procedure to identify trends or correlations in data. To set notation, we briefly revisit some elements of linear algebra of Chap. 3.

5.1 Vectors and Distances

We shall work in a three dimensional Cartesian coordinate system (x, y, z) with unit vectors $\{\mathbf{i}, \mathbf{j}, \mathbf{k}\}$, wherein a vector $\mathbf{a}$ can be expanded in terms of its coefficients (a_1, a_2, a_3) as

$$\mathbf{a} = a_1\mathbf{i} + a_2\mathbf{j} + a_3\mathbf{k} \equiv \begin{pmatrix} a_1 \\ a_2 \\ a_3 \end{pmatrix}. \tag{5.1}$$

In Euclidean space, the length of a vector is defined as the square root of the sum of the squares of its coefficients, $|\mathbf{a}| = \sqrt{a_1^2 + a_2^2 + a_3^3}$. Two vectors $\mathbf{a}$ and $\mathbf{b}$ are linearly independent if they span a plane, given by linear combinations $\mathbf{c} = \lambda\mathbf{a} + \mu\mathbf{b}$, where λ and μ are real numbers, and defined as $\mathbf{c} = (\lambda a_1 + \mu b_1)\mathbf{i} + (\lambda a_2 + \mu b_2)\mathbf{j} + (\lambda a_3 + \mu b_3)\mathbf{k}$. Conversely, we say that $\mathbf{a}$ and $\mathbf{b}$ are linearly dependent if $\lambda\mathbf{a} + \mu\mathbf{b} = \mathbf{0}$ for some nonvanishing λ and μ.

A corollary to (3.10) is an expression for the distance between (the endpoints of) vectors $\mathbf{a}$ and $\mathbf{b}$, given by $\mathbf{r} = \mathbf{b} - \mathbf{a}$,

M.H.P.M. van Putten, *Introduction to Methods of Approximation in Physics and Astronomy*, Undergraduate Lecture Notes in Physics,
DOI 10.1007/978-981-10-2932-5_5

$$|\mathbf{r}|^2 = (\mathbf{b} - \mathbf{a}) \cdot (\mathbf{b} - \mathbf{a}) = |\mathbf{a}|^2 + |\mathbf{b}|^2 - 2\mathbf{a} \cdot \mathbf{b} = a^2 + b^2 - 2ab\cos\theta. \quad (5.2)$$

The reciprocal, familiar from electrostatics and Newtonian gravitational binding energy, satisfies

$$\frac{1}{|\mathbf{r}|} = \frac{1}{\sqrt{a^2 + b^2 - 2ab\cos\theta}} = \frac{1}{a}\sum_{l=0}^{\infty}\left(\frac{a}{b}\right)^{l+1} P_l(\cos\theta) \quad (5.3)$$

when $b > a$, i.e.,

$$\frac{1}{|\mathbf{r}|} = \frac{1}{b}\left(1 + \frac{a}{b}P_1(\cos\theta) + \frac{a^2}{b^2}P_2(\cos\theta) + \cdots\right) \quad (5.4)$$

in terms of the Legendre polynomials $P_l(x)$,

$$\begin{aligned} &P_0(x) = 1, \quad P_1(x) = x, \quad P_2(x) = \tfrac{1}{2}(3x^2 - 1), \\ &P_3(x) = \tfrac{1}{2}(5x^3 - 3x), \quad P_4 = \tfrac{1}{8}(35x^4 - 30x^2 + 3), \ldots \end{aligned} \quad (5.5)$$

The Legendre polynomials can be expressed by Rodrigue's formula

$$P_n(x) = \frac{1}{2^n n!}\frac{d^n}{dx^n}\left(x^2 - 1\right)^n, \quad (5.6)$$

and satisfy the Legendre equation

$$\frac{d}{dx}\left[(1 - x^2)\frac{du(x)}{dx}\right] + n(n+1)u(x) = 0 \quad (n = 0, 1, 2, \ldots), \quad (5.7)$$

representing a Sturm-Liouville problem with eigenvalues $-n(n+1)$. They may be computed from (5.6) or by recursion,

$$(n+1)P_{n+1}(x) = (2n+1)xP_n(x) - nP_{n-1}(x). \quad (5.8)$$

What makes (5.4) relevant is the fact that the Legendre expansion is extremely efficient—the first few terms being sufficient for an accurate expansion—by virtue of the orthogonality property

$$\int_{-1}^{1} P_n(x)P_m(x) = \frac{2}{2n+1}\delta_{nm} \quad (5.9)$$

where δ_{nm} denotes the Kronecker delta function, $\delta_{nm} = 1$ for $n = m$ and $\delta_{nm} = 0$ otherwise.

5.2 Projections of Vectors

Finding a suitable reduction to a finite number of degrees of freedom is a starting point for computation on a computer. While this can take many forms, an efficient approach aims for a small number of dimensions. In the language of linear vector spaces, this typically involves *projections onto sub-spaces* of reduced dimension.

Following Chap. 3, consider the two-dimensional Euclidean plane $\mathbb{R}^2$ with vectors of the form

$$\mathbf{a} = x\mathbf{i} + y\mathbf{j} = \begin{pmatrix} x \\ y \end{pmatrix} \quad (x, y \in \mathbb{R}) \tag{5.10}$$

in a Cartesian coordinate system (x, y) with orthonormal basis (ONB) of $\mathbf{i}$ pointing along the x-axis and $\mathbf{j}$ pointing along the y-axis, each of unit length. As such, $\mathbb{R}^2$ is a linear subspace of $\mathbb{R}^3$. In particular, $\mathbb{R}^2$ has the Euclidean norm $|\mathbf{a}| = \sqrt{x^2 + y^2}$, wherein

$$\mathbf{i} = \begin{pmatrix} 1 \\ 0 \end{pmatrix}, \ \mathbf{j} = \begin{pmatrix} 0 \\ 1 \end{pmatrix} \tag{5.11}$$

have unit length $|\mathbf{i}| = 1$, $|\mathbf{j}| = 1$ and $\angle(\mathbf{i}, \mathbf{j}) = \frac{\pi}{2}$.

Any two non-zero vectors $\mathbf{a}$ and $\mathbf{b}$ can be normalized to elements from S^1,

$$\hat{\mathbf{a}} = \frac{\mathbf{a}}{||\mathbf{a}||} = \begin{pmatrix} \cos\alpha \\ \sin\alpha \end{pmatrix}, \ \hat{\mathbf{b}} = \frac{\mathbf{b}}{||\mathbf{b}||} = \begin{pmatrix} \cos\beta \\ \sin\beta \end{pmatrix}, \tag{5.12}$$

satisfying $\hat{\mathbf{a}} \cdot \hat{\mathbf{b}} = \cos(\alpha - \beta)$.

Geometrically, $x\mathbf{i}$ and $y\mathbf{j}$ in (5.10) represent the projections of $\mathbf{a}$ onto the x- and y-axis (the span of the basis vectors (5.11)), illustrated in Fig. 5.1. These are the *orthogonal projections* Π_x along the y-axis and, respectively, Π_y along the y-axis, that are linear:

$$\Pi_i\,(\lambda\mathbf{a} + \mu\mathbf{b}) = \lambda\Pi_i\,(\mathbf{a}) + \mu\Pi_i\,(\mathbf{b}) \quad (i = x, y) \tag{5.13}$$

satisfying

$$\Pi_i^2 = \Pi_i. \tag{5.14}$$

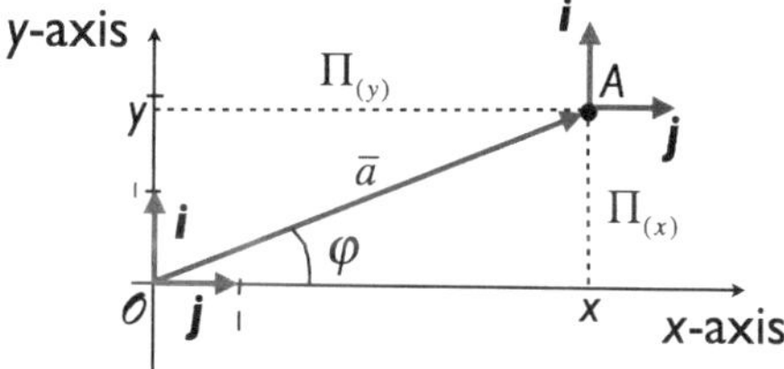

Fig. 5.1 $A = (x, y)$ has a support vector $\mathbf{a}$ satisfying $\Pi_x\mathbf{a}$ and $\Pi_y\mathbf{a}$, i.e., $x\mathbf{i} = \Pi_{(x)}\,\mathbf{a}$ and $y\mathbf{j} = \Pi_{(y)}\,\mathbf{a}$

This property (5.14) shows that the sub-space of projection is invariant under the projection. For Π_x, we have, for instance,

$$\Pi_x^2 \mathbf{a} = \Pi_x (\Pi_x \mathbf{a}) = \Pi_x x\mathbf{i} = x\Pi_x \mathbf{i} = x\mathbf{i}. \tag{5.15}$$

The idea of projections onto linear sub-spaces readily generalizes to higher dimensions. Consider, for instance, the Euclidean three-space $\mathbb{R}^3$ of Sect. 5.1. Let V denote the two-dimensional xy-coordinate plane described at the beginning of this section. For a given $\mathbf{a} \in \mathbb{R}^3$ in (5.1), let $\mathbf{a}'$ denote its orthogonal projection by Π_{xy} onto V. Algebraically, we have

$$\mathbf{a} = (x\mathbf{i} + y\mathbf{j}) + z\mathbf{k} = \Pi\mathbf{a} + z\mathbf{k} = \mathbf{a}' + \mathbf{e}, \tag{5.16}$$

where $\mathbf{e} = \mathbf{a} - \mathbf{a}' = z\mathbf{k}$ denotes a remainder, orthogonal to V that points to $\mathbf{a}$ away from $\mathbf{a}'$ in V. Then $\mathbf{a}' \in V$ is an approximation to $\mathbf{a} \in \mathbb{R}^3$ with error $\mathbf{e}$.

Example 5.1. Consider

$$\mathbf{a} = 2\mathbf{i} + 4\mathbf{j} + \mathbf{k} \tag{5.17}$$

and V given by the xy-coordinate plane as before. Then $\mathbf{a}' = \Pi_{(xy)}\, \mathbf{a} = 2\mathbf{i} + 4\mathbf{j}$ and $\mathbf{e} = \mathbf{k}$. In taking $\mathbf{a}'$ as an approximation to $\mathbf{a}$, how bad is $\mathbf{e}$? In absolute terms, we have

$$|\mathbf{e}| = 1. \tag{5.18}$$

In relative terms, however, we have

$$\frac{|\mathbf{e}|}{|\mathbf{a}|} = \frac{1}{\sqrt{21}} \simeq 0.22, \tag{5.19}$$

i.e., $\mathbf{a}'$ is possibly a useful first-order approximation to $\mathbf{a}$. As an alternative to (5.19), we can calculate the angle between $\mathbf{a}$ and $\mathbf{a}'$,

$$\cos\varphi = \frac{\mathbf{a} \cdot \mathbf{a}'}{|\mathbf{a}|\,|\mathbf{a}'|} = \frac{20}{\sqrt{20}\sqrt{21}} = \sqrt{\frac{20}{21}}. \tag{5.20}$$

With $\cos\varphi \simeq 1 - (1/2)\varphi^2$ for small angles φ, we hereby have

$$\varphi \simeq \sqrt{2}\,(1 - \cos\varphi)^{\frac{1}{2}} = \sqrt{2}\left(1 - \sqrt{\frac{20}{21}}\right)^{\frac{1}{2}} \simeq 0.22, \tag{5.21}$$

equivalent to (5.19); see Fig. 5.2.

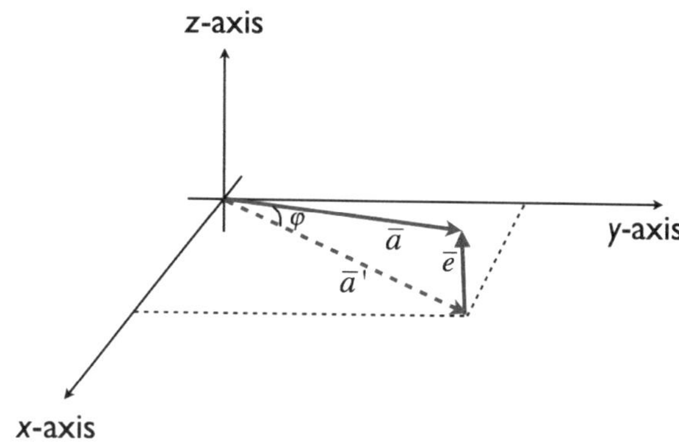

Fig. 5.2 Projection $\mathbf{a}'$ of a vector $\mathbf{a}$ onto the xy-plane V. The discrepancy $\mathbf{e} = \mathbf{a} - \mathbf{a}'$ indicates that $\mathbf{a}'$ serves as a leading order approximation to $\mathbf{a}$ provided that $\mathbf{e}$ has relatively small length or the angle of inclination φ is small

Following Sect. 3.3, we write our projections as

$$\Pi_x = \begin{pmatrix} 1 & 0 \\ 0 & 0 \end{pmatrix}, \quad \Pi_y = \begin{pmatrix} 0 & 0 \\ 0 & 1 \end{pmatrix}. \tag{5.22}$$

Evidently, we have

$$\begin{pmatrix} 1 & 0 \\ 0 & 0 \end{pmatrix}\begin{pmatrix} x \\ y \end{pmatrix} = \begin{pmatrix} x \\ 0 \end{pmatrix} = x\mathbf{i}, \quad \begin{pmatrix} 0 & 1 \\ 0 & 0 \end{pmatrix}\begin{pmatrix} x \\ y \end{pmatrix} = \begin{pmatrix} 0 \\ y \end{pmatrix} = x\mathbf{j} \tag{5.23}$$

as desired, and they can be seen to satisfy (5.14) by explicit calculation. Likewise, the orthogonal projection onto the xy-plane in $\mathbb{R}^3$ is

$$\Pi_{xy} = \begin{pmatrix} 1 & 0 & 0 \\ 0 & 1 & 0 \\ 0 & 0 & 0 \end{pmatrix} = \Pi_x + \Pi_y, \tag{5.24}$$

where Π_i are the projections onto the x- and y-axis analogous to (5.22).

Example 5.2. Projections can be used to derive the matrix of rotations as follows. A rotation of the orthonormal basis while keeping a vector $\mathbf{a}$ fixed satisfies

$$\mathbf{a} = x\mathbf{i} + y\mathbf{j} = x'\mathbf{i}' + y'\mathbf{j}'. \tag{5.25}$$

If $\{\mathbf{i}', \mathbf{j}'\}$ is a rotation of $\{\mathbf{i}, \mathbf{j}\}$ over φ (anti-clockwise), then

$$\mathbf{i}' = \mathbf{i}\cos\varphi + \mathbf{j}\sin\varphi, \quad \mathbf{j}' = -\mathbf{i}\sin\varphi + \mathbf{j}\cos\varphi, \tag{5.26}$$

where (Fig. 5.3)

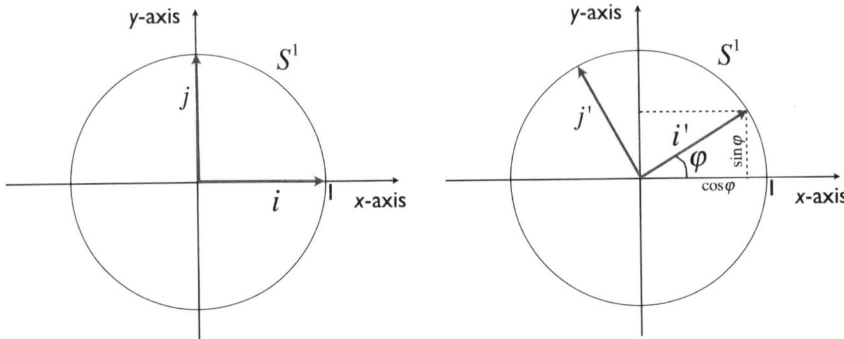

Fig. 5.3 Rotation of an orthonormal basis over an angle φ. These unit vectors rotate over the unit circle S^1

$$\cos\varphi = \mathbf{i}' \cdot \mathbf{i} = \mathbf{j}' \cdot \mathbf{j}, \quad \sin\varphi = \mathbf{i}' \cdot \mathbf{j} = -\mathbf{j}' \cdot \mathbf{i}. \tag{5.27}$$

Then

$$x'\mathbf{i}' + y'\mathbf{j}' = \left(x'\cos\varphi - y'\sin\varphi\right)\mathbf{i} + \left(x'\sin\varphi + y'\cos\varphi\right)\mathbf{j} = x\mathbf{i} + y\mathbf{j}. \tag{5.28}$$

In matrix form, the result is

$$\begin{pmatrix} x \\ y \end{pmatrix} = R(\varphi)\begin{pmatrix} x' \\ y' \end{pmatrix}, \tag{5.29}$$

in terms of the rotation matrix (3.76).

Alternatively, consider the rotation of **a** keeping the orthonormal basis fixed,

$$\mathbf{a} = x\mathbf{i} + y\mathbf{j} \to \mathbf{a}' = x'\mathbf{i} + y'\mathbf{j}. \tag{5.30}$$

Rotation of **a** over φ is equivalent to rotation of the ONB over $-\varphi$. Evidently, the rotation matrix satisfies

$$R(-\varphi) = R^{-1}(\varphi) = R^T(\varphi), \tag{5.31}$$

where R^{-1} refers to the inverse of R and R^T refers to the transpose. It follows that

$$\begin{pmatrix} x' \\ y' \end{pmatrix} = [R(-\varphi)]^{-1}\begin{pmatrix} x \\ y \end{pmatrix} = R(\varphi)\begin{pmatrix} x \\ y \end{pmatrix}. \tag{5.32}$$

This transformation may be verified by considering its action on the basis vectors:

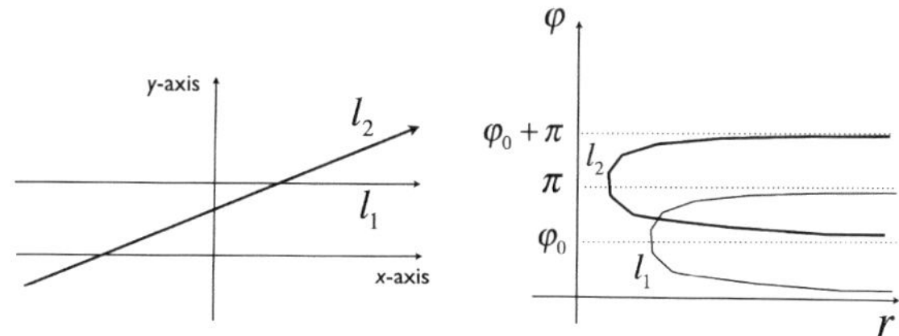

Fig. 5.4 Straight lines in $\mathbb{R}^2$ expressed in Cartesian coordinates (x, y) (*left*) and in polar coordinates (r, φ) (*right*)

$$\mathbf{i}' = R(\varphi)\mathbf{i} = \begin{pmatrix} \cos\varphi \\ \sin\varphi \end{pmatrix}, \quad \mathbf{j}' = R(\varphi)\mathbf{j} = \begin{pmatrix} -\sin\varphi \\ \cos\varphi \end{pmatrix}, \tag{5.33}$$

which recovers (5.26). We leave it as an exercise to show that $R(\varphi)$ is unitary, leaving invariant norms and angles:

$$|\mathbf{a}'| = |\mathbf{a}|, \quad \mathbf{a}' \cdot \mathbf{b}' = \mathbf{a} \cdot \mathbf{b}. \tag{5.34}$$

Example 5.3. Consider the problem of a point constraint to straight line

$$l: \quad y = a + bx \tag{5.35}$$

with intercept a at the y-axis and slope b. What point on this *constraint surface* l is closest to the origin? Geometrically, this happens at the point, whose support vector is orthogonal to l, as in Fig. 5.2,

$$(x\mathbf{i} + (a + bx)\mathbf{j}) \cdot (\mathbf{i} + b\,\mathbf{j}) = 0: \quad x = -\frac{ab}{1+b^2}, \tag{5.36}$$

giving a separation $\rho_0 = \sqrt{x^2+y^2} = a/\sqrt{1+b^2}$. The distance of points on l to the origin satisfy

$$\rho(x) = \sqrt{x^2 + (a+bx)^2} = \sqrt{(1+b^2)x^2 + 2abx + a^2}. \tag{5.37}$$

Now minimize $\rho(x)$,

$$\frac{d\rho(x)}{dx} = \frac{(1+b^2)x + ab}{\rho(x)} = 0: \quad x = -\frac{ab}{1+b^2}, \tag{5.38}$$

giving ρ_0 once again.

Alternatively, we may set up l in polar coordinates (Fig. 5.4). Projections of vectors in polar coordinates (r, φ) give us the vector components in Cartesian coordinates (x, y) according to

$$\mathbf{a} = x\mathbf{i} + y\mathbf{j}: \quad x = r\cos\varphi, \quad y = r\sin\varphi, \tag{5.39}$$

where $r = |\mathbf{a}|$ and $\cos\varphi = \hat{\mathbf{a}} \cdot \mathbf{i}$ is the angle of $\mathbf{a}$ to the x-axis. A special case is the horizontal line with $b = 0$, i.e., $y = y_0$ is a constant. In this event, (5.39) implies $r \sin\varphi = y_0 : r = y_0/\sin\varphi$ $(0 < \varphi < \pi)$ in case of $y_0 \neq 0$. The general case (5.35) obtains by rotation over ϕ_0, $b = \tan\varphi_0$,

$$r(\varphi) = \frac{\rho_0}{\sin(\varphi - \varphi_0)} \quad (\varphi_0 < \varphi < \varphi_0 + \pi), \tag{5.40}$$

and we read off that $r(\varphi)$ attains ρ_0 for $\varphi = \varphi_0$, i.e., whereby $\rho_0 = (a/b)\sin\varphi_0 = a\cos\varphi_0 = a/\sqrt{1+b^2}$ as before.

5.3 Snell's Law and Fermat's Principle

Snell's law describes the propagation of light encountering an interface between two media with possibly different indices of refraction n, that defines the phase velocity $c' = c/n$, where c denote the velocity of light in vacuum. We here discuss some of its derivations, that are illustrative for the relation between orthogonal projection and *minimal distances*.

For instance, in glass, n may assume values around $n = 1.5$ with small dependence on color. (With n of red light smaller than n of blue light.) Let $\mathbf{p_i}$ ($\mathbf{p}_r$) denote the momentum of a photon with angle of incidence θ_i (θ_r) to the normal of such interface. Snell's law states (Fig. 5.5)

$$n_i \sin\theta_i = n_r \sin\theta_r, \tag{5.41}$$

where θ_r denotes the angle of refraction of $\mathbf{p_r}$. The recordings of Claudius Ptolemaeus satisfy (5.41) rather well (Fig. 5.6).

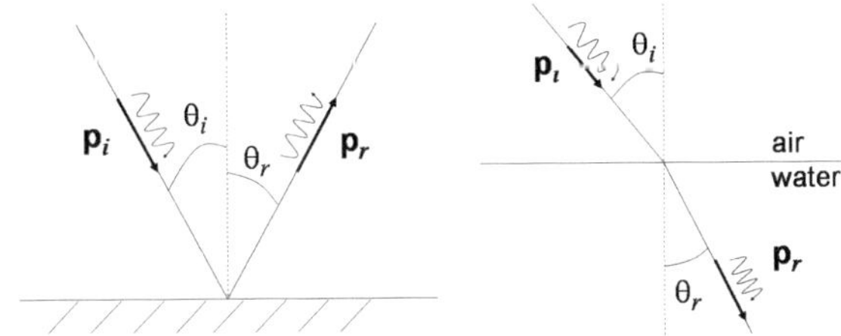

Fig. 5.5 (*Left*) Reflection of light in a mirror. Here, the normal component $p_\perp$ of the momentum $\mathbf{p}_i$ of the incident photon is reversed in sign in the momentum $\mathbf{p}_r$ of the reflected photon, while the tangential component $p_{||}$ parallel to the mirror is conserved. Total momentum is conserved in this process. (*Right*) Refraction of light from air to water due to a change of the velocity of light by a factor *1/n*. By conservation of energy and the relation $p = E/c'$, $c' = c/n$, where c is the velocity of light in vacuum, $p_r = np_i$, while $p_{||,r} = p_{||,i}$

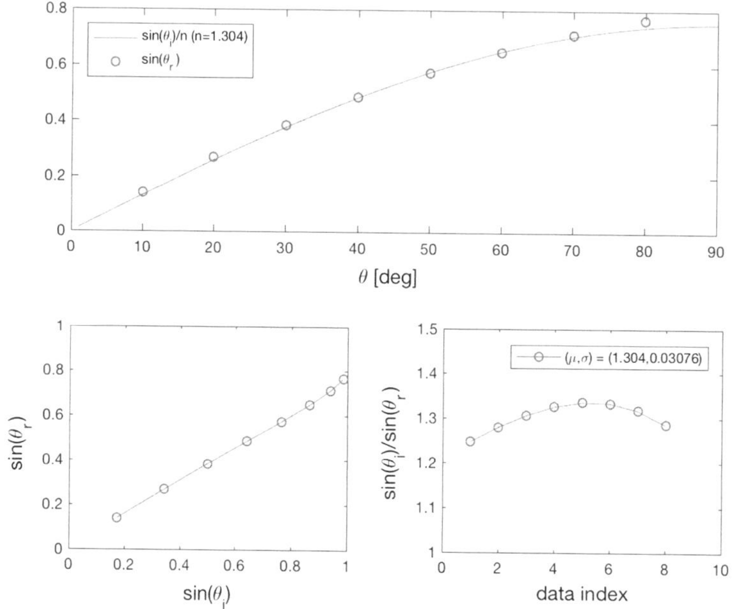

Fig. 5.6 A plot of Ptolemaeus' data and Snell's law showing an index of refraction $n \simeq 1.3$ of water. Ptolemaeus' data vary about Snell's law with a standard deviation of 2.3%

Let (x, y) denote a Cartesian coordinate system with the y-axis normal and the x-axis tangential to the interface with unit vectors $\mathbf{i}, \mathbf{j}$. The incident and refracted momenta can be decomposed in terms of their normal and tangential components

$$\mathbf{p} = p_\perp \mathbf{i}_y + p_{||}\mathbf{i}_x, \quad p_\perp = p\sin\theta, \quad p_{||} = p\cos\theta, \tag{5.42}$$

where $p = |\mathbf{p}|$.

In case of reflection (Fig. 5.5), p is conserved and $n_r = n_i$ and (5.41) reduces to $\theta_r = \theta_i$. In this event, reflection turns $p_\perp$ into $-p_\perp$, i.e., a change of $-2p_\perp$ per photon, much like bouncing off a ping pong ball on a hard surface.

Example 5.4. Consider the index of refraction of air,

$$n - 1 = 2.7 \times 10^{-4}\left(\frac{\rho}{\rho_0}\right), \tag{5.43}$$

where ρ_0 denotes the mass density under standard atmospheric conditions. Snell's llaw expresses a relation between the angles of incidence θ_i and refraction θ_r as light passes from a medium with index of refraction n_1 to a medium with index of refraction n_2,

$$n_1 \sin\theta_1 = n_2 \sin\theta_r. \tag{5.44}$$

Recall that both θ_i and θ_r are relative to the normal to the interface. In case of hot above cold air closer to the ground, n_1 exceeds n_2 higher up by (7.40), i.e., $n_1 > n_2 > 1$.

- In the propagation of light from the ground to the interface above between the two layers, there is a critical angle θ_i for *total reflection*.
- For an angle θ_i close to $\pi/2$, we can write $\theta_i = \pi/2 - \epsilon$, where ϵ is small. Then $\sin\theta_i = \cos\epsilon$. The result is conveniently expressed by a Taylor series expansion of $\cos\epsilon$ for small ϵ.
- For a temperature difference 30 K, this leads to an estimate of θ_i.[1]

To address these three points, we note that total reflection occurs for $\sin\theta_r = \pi/2$, i.e.,

$$\sin\theta_i = \frac{n_2}{n_1} = \frac{n_2}{n_2 + \Delta n} \simeq 1 - \frac{\Delta n}{n_2} \simeq 1 - \Delta n. \tag{5.45}$$

Here, $n_1 = n_2 + \Delta n$, $\Delta n = n_1 - n_2 = 2.7 \times 10^{-4}\,(\Delta\rho/\rho_0)$. According to the ideal gas law, changes in temperature (at constant pressure) satisfy $\rho T =$ const., i.e.,

$$\frac{\Delta\rho}{\rho_0} = -\frac{\Delta T}{T_0} \simeq \frac{30\,\mathrm{K}}{300\,\mathrm{K}} = 0.1. \tag{5.46}$$

Consequently, $\Delta n \simeq 2.7 \times 10^{-5}$. Also,

$$\sin\left(\frac{\pi}{2} - \epsilon\right) = \cos\epsilon = 1 - \frac{1}{2}\epsilon^2 + O(\epsilon^2). \tag{5.47}$$

Combining the above, we solve (5.45)–(5.47),

$$1 - \Delta n = 1 - \frac{1}{2}\epsilon^2 \tag{5.48}$$

to obtain

$$\epsilon = \sqrt{2\Delta n} \simeq 0.0075\,\mathrm{rad} = 0.42^\circ. \tag{5.49}$$

[1] This defines the critical angle for a *Fata Morgana* to occur.

Radiation reflected by a mirror exerts a net normal pressure with vanishing tangential pressure. Since $p = E/c$ for a single photon of energy E, incoming radiation with intensity I carries a pressure

$$P = \frac{I}{c}\cos^2\theta_i, \tag{5.50}$$

where $[I] = \text{erg s}^{-1}\text{ cm}^{-2}$. Daylight intensity from the Sun reaches about 1400 W per square meter ($I = 1.4\times10^6\text{ erg s}^{-1}$) at perihelion, corresponding to a normal pressure of about 5×10^{-5} dyn cm^{-2} (5 μPa). These pressures are generally imperceptibly small but can be measured in table top experiments and may possibly be used for propulsion of small satellites equipped with solar sails (e.g., [1]).

In light propagation through an interface from, say, air ($n = 1$) to water ($n = 1.33$), there are two invariants: energy (color) and tangential momentum (no shear stress), whereby

$$\text{(a)}p_r = np_i, \quad \text{(b)}p_{||,r} = p_{||,i}. \tag{5.51}$$

With the projection $p_{||} = p\sin\theta$, Snell's law immediately follows,

$$\sin\theta_i = n\sin\theta_r. \tag{5.52}$$

Reflection of light in a mirror and propagation of light across two media according to Snell's law illustrate the *geometric-optics* approximation by ray-tracing, in which light propagates along piecewise linear trajectories.

In generalizing to stratified media described by a gradient in n, propagation is found to be along curved trajectories. In air, n increases with density and hence it decreases with height in the Earth's atmosphere, $n - 1 \simeq 2.7\times10^{-4}$ at standard atmospheric conditions down to essentially zero in the stratosphere and beyond. Light entering the atmosphere at a slight angle above the horizon (θ_i close to $\pi/2$) hereby bends gently over towards the surface of the Earth ($\theta_r < \theta_i$), allows to see the Sun for a few more minutes at sunset, even as it has already gone down under the visible horizon.

Remarkably, Snell's law also describes a minimum path length in light propagation from A to B in different media separated by an interface. This is commonly expressed in therms of travel time $\Delta t = t_B - t_A$, the result of which is *Fermat's principle*.[2] Fermat's principle posits that the travel time of the actual path is extremal relative to that of alternative, neighboring paths.

[2]In conserving photon energy, E of the incident and refracted photons being the same, this is equivalent to extremizing the action $S = E\Delta t$, i.e., distance measured in total phase, relevant to covariant formulations of Fermat's principle.

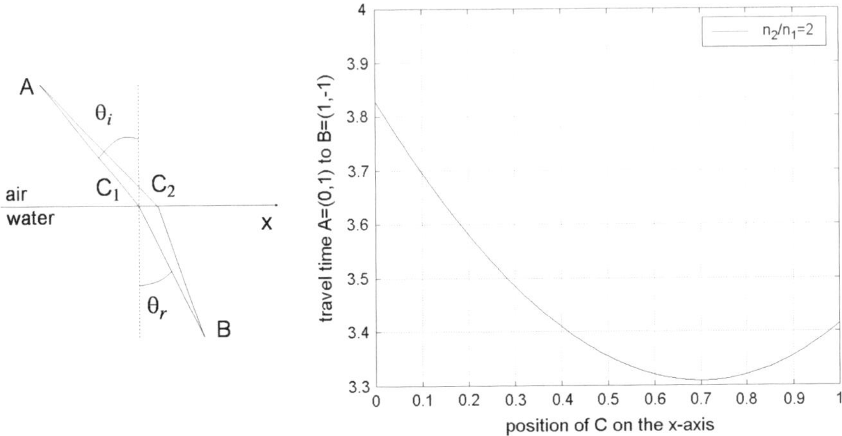

Fig. 5.7 (*Left*) Schematic overview of two neighboring piecewise linear paths from A in air to B in water with corresponding different intermediate points C_1 and C_2 at the interface on the x-axis. (*Right*) Light travel time from $A = (0, 1)$ in medium 1 to $B = (1, -1)$ in medium 2 via straight lines AC and CB with $C = (0, x)$ on the interface between with index of refraction $n_1 = 1$ and n_2. By Fermat's principle, the minimum in the curve shown identifies C for the optimal piece wise linear path from A to B

Example 5.5. In a homogeneous isotropic medium, points A, B and C satisfy the triangle inequality[3]

$$|\mathbf{r}_A - \mathbf{r}_B| \leq |\mathbf{r}_A - \mathbf{r}_C| + |\mathbf{r}_C - \mathbf{r}_B|, \tag{5.53}$$

where the $\mathbf{r}_i$ refer to the position vectors to the points i = A, B and C. Fermat's principle implies that the true photon path from A to B is the straight line from A to B.

Consider photon paths by piecewise linear trajectories from A $= (0, y)$ $(y > 0)$ in air to B $= (b, c)$ $(c < 0)$ in water via a point C $= (0, x)$ at the air-water interface along the x-axis (Fig. 5.7). Thus, $n = 1$ in air $(y > 0)$ above and $n > 1$ in water $(y < 0)$ below the interface, and hence

$$\Delta t(x) = AC + nBC = \sqrt{x^2 + y^2} + n\sqrt{(x - b)^2 + c^2}, \tag{5.54}$$

where n is the index of refraction of water. Figure 5.7 illustrates $cT(x)$ scaled to $n = 2$. The extremal value in $\Delta t(x)$ attains at

[3]In flat space with Euclidean metric, the triangle inequality holds locally and globally. In curved space with Euclidean signature, it holds locally.

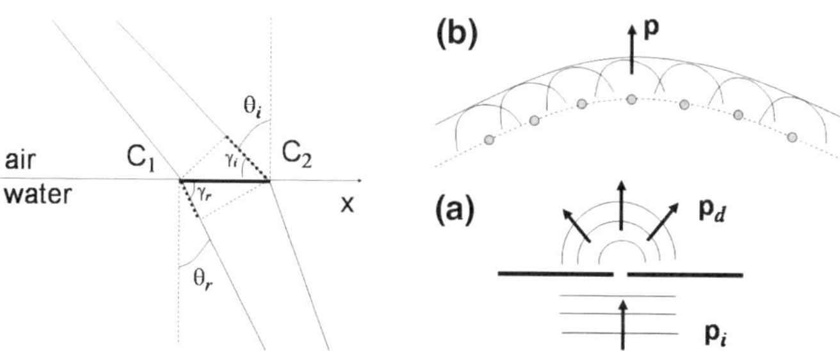

Fig. 5.8 (*Left*) The net change in travel time results from ΔAC $= \Delta x \cos\gamma_i$ and ΔCB $= -\Delta x \cos\gamma_r$. (*Right*) **a** Schematic overview of wave *diffraction* through a pinhole with isotropic dispersion of momentum in the incoming wave with all directions in accord with Fermat's principle. **b** Huygens attributes the same to waves creating by an elementary point source. Repeating Huygens argument describes the propagation of a wave front (*solid curve*) of synchronous point sources (*solid dots*) at a common distance to the pinhole, excited by prior wave front (*dashed curve*)

$$0 = \frac{d\Delta t(x)}{dx} = \frac{x}{\sqrt{x^2+y^2}} + n\frac{x-b}{\sqrt{(x-b)^2+c^2}} = \sin\theta_i - n\sin\theta_r, \quad (5.55)$$

which recovers (5.52).

Feynman gives an interesting local interpretation of (5.55) [2] shown in Fig. 5.8. The extremum in $s(x)$ attains when the displacement Δx (thick continuous line) in C (neighboring paths C_1 and C_2) is such that the lengthening in AC is canceled by the shortening of BC in regards to travel times, i.e.,

$$\Delta AC + n\Delta CB = 0. \quad (5.56)$$

By inspection, it corresponds to the thick dashed line-segments,

$$\Delta AC = \Delta x \cos\gamma_i, \quad \Delta CB = -\Delta x \cos\gamma_r. \quad (5.57)$$

With $\gamma_i = \pi/2 - \theta_i$ and $\gamma_r = \pi/2 - \theta_r$, Snell's law (5.52) follows once more.

Example 5.6. In the propagation of light in the Earth's atmosphere, consider two alternative paths from A to B on the Earth's surface, e.g., straight versus bend over an arc over an angle α_0 with radius of curvature R. If $\rho = \rho_0(1-z/H)$ is the local density with scale height $H \simeq 6$ km, then by (5.43)

$$n(z) = n_0\left(1 - \epsilon z/H\right), \quad (5.58)$$

where $n_0 = 1+\eta \simeq 1$, $\epsilon = \frac{\eta}{1+\eta} \simeq \eta$, $\eta = 2.7 \times 10^{-4}$. By Fermat's principle, the path of shortest travel time defined by the phase velocity $v = c/n$ is closest to the true path. In what follows, we put $c = 1$. Thus, compare $T_0 = 2R\sin\alpha_0$

with the travel time over the arc defined by the height

$$z = h_0 - R(1 - \cos\alpha) \simeq h_0\left(1 - \frac{R\alpha^2}{2h_0}\right) \quad (-\alpha_0 \le \alpha \le \alpha_0). \tag{5.59}$$

Note that the mean height $\bar{z} = \alpha_0^{-1}\int_0^{\alpha_0} z(\alpha)d\alpha = (2/3)h_0$, $h_0 = (1/2)R\alpha_0^2$ with an associated reduction in travel time by $(2/3)\epsilon h_0/H$, that appears in the difference

$$T - T_0 = \int_{-\alpha_0}^{\alpha_0} n(z)Rd\alpha - T_0 \simeq \frac{1}{3}\alpha_0 R\left(\alpha_0^2 - 4\epsilon'\right), \tag{5.60}$$

where $\epsilon' = \epsilon h_0/H$ and using $\sin\alpha_0 = \alpha_0 - (1/6)\alpha_0^3 + O(\alpha_0^5)$. So, light travels along curved trajectories whenever $2\epsilon R/H > 1$, i.e.,

$$\epsilon > \frac{H}{2R} \ge \frac{H}{2R_\oplus} \simeq 5 \times 10^{-4}. \tag{5.61}$$

Since $\epsilon = 2.7 \times 10^{-4}$, light between A and B has no tendency take a path with radius of curvature less that the radius of the Earth. Nevertheless, bending at $R > R_\oplus$ does exist, notably for incoming light allowing us to see the Sun for a prolonged time, when it has already dropped below the Earth's horizon.

When a plane wave of light hits a small pinhole of a size less than the wave length, Fermat's principle predicts that a wave front disperses isotropically in all directions, as Δt is the same for propagation to a hemisphere centered about the pinhole independent of direction. In Huygen's interpretation, the pinhole contains an elementary point source creating waves in all directions (Fig. 5.8).

5.4 Fitting Data by Least Squares

Projections give a method of fitting curves to observational data, common to the understanding and interpretation of data. A *best-fit* refers to a compromise that minimizes a pre-defined error. Consider two sets of observations A and B in (x_i, y_i):

$$(A)\ \begin{array}{cc} x_i & y_i \\ \hline 1 & k \\ 2 & 2+\epsilon \\ 3 & 3 \end{array} \qquad (B)\ \begin{array}{cc} x_i & y_i \\ \hline 1 & k+\epsilon \\ 2 & 2 \\ 3 & 3 \end{array} \tag{5.62}$$

Here, k shall refer to 1 or 2 and ϵ represents an error. These data can refer to a short time series, e.g., observations on the height $z = y_i$ of our proverbial apple falling from the tree at instances $t_i = x_i$, $i = 1, 2, 3$. Or it can be the pricing of a security on the stock market in three consecutive days. Either way, we wish to look at the data with the support of a line that most closely resembles running the three points—which will be approximate whenever $(k, \epsilon) \neq (1, 0)$.

The problem of fitting can be approached as the minimization of squares of errors $e_i = y_i - f_i$, expressed by

$$\rho(a, b) = |\mathbf{e}|^2 = \sum_{i=1}^{3} (y_i - f_i)^2 \tag{5.63}$$

where $f_i = f(x_i)$ is our choice of fitting function, here the linear function

$$f(x) = a + bx. \tag{5.64}$$

Thus, our error is a function of the unknown coefficients (a, b). Best-fit in terms of *least squares* aims at finding (a, b) such that $\rho(a, b)$ is minimal,

$$d\rho = \sum_{i=1}^{3} e_i df_i = \mathbf{e}^T d\mathbf{f} = 0 \tag{5.65}$$

showing that $\mathbf{e}$ and the constraint variations $d\mathbf{f}$ defined by (5.64) are mutually orthogonal as in Fig. 5.2.

According to (5.65), we require

$$\frac{\partial \rho(a, b)}{\partial a} = 0, \quad \frac{\partial \rho(a, b)}{\partial b} = 0. \tag{5.66}$$

Since (5.63) is quadratic in $f_i = f(x_i)$, it will be quadratic in (a, b), so that (5.66) defines two equations linear in the (a, b). In other words, (5.66) defines a system of two equations in the two unknowns (a, b), that we should be able to solve algebraically.

To be precise, (5.66) is

$$\sum_{i=1}^{n} (y_i - (a + bx_i)) = 0, \quad \sum_{i=1}^{n} (y_i - (a + bx_i)\, x_i = 0, \tag{5.67}$$

where we silently generalized to an arbitrary number of n data points. Writing it out gives

$$\sum_{i=1}^{n} y_i = na + b \sum_{i=1}^{n} x_i, \quad \sum_{i=1}^{n} x_i y_i = a \sum_{i=1}^{n} x_i + b \sum_{i=1}^{n} x_i^2. \tag{5.68}$$

This may be written somewhat cleaner by using the averages

$$\bar{x} = \frac{1}{n}\sum_{i=1}^{n} x_i, \quad \bar{y} = \frac{1}{n}\sum_{i=1}^{n} y_i, \tag{5.69}$$

so that (5.68) becomes

$$\begin{pmatrix} n & n\bar{x} \\ n\bar{x} & \sum_{i=1}^n x_i^2 \end{pmatrix} \begin{pmatrix} a \\ b \end{pmatrix} = \begin{pmatrix} n\bar{y} \\ \sum_{i=1}^n x_i y_i \end{pmatrix}. \tag{5.70}$$

Dividing the first equation by n, we have

$$\begin{pmatrix} 1 & \bar{x} \\ n\bar{x} & \sum_{i=1}^n x_i^2 \end{pmatrix} \begin{pmatrix} a \\ b \end{pmatrix} = \begin{pmatrix} \bar{y} \\ \sum_{i=1}^n x_i y_i \end{pmatrix}. \tag{5.71}$$

Inverting (5.71) gives

$$\begin{pmatrix} a \\ b \end{pmatrix} = D^{-1} \begin{pmatrix} \sum_{i=1}^n x_i^2 & -\bar{x} \\ -n\bar{x} & 1 \end{pmatrix} \begin{pmatrix} \bar{y} \\ \sum_{i=1}^n x_i y_i \end{pmatrix} \tag{5.72}$$

where

$$D = \sum_{i=1}^{n} x_i^2 - n\bar{x}^2. \tag{5.73}$$

It follows that

$$a = \frac{1}{D}\left[\bar{y}\sum_{i=1}^{n} x_i^2 - \bar{x}\sum_{i=1}^{n} x_i y_i\right], \quad b = \frac{1}{D}\left[\sum_{i=1}^{n} x_i y_i - n\bar{x}\bar{y}\right]. \tag{5.74}$$

Example 5.7. Let us apply (5.74) to our data set A in (5.62) for which $n = 3$. For $k = 1$, we find

$$a = \frac{\epsilon}{3}, \quad b = 1. \tag{5.75}$$

This result for the least square fit of a straight line to (5.62) brings about the balanced approach, in that the contribution of each data point is weighted evenly by *1/n* in a. In viewing the deviation from the straight line to be at the mid-point x_2, the slope is the same as that of the line through (x_1, y_1) and (x_3, y_3).

5.5 Gauss-Legendre Quadrature

Integration by finite summation allows for higher order schemes by considering non-uniform partitions of the nodes x_i of interpolation. Consider an approximation of a function $f(x)$ on $[-1, 1]$ by a polynomial $p(x)$ of degree $2n - 1$,

$$p(x) = c_0 + c_1 x + \cdots + c_{2n-1} x^{2n-1}. \tag{5.76}$$

Given its $2n$ coefficients $\{c_k\}_{k=0}^{2n-1}$, we can factor $p(x)$ over a Legendre polynomial $P_n(x)$ of order n and a polynomial $A(x)$ with a remainder $B(x)$, both of degree $n-1$,

$$p(x) = A(x)P_n(x) + B(x). \tag{5.77}$$

To see this, use the n coefficients in $A(x) = A_0 + A_1 x + \cdots A_{n-1}x^{n-1}$ to match the upper half c_k, $k \geq n$. With $P_n(x) = a_0 + a_1 x + \cdots + a_n x^n$, this matching obtains from

$$\begin{aligned}
&A_{n-1}a_n = c_{2n-1},\\
&A_{n-1}a_{n-1} + A_{n-2}a_n = c_{2n-2},\\
&A_{n-1}a_{n-2} + A_{n-2}a_{n-1} + A_{n-3}a_n = c_{2n-3},\\
&\cdots\\
&A_{n-1}a_1 + A_{n-2}a_2 + \cdots + A_0 a_n = c_n.
\end{aligned} \tag{5.78}$$

The polynomial $B(x)$ of degree $n - 1$ can now be used to match the remaining c_k, $0 \leq k \leq n - 1$.

Next, we define Gauss-Legendre quadrature by a weighted sum (e.g., [3])

$$\int_{-1}^{1} p(x)dx = \sum_{i=1}^{n} w_i p(x_i) = \sum_{i=1}^{n} w_i B(x_i) \tag{5.79}$$

of function values at the zeros $\{x_i\}_{i=1}^{n}$ of

$$P_n(x) = a_n \Pi_{i=1}^{n}(x - x_i). \tag{5.80}$$

The n coefficients $\{w_i\}_{i=1}^{n}$ are the weights at x_i, that sum up to the size of the domain,

$$\sum_{i=1}^{n} w_i = 2. \tag{5.81}$$

Integration of $p(x)$ of degree $2n-1$ hereby reduces to the problem of exact integration of a polynomial of degree $n-1$ set by lower half c_k, $0 \leq k \leq n-1$, in the interpolating expression

$$B(x) = B_0 + B_1 x + \cdots B_{n-1} x^{n-1} = \sum_{i=1}^{n} B(x_i) \Pi_{j\neq i} \frac{x - x_j}{x_i - x_j}. \tag{5.82}$$

Requiring

$$\sum_{i=1}^{n} w_i B(x_i) = \int_{-1}^{1} B(x)dx = \sum_{i=1}^{n} B(x_i) \int_{-1}^{1} \Pi_{j\neq i} \frac{x - x_j}{x_i - x_j} dx, \tag{5.83}$$

we shall thus insist

$$w_i = \int_{-1}^{1} \Pi_{j\neq i} \frac{x - x_j}{x_i - x_j} dx = \frac{1}{P_n'(x_i)} \int_{-1}^{1} \frac{P_n(x)}{x - x_i} dx. \tag{5.84}$$

Let $Q(x)$ denote a polynomial up to degree n and write $Q(x) = Q(x) - Q(x_i) + Q(x_i)$. The expression on the right hand side satisfies

$$\int_{-1}^{1} Q(x) \frac{P_n(x)}{x - x_i} dx = \int_{-1}^{1} \frac{Q(x) - Q(x_i)}{x - x_i} P_n(x) dx + Q(x_i) \int_{-1}^{1} \frac{P_n(x)}{x - x_i} dx. \tag{5.85}$$

Since $(Q(x) - Q(x_i))/(x - x_i)$ is of degree $n - 1$ (or less), the first integral on the right hand side vanishes by orthogonality to $P_n(x)$, leaving

$$\int_{-1}^{1} \frac{P_n(x)}{x - x_i} dx = Q(x_i)^{-1} \int_{-1}^{1} Q(x) \frac{P_n(x)}{x - x_i} dx. \tag{5.86}$$

To proceed, we apply (5.86) to

$$\int_{-1}^{1} \frac{P_n(x)}{x - x_i} dx = \frac{1}{P_n'(x_i)} \int_{-1}^{1} P_n'(x) \frac{P_n(x)}{x - x_i} dx, \tag{5.87}$$

whereby

$$\int_{-1}^{1} \frac{P_n(x)}{x - x_i} dx = \frac{1}{2P_n'(x)} \left(\left[\frac{P_n^2(x)}{x - x_i} \right]_{-1}^{1} + \int_{-1}^{1} \frac{P_n^2(x)}{(x - x_i)^2} dx \right). \tag{5.88}$$

Since $P_n^2(\pm 1) = 1$ and $P_n(x_i) = 0$,

$$\int_{-1}^{1} \frac{P_n^2(x)}{(x - x_i)^2} dx = \int_{-1}^{1} Q(x) \frac{P_n(x)}{x - x_i} dx \tag{5.89}$$

with a polynomial $Q(x)$ of degree $n - 1$,

$$Q(x) = \frac{P_n(x) - P_n(x_i)}{x - x_i} = a_n \Pi_{j\neq i} (x - x_j) \to P_n'(x_i) \tag{5.90}$$

Table 5.1 Nodes and weights of Gauss-Legendre quadrature

$2n-1=5$		$2n-1=9$		$2n-1=19$	
$\pm x_i$	w_i	$\pm x_i$	w_i	$\pm x_i$	w_i
0	0.56888	0.14887	0.29552	0.07652	0.15275
0.53846	0.47862	0.43339	0.26926	0.22778	0.14917
0.90617	0.23692	0.67940	0.21908	0.37370	0.14209
–	–	0.86506	0.14945	0.51086	0.13168
–	–	0.97390	0.06667	0.63605	0.11819
–	–	–	–	0.74633	0.10193
–	–	–	–	0.83911	0.08327
–	–	–	–	0.91223	0.06267
–	–	–	–	0.96397	0.04060
–	–	–	–	0.99312	0.01761

as x approaches x_i. Applying (5.86) once more, it follows from (5.88) that

$$\int_{-1}^{1} \frac{P_n(x)}{x-x_i}dx = \frac{1}{2P_n'(x_i)}\left(\frac{1}{1-x_i}+\frac{1}{1+x_i}+P_n'(x_i)\int_{-1}^{1}\frac{P_n(x)}{x-x_i}dx\right), \tag{5.91}$$

and hence

$$\frac{1}{2}\int_{-1}^{1}\frac{P_n(x)}{x-x_i}dx = \frac{1}{(1-x_i^2)P_n'(x_i)}. \tag{5.92}$$

Accordingly, (5.84) becomes (Table 5.1)

$$w_i = \frac{2}{(1-x_i^2)\left[P_n'(x_i)\right]^2}. \tag{5.93}$$

Example 5.8. The Lorentzian distribution,[4]

$$I(s) = \int_{-1}^{1} f(x)dx, \quad f(x) = \frac{1}{s+x^2} \tag{5.94}$$

can be integrated by Gauss-Legendre quadrature. Fig. 5.9 shows the results as a function of the number of nodes Fig. 5.9.

Table 5.2 gives a summary of this discussion.

[4]For numerical integration over the real line, see, e.g., [4].

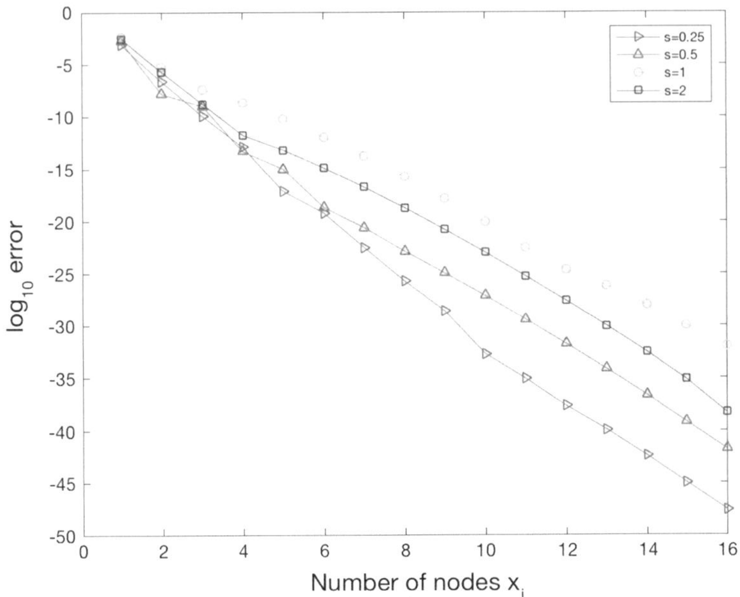

Fig. 5.9 Gauss-Legendre quadrature applied to (5.94). The results show rapid exponential convergence as a function of the number of nodes, especially so when the poles of $z = \pm i\sqrt{s}$ of $f(x)$ move further away from the real line

5.6 Exercises

5.1. Recall the first few Legendre polynomials $P_0 = 1$, $P_1 = x$ and $P_2 = \frac{1}{2}(3x^2 - 1)$. Verify by explicit calculation that these functions are orthogonal in the sense of the inner product

$$(f, g) = \int_{-1}^{1} f(x)g(x)dx \tag{5.95}$$

between two functions $f(x)$ and $g(x)$ on $[-1, 1]$.

5.2. Consider the matrix

$$A = \begin{pmatrix} 1 & 2 \\ 2 & 0 \end{pmatrix}. \tag{5.96}$$

Is A a projection?

5.3. For a unit vector $\mathbf{u}$ in $\mathbb{R}^n$, consider

$$\Pi = I - k\mathbf{u}\mathbf{u}^T, \tag{5.97}$$

Table 5.2 Overview projections and minimal distances

1. Vector fields $\mathbf{a} = \mathbf{a}(t, x, y, z)$ are elements of a linear vector space defined over time and space. The can be expressed by orthogonal projections on basis vectors, tangent to curves of constant coordinates. Different choices of coordinates generally bring along different basis vectors.

2. Integration by summation can be extended to quadratures with exponential convergence for smooth functions, such as Gauss-Legendre quadrature.

3. Orthogonal projections Π are linear maps onto sub-spaces satisfying $\Pi^2 = \Pi$. Projections provide approximations to vectors in lower dimensions and provide a starting point for data fitting, that minimize errors in terms of least square errors. Least squares gives a model-independent and unbiased fit, wherein all points are weighted equally (canonical formulation).

4. The geometric-optics approximation describes ray tracing, applicable when the wave length light is much smaller than the length scale of the geometry at hand. In homogeneous media, light rays propagate as straight lines with conservation of photon energy and total momentum.

5. Light propagation across interface between different media is described by Snell's law, that describes conservation of tangential photon momenta or, equivalently, minimization of total travel time (Fermat's principle). In the isotropic diffraction of light passing through a small pinhole, Fermat's principle reduces to Huygen's principle.

where the transpose T takes the column vector $\mathbf{u}$ into a row vector $\mathbf{u}^T$.

(i) For $k = 1$, show Π is a projection. What is the plane of the projection?
(ii) Consider $\mathbf{u}^T = (0, 0, 1)$ in $\mathbb{R}^3$. If $\mathbf{a}^T = (4\cos\alpha, 4\sin\alpha, 2 + \sin\alpha)$, compute $\mathbf{a}' = \Pi\,\mathbf{a}$ and determine $\alpha = \alpha_0$ when $\mathbf{e} = \mathbf{a} - \mathbf{a}'$ is smallest in norm. What is the error $e = ||\mathbf{e}||$ and the associated angle between $\mathbf{a}$ and $\mathbf{a}'$? Similar for $\mathbf{a}^T = (4, 3, 1 + x^2)$.
(iii) For $k = 2$ with $\mathbf{u}^T = (0\ \ 0\ \ 1)$, describe Π for $k = 2$.

5.4. Derive the best fit of a linear polynomial (5.64) to the first set A with $k = 2$. Sketch the result including the error $\delta = (y_1 - f_1, y_2 - f_2, y_3 - f_3)$.

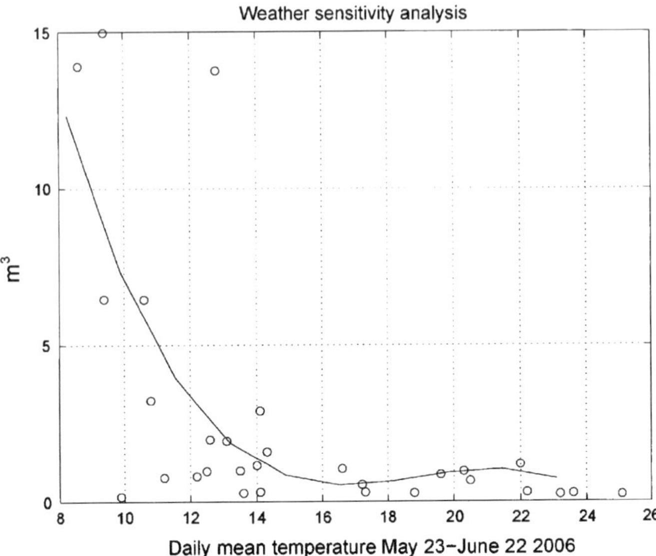

Fig. 5.10 A distribution of heating measured in m^3 of natural gas usage per day as a function of mean outside temperature over the course of one month. The *continuous curve* is a best-fit cubic interpolation (Reprinted from van Putten, M.H.P.M., & van Putten, A.F.P., 2007, Proc. Roy. Soc. London A., 463, 2495, US Pat. 7636666, 2009)

5.5. Derive the best fit of a linear polynomial (5.64) to the second data set B in (5.62) for $k = 1, 2$. Sketch the results including the error $\delta = (y_1 - f_1, y_2 - f_2, y_3 - f_3)$.

5.6. Consider estimating home energy efficiency based on sensitivity $\Delta h_i/\Delta T_i$ of heating h_i in natural gas usage [m^3] to fluctuations in outside temperature T_i (Fig. 5.10) about a mean T_0. Determine $\Delta h_i/\Delta T_i$ from the slope of a best-fit linear approximation to the following (approximate) data:

T_i	h_i	T_i	h_i
8.5	14	12.5	2
9.5	15	13	14
9.5	6.5	13	2
10	0	13.5	1
10.5	6.5	13.5	0
10.5	3.5	14	0
11	1	14	1.5
12	1	14	2
12.5	1	14	3

(5.98)

What is T_0? How does this efficiency change for a choice of $T_0' < T_0$? For the one-month observation shown in Fig. 5.10, the scatter in data about the best-fit linear approximation is considerable. What does this say about the estimated efficiency and how can this limitation be ameliorated?

5.7. Show that the n-th coefficient a_n in the Legendre polynomial $P_n(x) = a_0 + a_1 x + \cdots + a_n x^n$ of degree n satisfies

$$a_n = \frac{(2n)!}{2^n (n!)^2}. \tag{5.99}$$

5.8. For $f(x) = e^{-x^2}$ on $[-1, 1]$, numerically evaluate the errors in the integration of by 3 and 5 point Gauss-Legendre quadrature. Determine the errors and compare with those in the trapezoidal rule above.

5.9. Based on (5.40), identify the trigonometric relations giving $\rho_0 = a/\sqrt{1+b^2}$.

5.10. Elaborating further on straight lines, we can express straight lines to be shortest paths $(r(\lambda), \varphi(\lambda))$ between two points A and B by the minimum of the Euclidean distance

$$S = \int_A^B \sqrt{\dot{r}^2 + r^2\dot{\varphi}^2}\, d\lambda. \tag{5.100}$$

If $d\lambda = ds$ along the extremal path $(\sqrt{\dot{r}^2 + r^2\dot{\varphi}^2} = 1)$, work out the extremal condition $\delta S = 0$ with respect to variations δr and $\delta\varphi$ subject to $\delta r = \delta\varphi = 0$ at A and B.[5] Identify the two resulting integrals of motion with conservation of energy and angular momentum along straight lines.

5.11. Show that light rays passing through a slab of glass propagate with a parallel displacement.

5.12. Consider a an ellipse with semi-major axis a and semi-minor axis b, $a \geq b$, described by

$$\frac{x^2}{b^2} + \frac{(y-a)^2}{a^2} = 1. \tag{5.101}$$

1. Derive (5.101) from the definition of an ellipse as a closed curve of points with constant sum of distances l_1 and l_2 to two focal points, say, at $(-p, 0)$ and $(p, 0)$ along the x-axis, $p = \sqrt{a^2 - b^2}$.
2. Show that the minor semi-axis satisfies $b = a\sqrt{1-e^2}$ in terms of the ellipticity e.
3. For small x, y about the origin, show that (5.101) reduces to a parabola. Identify the focal point of reflected light rays along the y-axis of light rays coming in parallel to the y-axis. How does it relate to p?

5.13. Light refraction in Newton's prism experiments emulates a rainbow: blue light is refracted relatively more than red light. Sketch the associated *dispersion relation*

[5] These are the Euler-Lagrange equations of a free particle in polar coordinates.

$\omega = \omega(k)$ for glass, $E = \hbar\omega$, $p = \hbar k$, where $\hbar$ denotes Planck's constant. For reference, some particular glass used in prisms has $n \simeq 1.54$ for blue (400 nm) and $n \simeq 1.51$ for red light (700 nm). Compare your result with the dispersion relation of vacuum.

5.14. Light *can* reflect off water when its rays are almost skimming the surface. Derive the critical angle θ_* that defines the window $\theta_* < \theta_i < \pi/2$ of complete reflection.

References

1. Demir, D., 2011, *A table top demonstration of radiation pressure* (Diplomarbeit, University of Vienna).
2. Feynman, R.P., Leighton, R.B., & Sands, M., Lectures on Physics, Vol. I, 1963, Ch. 26.
3. Gautschi, W., 2004, *Orthogonal Polynomials: Computation and Approximation*. Numerical Mathematics and Scientific Computation (Oxford University Press, New York).
4. Takahasi, H., & Mori, M., 1974, "Double Exponential Formulas for Numerical Integration," Publ. RIMS, Kyoto University, vol. 9, 721–741.

Chapter 6
Spectral Methods and Signal Analysis

Efficient and accurate representations of functions is frequently approached by *spectral methods*. In some sense as a converse, these methods also provide a starting point for signal analysis in noisy data. We begin with an example comparing Taylor series and expansions in Legendre polynomials. Further, the Fourier transform provides a general method to represent analytic functions with exponential convergence. It enables efficient data-analysis by correlations.

6.1 Basis Functions

Consider a function $f(x)$ on some sub-domain D of the real line. We set out to expand $f(x)$ in terms of an ensemble of *basis* functions $\{\phi_n(x)\}_{n=0}^{\infty}$. Quite generally, we seek expansions

$$f(x) = \Sigma_{n=0}^{\infty} a_n \phi_n(x), \tag{6.1}$$

using a suitable choice of coefficients a_n. The class of basis functions one might consider is quite broad. Perhaps the most familiar are $\phi_n(x) = (x-a)^n$ encountered in Taylor series expansions about $x = a$. Upon analytic continuation into the complex plane, the Taylor expansion functions $\{z^n\}_{n=0}^{\infty}$ define an orthogonal basis set of analytic functions about the origin, e.g., in the unit disk with continuity on S^1 in $\mathbb{C}$ with inner product

$$\langle f(z), g(z)\rangle = \int_0^{2\pi} f(z)\bar{g}(z)d\varphi. \tag{6.2}$$

M.H.P.M. van Putten, *Introduction to Methods of Approximation in Physics and Astronomy*, Undergraduate Lecture Notes in Physics,
DOI 10.1007/978-981-10-2932-5_6

However, they fail to be orthogonal on intervals $D = [-1, 1]$ on the real line with

$$\langle f(x), g(x)\rangle = \int_{-1}^{1} f(x)\bar{g}(x)dx. \tag{6.3}$$

A suitable basis set of functions, therefore, depends on the choice of domain in the associated inner product.

Example 6.1. Let us illustrate (6.1) for

$$f(x) = \frac{1}{2-x^2} \quad (x \in D). \tag{6.4}$$

The Taylor series expansion of $f(x)$ about $x = 0$ is readily obtained as a Neumann series,

$$f(x) = \frac{1/2}{1-(x/\sqrt{2})^2} = \frac{1}{2}\left(1+\frac{1}{2}x^2+\frac{1}{4}x^4+\cdots+2^{-n/2}x^n\right)+O\left(2^{-\frac{n}{2}-1}\right), \tag{6.5}$$

where n is even. From this, we can read off the Taylor series coefficients

$$a_n = \left[1+(-1)^n\right]2^{-\frac{n}{2}-2} = O\left(\left[\frac{1}{\sqrt{2}}\right]^n\right) \tag{6.6}$$

with vanishing coefficients for n odd.

Since (6.6) shows exponential convergence,[1] our Taylor series expansion might be considered efficient on D. Improved efficiency, however, obtains with basis functions that form an *orthogonal basis set on D* as follows.

6.2 Expansion in Legendre Polynomials

The Legendre polynomials $P_n(x)$ are alternatingly even and odd polynomials on D satisfying the boundary conditions $P_n(\pm 1) = (\pm 1)^n$, the first few of which are given by (cf. Sect. 5.1)

$$P_0(x) = 1, \quad P_1(x), \quad P_2(x) = \frac{1}{2}\left(3x^2-1\right), \quad P_3(x) = \frac{1}{2}\left(5x^3-3x\right). \tag{6.7}$$

[1]The $D = [-1, 1]$ is entirely within the complex domain of convergence $|z| < \sqrt{2}$ of the analytic extension $f(z)$ of $f(x)$.

Applying $\phi_n(x) = P_n(x)$ to (6.1), we obtain the *Legendre expansion* for (6.4),

$$f(x) = b_0 + b_1 x + b_2 P_2(x) + b_3 P_3(x) + \cdots . \tag{6.8}$$

The expansion coefficients b_n derive from orthogonality of the Legendre polynomials on $D = [-1, 1]$, defined by the Euclidean inner product (6.3). Specifically, we have

$$\langle P_n(x), P_m(x)\rangle = \frac{2}{2n+1}\delta_{mn}, \tag{6.9}$$

where δ_{nm} denotes the Kronecker delta function

$$\delta_{mn} = \begin{cases} 1 \ (n = m) \\ 0 \ (n \neq m) \end{cases} \tag{6.10}$$

Example 6.2. The orthogonality condition (6.9) can be used to successively compute the Legendre polynomials, starting with $P_0 = 1$. Thus, $P_1(x) = a_1 x + a_0$ satisfies

$$\int_{-1}^{1} P_0(x)P_1(x)dx = A\int_{-1}^{1} x dx = 2a_0 = 0; \tag{6.11}$$

$$\int_{-1}^{1} P_1(x)P_1(x)dx = A^2\int_{-1}^{1} x^2 dx = \frac{2}{3}A^2. \tag{6.12}$$

For $n = 1$, equating the latter to the normalization condition $2/(2n+1)$ requires $a_1 = 1$. For $P_2(x) = a_2 x^2 + a_1 x + a_0$, we have

$$\int_{-1}^{1} P_0(x)P_2(x)dx = \frac{2}{3}a_2 + 2a_0 = 0, \tag{6.13}$$

$$\int_{-1}^{1} P_1(x)P_2(x)dx = \frac{2}{3}a_1 = 0;$$

$$\int_{-1}^{1} P_2(x)P_2(x)dx = \frac{2}{5}a_2^2 + \frac{4}{3}a_2 a_0 + 2a_0^2 = \frac{2}{5}, \tag{6.14}$$

from which it follows that $a_2 = 3/2$ and $a_0 = -1/2$.

By these properties, the Legendre expansion of a function $f(x)$ on $[-1, 1]$ obtains with

$$b_n = \frac{2n+1}{2}\int_{-1}^{1} f(x)P_n(x)dx. \tag{6.15}$$

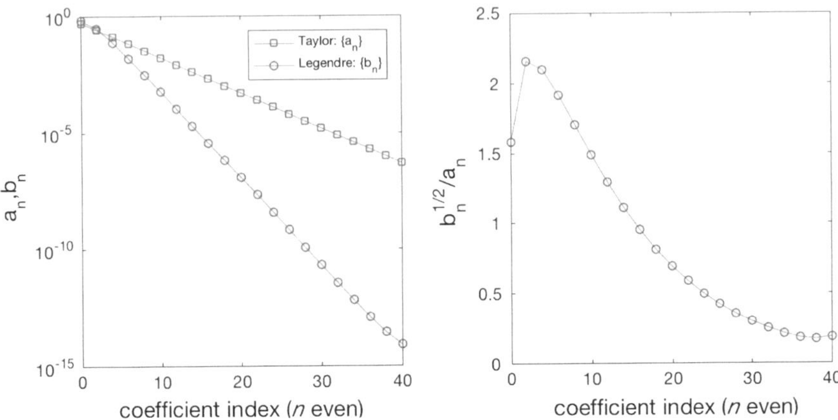

Fig. 6.1 Shown are the expansion coefficients a_n and b_n (n even) in the Taylor and, respectively, Legendre expansions (6.6) and (6.8) of $f(x)$ in (6.4) on $D = [-1, 1]$. Though both series show exponential decay by analyticity of $f(x)$ on D, the Legendre coefficients decay about twice as fast

Example 6.1 (Cont'd). Applied to $f(x)$ in (6.4), we find

$$b_0 = 2^{-\frac{5}{2}} \log\left(\frac{3+2^{\frac{3}{2}}}{3-2^{\frac{3}{2}}}\right),\ b_1 = 0,\ b_2 = \frac{25}{2^{\frac{7}{2}}} \log\left(\frac{3+2^{\frac{3}{2}}}{3-2^{\frac{3}{2}}}\right) - \frac{15}{2}, \ldots \quad (6.16)$$

the results of which are shown in Fig. 6.1.

As Fig. 6.1 shows, for a given N in the partial sum of (6.1), a truncated Legendre expansion will be more accurate than a Taylor series (6.5) truncated at the same power x^N. Why do the Legendre polynomials work so well? In (6.5) and (6.8), adding a term $a_{N+1}\phi_{N+1}$ to the partial sum $S_N(x)$ serves to approximate the residual $f(x) - S_N(x)$. This is done most efficiently, provided the new term is orthogonal to $S_N(x)$. This happens when $\phi_{N+1}(x)$ is linearly independent of the $\{\phi_n\}_{n=0}^N$, i.e., $\phi_{N+1}(x)$ is orthogonal to all of $\phi_n(x)$ ($n = 0, 1, 2, \ldots, N$) (cf. Sect. 3.8). Orthogonality (6.9) of Legendre functions on $D = [-1, 1]$ should be contrasted with the complete lack thereof in the basis functions x^n of a Taylor series, apparent in the cosines

$$\frac{< x^n, x^m >}{|x^n||x^m|} = \begin{cases} 2\frac{\sqrt{(2n+1)(2m+1)}}{n+m+1} & (n+m \text{ even}) \\ 0 & (n+m \text{ odd}) \end{cases}, \quad (6.17)$$

where $|f(x)| = \sqrt{< f(x), f(x) >}$ denotes the norm of $f(x)$.

Example 6.3. The function $f(x) = \tan\left(\frac{\pi}{4}x\right)$ on $[-1, 1]$ has a Taylor series expansion $f(x) = a_0 + a_1x + a_2x^2 + a_3x^3 + \cdots$ about the origin

$$f(x) = \frac{\sin y}{\cos y} = \frac{y - \frac{1}{6}y^3 + O\left(y^5\right)}{1 - \frac{1}{2}y^2 + O\left(y^4\right)} = y + \frac{1}{3}y^3 + O\left(y^5\right) \tag{6.18}$$

with $y = \pi x/2$, i.e.,

$$f(x) = \frac{\pi}{4}x + \frac{\pi^3}{192}x^3 + O\left(x^5\right) = 0.7854x + 0.1615x^3 + O\left(x^5\right). \tag{6.19}$$

A Legendre expansion of $f(x)$ gives $f(x) = b_1x + b_3P_3(x) + O\left(x^5\right)$ with

$$\begin{aligned} b_1 &= \tfrac{6}{\pi^2}\left(4K - \pi\ln 2\right), \\ b_3 &= \tfrac{7}{24\pi^4}\left(1152\pi^2K - 48\pi^3\ln 2 + 540\,\zeta(3) + 11520i\mathrm{Li}_4(i) + 7i\pi^4\right) \end{aligned} \tag{6.20}$$

in terms of the Catalan constant, the Riemann-zeta function (Appendix B) and the polylogarithm defined by, respectively,

$$\begin{aligned} K &= \textstyle\sum_{k=0}^{\infty}(-1)^k(2k+1)^{-2} = 0.91597..., \\ \zeta(z) &= \textstyle\sum_{k=0}^{\infty} n^{-z}, \quad \mathrm{Li}_s(z) = \sum_{k=1} z^k n^{-s}, \end{aligned} \tag{6.21}$$

i.e.,

$$f(x) = 0.90354x + 0.08779P_3(x) + O\left(x^5\right). \tag{6.22}$$

Note the discrepancy by a factor of about two in Taylor and Legendre coefficients a_3 and, respectively, b_3. By the orthogonality property of the Legendre functions, such is expected to continue to subsequent coefficients. Indeed, $a_5 = 0.03985$, $a_7 = 0.009949$, ... and $b_5 = 0.007933$, $b_7 = 0.000069556$, .., whereby the expansion in a Legendre series provides relatively uniform approximation of the function at hand (Fig. 6.2).

6.2.1 Symbolic Computation in Maxima

We use a free software package *Maxima*[2] for some illustrative symbolic and numerical experiments.[3]

[2] Distribution under GPL licence of *Macsyma* (Project MAC's SYmbolic MAnipulation), see further [1].

[3] For a brief summary of frequently used *Maxima* commands, see, e.g., http://www.math.harvard.edu/computing/maxima/.

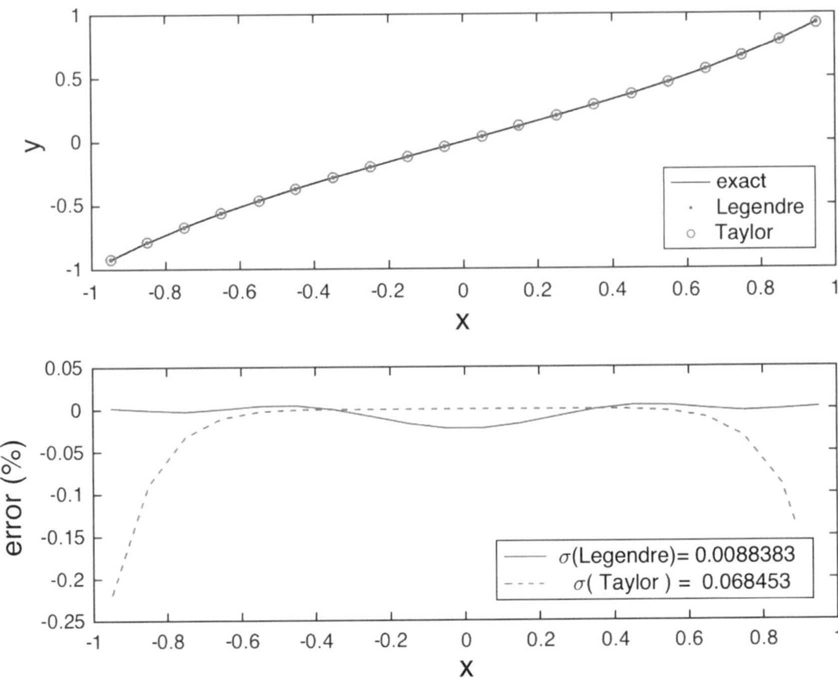

Fig. 6.2 Shown is the approximation of $y(x) = \tan\left(\frac{\pi}{4}x\right)$ by an expansion in Legendre and Taylor series (about $x = 0$) over the domain $[-1, 1]$. The errors in the Legendre series are relatively uniform, whereas the Taylor series provides superior approximation only in a neighborhood of the origin

To begin, we invoke the build-in Legendre polynomials and set the environment variable "orthopoly_returns_intervals" to $false$. *Maxima* hereby returns function values to numerical evaluations such as $P_2(0.5)$ of $P_2(x)$ at $x = 0.5$. The following commands reproduce (6.7) above,

```
orthopoly_returns_intervals:false;
legendre_p(0,x); legendre_p(1,x); legendre_p(3,x);
```

and

```
factor(legendre_p(2,x));
factor(legendre_p(10,x));
```

produces expressions with a slightly more sanitized look and feel. Specific function evaluations can be produced by substituting a value for x, e.g.,

```
ev(legendre_p(2,0.5));
```

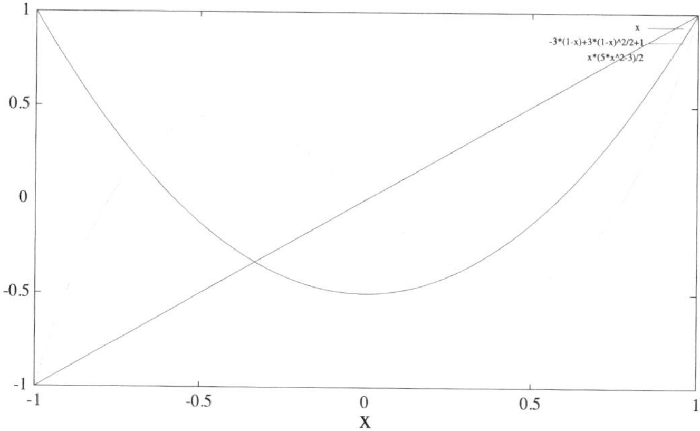

Fig. 6.3 A plot of the Legendre polynomials $P_i(x)$ $(i = 1, 2, 3)$ in *Maxima*

Figure 6.3 shows a plot of the first few Legendre polynomials, by

```
wxplot2d([legendre_p(1,x),legendre_p(2,x),factor(legendre_p(3,x)],
[x,-1,1]);
```

Maxima readily performs symbolic integration of polynomial expressions, for instance

```
integrate(x^2,x,0,1); integrate(x^2,x,a,b);
```

returns exact results $1/3$ and $b^3/3 - a^3/3$ for integration of x^2 over the interval $[0, 1]$ and, respectively, $[a, b]$.

Here, we seek explicit expressions for the partial sums $S_N(x)$ of an expansion of $f(x) = 1/(2-x^2)$ in (6.4) in $P_n(x)$. To this end, we define the expansion coefficients (6.15) as a *Maxima* function:

```
A(n):=(2*n+1)/2*integrate(1/(2-x^2)*legendre_p(n,x),x,-1,1);
```

The following commands give the exact expressions for some of the first few coefficients in (6.16):

```
A(0);A(1);expand(A(2));
```

We now put the first few of the numerical values of the A_m into an array:

```
N:10
for n:0 thru N do a[n]:ev(A(n),bfloat);
for n:0 thru N do display(a[n]);
```

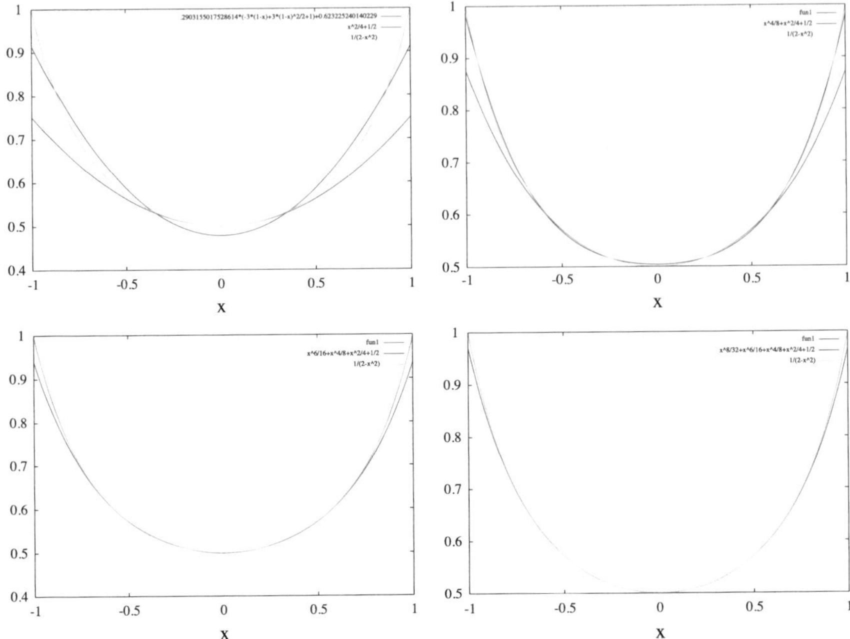

Fig. 6.4 Approximation of $f(x)$ in (6.4) on $D = [-1, 1]$ by its Taylor series about $x = 0$ and an expansion in Legendre polynomials shown for partial sums up to $n = 2, 4, 6, 8$. The Legendre expansion shows superior convergence

showing (ignoring the zero coefficients $b_{2k+1} = 0$)

$$\begin{array}{ll} b_0 = 6.2322 \times 10^{-1}, & b_2 = 2.9032 \times 10^{-1}, \\ b_4 = 6.8657 \times 10^{-2}, & b_6 = 1.431 \times 10^{-2}, \\ b_8 = 2.825 \times 10^{-3}, & b_{10} = 5.407 \times 10^{-4}. \end{array} \tag{6.23}$$

In the numerical evaluation by **ev(A(n),bfloat)** above, note the use of **bfloat**. By default, *Maxima*'s **ev** works in single (**float**(*expr*)). Here, double precision (**bfloat**(*expr*), **fpprec:16**) is required in evaluation of b_n with large n, e.g., $n \geq 10$.

We develop some partial sums as follows:

```
SN(x):=sum(a[n]*legendre_p(n,x),n,0,N);
```

and plot the graphs of $SN(x)$ along with the Taylor series $QN(x)$ of $f(x)$ up to the same degree in x,

```
QN(x):=sum(1/2*(x/sqrt(2))^(2*n),n,0,N/2);
f(x):=1/(2-x^2); wxplot2d([SN(x),QN(x),f(x)],[x,-1,1]);
```

Figure 6.4 shows the results of a re-run the above for $N = 4, 6, 8$, illustrating the extremely rapid convergence of the Legendre expansion to $f(x)$ with good approximations already for $N = 4$. In contrast, N must be much larger for the Taylor series to obtain a similar degree of approximation to $f(x)$. The expansion in Legendre polynomials is more efficient than the Taylor series expansion by the power of the mutually orthogonal basis functions $P_n(x)$.

6.3 Fourier Expansion

Smooth transient functions can be efficiently represented by a Fourier transform. Those that can be effectively truncated or those showing periodic or quasi-periodic behavior can be efficiently represented by Fourier series. This approach is key also to searches for signals in noisy data. Mathematically, Fourier series are import as orthogonal basis sets on intervals on the real line.

6.4 The Fourier Transform

The Fourier transform of a transient signal[4] $f(t)$ on the real line $\mathbb{R}$ is defined as

$$\tilde{f}(\omega) = \int_{-\infty}^{\infty} f(t)e^{-i\omega t}dt \;\; (-\infty < \omega < \infty). \tag{6.24}$$

If $f(t)$ has no discontinuities, it can be shown that (6.24) has an inverse pointwise on $\mathbb{R}$, given by

$$f(t) = \frac{1}{2\pi}\int_{-\infty}^{\infty} \tilde{f}(\omega)e^{i\omega t}d\omega. \tag{6.25}$$

If $f(t)$ is real, then $\overline{\tilde{f}(\omega)} = \tilde{f}(-\omega)$ and

$$f(t) = \frac{1}{2\pi}\int_{0}^{\infty} \left[\tilde{f}(\omega)e^{i\omega t} + \overline{\tilde{f}(\omega)}e^{-i\omega t}\right]dt. \tag{6.26}$$

If we put

$$\alpha(\omega) = \frac{\tilde{f}(\omega) + \overline{\tilde{f}(\omega)}}{2}, \;\; \beta(\omega) = i\frac{\tilde{f}(\omega) - \overline{\tilde{f}(\omega)}}{2}, \tag{6.27}$$

[4]We will assume these functions to be of finite energy, i.e., they are square integrable on the real line.

then we obtain an explicit Fourier representation of $f(t)$ in terms of cosines and sines,

$$f(t) = \frac{2}{\pi}\int_0^\infty [\alpha(\omega)\cos(\omega t) + \beta(\omega)\sin(\omega t)]\,dt. \tag{6.28}$$

Example 6.4. Consider the Gaussian $f(x) = e^{-at^2}$ $(a > 0)$ on the real axis, which is integrable and vanishes at infinity. Its Fourier transform (6.24) is therefore well-defined, which obtains again a Gaussian:

$$\tilde{f}(\omega) = \frac{1}{\sqrt{a}} e^{-\frac{\omega^2}{4a}} \int_{-\infty}^{\infty} e^{-(s-s_0)^2} ds = \sqrt{\frac{\pi}{a}} e^{-\frac{\omega^2}{4a}}, \tag{6.29}$$

where $s_0 = i\omega/(2\sqrt{a})$, obtained by contour integration.

The following example derives from scattering by *moving mirrors*, mapping wave fronts with uniform phase distribution at null-infinity (Y) backwards in time to null-infinity in the past (X), shown in Fig. 6.5.

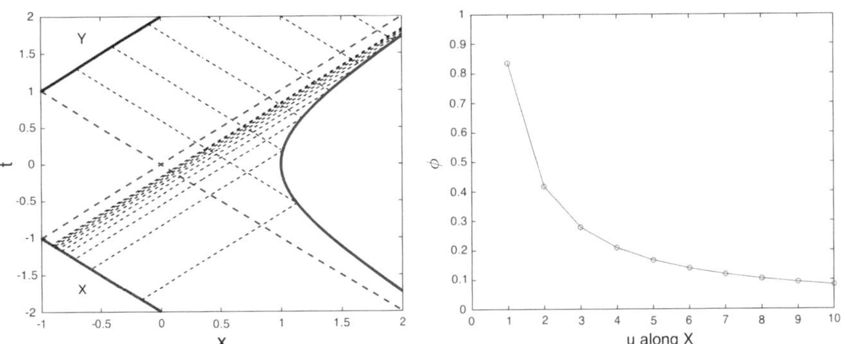

Fig. 6.5 A Rindler observer $\mathcal{O}$ moves at constant acceleration a in Minkowski spacetime (t, x), at constant distance $\xi = 1/a$ to an event horizon H (part of the null-cone at the origin) that precludes receiving signals coming from the left (Fig. 1.6). Acting as a mirror, $\mathcal{O}$ maps a uniform distribution of phase over a future null-direction Y onto a logarithmic distribution of phase over a past null-direction X. (After Birrell, N.D., & Davies, P.C.W., 1982, *Quantum Fields in Curved Space*, Cambridge University Press, Cambridge.)

Example 6.5. Consider the Fourier transform of a logarithmic distribution of phase $\phi(x) = \alpha \ln x$ on $X = (0, \infty)$

$$\tilde{f}(y) = \int_0^\infty e^{i\alpha \ln x} e^{-iyx} dx. \tag{6.30}$$

Then the Fourier spectrum is mixed with positive and negative frequencies satisfying

$$\tilde{f}(-y) = -e^{-\alpha\pi} \tilde{f}(y) \ \ (y > 0); \tag{6.31}$$

To see this, consider $y > 0$, allowing a contour deformation $y = -it$ to

$$\tilde{f}(y) = e^{\frac{1}{2}\alpha\pi} \int_0^\infty e^{i\alpha \ln t} e^{-yt} dt \ \ (y > 0). \tag{6.32}$$

Likewise, $y < 0$ allows a contour deformation $y = it$ to

$$\tilde{f}(y) = e^{-\frac{1}{2}\alpha\pi} \int_0^\infty e^{i\alpha \ln t} e^{yt} dt \ \ (y < 0). \tag{6.33}$$

By (6.32) and (6.33), the result (6.31) follows. More explicitly, we have

$$\tilde{f}(y) = e^{\frac{1}{2}\alpha\pi} y^{-i\alpha-1} \Gamma(1 + i\alpha) \ \ (y > 0) \tag{6.34}$$

in terms of the special function (Appendix B)

$$\Gamma(z) = \int_0^\infty t^{z-1} e^{-t} dt. \tag{6.35}$$

6.4.1 *The Sinc Function*

The overall shape of the Fourier transform $\tilde{f}(\omega)$ is directly related to the behavior of $f(t)$ in the time domain. Quite generally, $\tilde{f}(\omega)$ shows steep (exponential) decay whenever $f(t)$ is smooth (analytic). Conversely, when $f(t)$ is non-smooth, e.g., $f(t)$ shows jumps in its derivatives or even jumps in the function itself, $\tilde{f}(t)$ shows algebraic decay in the limit as ω becomes large (positive and negative).

To illustrate this, consider the block function

$$f(t) = K \ (-a < t < a), \ \ f(t) = 0 \ \ (|t| > a) \tag{6.36}$$

with total surface area

$$A = \int_{-\infty}^{\infty} f(t)dt = 2aK. \tag{6.37}$$

Its Fourier transform satisfies

$$\tilde{f}(\omega) = K \int_{-a}^{a} e^{i\omega t} dt = K \frac{1}{i\omega} \left[e^{i\omega t} \right]_{t=-a}^{a} \equiv A \ \mathrm{sinc}(\omega a) \tag{6.38}$$

with $A = 2aK$ and

$$\mathrm{sinc(x)} = \frac{\sin x}{x}. \tag{6.39}$$

It has a removable singularity at $x = 0$, whereby $\mathrm{sinc}(0) = 1$ in view of $\lim_{x \to 0} x^{-1} \sin x = 1$. In particular,

$$\tilde{f}(0) = A \tag{6.40}$$

is the integral of $f(t)$ over $\mathbb{R}$.

Upon varying a keeping the total area fixed, $\tilde{f}(\omega)$ shows horizontal stretching (scaling in ω). Specifically, a narrow spike by small a gives rise to a broad (*hard*) spectrum $\tilde{f}(\omega)$. Conversely, a broad pulse with large a has a steep (*soft*) spectrum.

6.4.2 Plancherel's Theorem

Square integrable functions are such that

$$E = \int_{-\infty}^{\infty} f(t)\bar{f}(t)dt < \infty. \tag{6.41}$$

For these functions, the *energy* E can alternatively be expressed in terms of an equivalent integral of $\tilde{f}(\omega)$, since

$$E = \frac{1}{4\pi^2} \int_{-\infty}^{\infty} \left[\int_{-\infty}^{\infty} \tilde{f}(\omega) e^{i\omega t} d\omega \overline{\int_{-\infty}^{\infty} \tilde{f}(\omega) e^{i\omega' t} d\omega'} \right] dt, \tag{6.42}$$

i.e.,

$$E = \frac{1}{2\pi}\int_{-\infty}^{\infty}\int_{-\infty}^{\infty}\tilde{f}(\omega)\overline{\tilde{f}(\omega')}\left[\frac{1}{2\pi}\int_{-\infty}^{\infty}e^{i(\omega-\omega')t}dt\right]d\omega d\omega'. \tag{6.43}$$

According to (6.24) and (6.25), we have the distribution function

$$\frac{1}{2\pi}\int_{-\infty}^{\infty}e^{i(\omega-\omega')t}dt = \delta(\omega-\omega'). \tag{6.44}$$

Hence, (6.43) reduces to *Plancherel's Theorem*

$$E = \frac{1}{2\pi}\int_{-\infty}^{\infty}\left|\tilde{f}(\omega)\right|^2 d\omega. \tag{6.45}$$

For our block function (6.36), we have by explicit calculation $E = AK$ and

$$\int_{-\infty}^{\infty}\left|\tilde{f}(\omega)\right|^2 d\omega = 2E\int_{-\infty}^{\infty}\text{sinc}^2(s)ds. \tag{6.46}$$

It can be shown, e.g., by contour integration in the complex plane, we have

$$\int_{-\infty}^{\infty}\left(\frac{\sin s}{s}\right)^2 ds = \pi, \tag{6.47}$$

so that

$$\frac{1}{2\pi}\int_{-\infty}^{\infty}\left|\tilde{f}(\omega)\right|^2 d\omega - E \tag{6.48}$$

in accord with (6.45).

6.4.3 Fourier Series by Sampling $\tilde{f}(\omega)$

In formulating problems in signal analysis for numerical evaluation by a computer, we inevitably work on a finite number of data points.

Consider a finite, uniformly sampled spectrum, e.g., in the evaluation of the inverse shown in Fig. 6.6. We consider the discrete angular frequencies

$$\omega_n = \frac{2\pi}{T}n \tag{6.49}$$

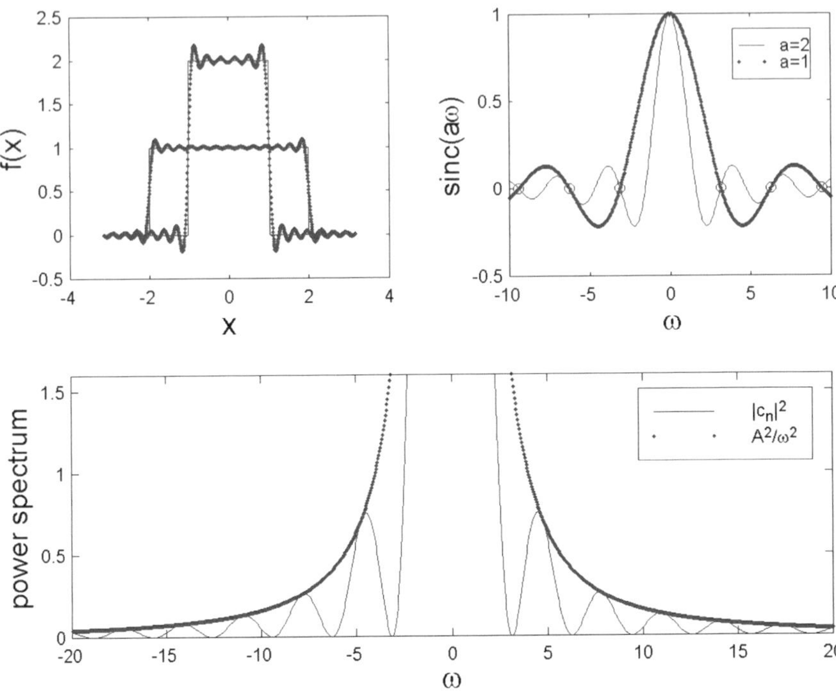

Fig. 6.6 Two block functions with widths $a = 2$ and $a = 1$ with the same total surface area $A = 4$ (*left*) a reconstructions by the Fourier inverse $A\text{sinc}(a\omega)$, here computed by a truncated Riemann sum with a uniform partition over $N = 1000$ frequencies ω in $[-20, 20]$. The normalized Fourier transforms (*right*) show the relatively broad spectrum for $a = 1$ compared to that of $a = 2$. Because of the discontinuities, the power spectrum (shown for $a = 1$) has an algebraic ω^{-1} decay defined by the sinc function (*bottom*)

for some time T, that defines $2\pi/T$ to be a unit of frequency. In this event, the exponential $e^{-i\omega_n t}$ becomes periodic in t with period T.

Taking $T = 2\pi$, the Fourier transform of a transient $f(t)$ on $\mathbb{R}$ can now be re-expressed by using a trick, also known as the Poisson summation formula:

$$\tilde{f}(\omega_n) = \int_{-\infty}^{\infty} f(t)e^{-i\omega_n t}dt = \sum_{m\epsilon\mathbb{Z}} \int_{2\pi m}^{2\pi(m+1)} f(t)e^{-i\omega_n t}dt \tag{6.50}$$

that is

$$\tilde{f}(\omega_n) = \sum_{n\epsilon\mathbb{Z}} \int_0^{2\pi} \left[f(t + 2\pi m)e^{-i\omega_n(t+2\pi m)} \right] dt = \int_0^{2\pi} f_{2\pi}(t)e^{-i\omega_n t}dt, \tag{6.51}$$

where we define the 2π periodic sum

$$f_{2\pi}(t) = \sum_{m\epsilon\mathbb{Z}} f(t + 2\pi m). \tag{6.52}$$

So, we have

$$\tilde{f}(\omega_n) = \int_{-\infty}^{\infty} f(t)e^{-i\omega_n t}dt = \int_0^{2\pi} f_{2\pi}(t)e^{-i\omega_n t}dt. \tag{6.53}$$

This identity shows that the spectrum $\tilde{f}(\omega_n)$ derives both by sampling of the continuous spectrum of $f(t)$ (with $D = \mathbb{R}$) or, equivalently, of as the spectrum of the induced periodic sum $f_{2\pi}(t)$ (with $D = [0, 2\pi]$). We thus arrive at the *Fourier coefficients* $a_n = (2\pi)^{-1}\tilde{f}(\omega_n)$,

$$a_n = \frac{1}{2\pi}\int_0^{2\pi} f_{2\pi}(t)e^{-i\omega_n t}dt. \tag{6.54}$$

The Fourier spectrum of the periodic function $f_{2\pi}(t)$ is discrete. Since, $f_{2\pi}(t) = f_{2\pi}(t+T)$, the inverse Fourier transform satisfies

$$\int_{-\infty}^{\infty} \tilde{f}_{2\pi}(\omega)e^{i\omega t}d\omega = \int_{-\infty}^{\infty} \tilde{f}_{2\pi}(\omega)e^{i\omega(t+T)}d\omega. \tag{6.55}$$

It follows that

$$\int_{-\infty}^{\infty} \tilde{f}_{2\pi}(\omega)\left[1 - e^{i\omega T}\right]e^{i\omega t}d\omega \equiv 0. \tag{6.56}$$

A non-trivial spectrum $\tilde{f}_{2\pi}(\omega)$ hereby reduces to

$$\tilde{f}_{2\pi}(\omega) = 2\pi\sum_{n\epsilon\mathbb{Z}} a_n\delta\left(\omega - \omega_n\right), \tag{6.57}$$

whereby

$$f_{2\pi}(t) = \frac{1}{2\pi}\int_{-\infty}^{\infty} \tilde{f}_{2\pi}(\omega)e^{i\omega t}d\omega = \sum_{n\epsilon\mathbb{Z}} a_n e^{i\omega_n t} \tag{6.58}$$

is the inverse of (6.54).

For periodic functions, therefore, we can use the Fourier series (6.54)–(6.58) directly. In case of the Fourier transform (6.24) and (6.25) applied to a transient function $f(t)$ on $\mathbb{R}$, we consider the induced periodic function

$$f_T(t) = \sum_{\mathbb{Z}} f(t + nT) = \cdots + f(t - T) + f(t) + f(t + T) + \cdots \quad (6.59)$$

by choice of T, where T is to be chosen sufficiently large to ensure that

$$f_T(t) \simeq f(t) \;\; (t\epsilon[-T/2, T/2]). \quad (6.60)$$

The discrepancy $f_T(t) - f(t)$ in $[-T/2, T/2]$ is commonly referred to as the *aliasing error*.

6.4.4 Discrete Fourier Transform (DFT)

Our final step in moving (6.24) and (6.25) to a numerical implementation is by considering $f_{2\pi}(t)$ in (6.54)–(6.58) for a finite number of data points, obtained by sampling. The infinite series $\{a_n\}$ in (6.54)–(6.58) hereby leads to a finite series. This final step is remarkable simple yet powerful.

Consider the uniform sampling of the working domain $D = [0, T]$ with $T = 2\pi$ by N points, i.e.,

$$t_m = \frac{2\pi}{N} m, \quad (6.61)$$

where $2\pi/N$ defines a unit step size in the time domain. Then

$$e^{int_m} = e^{2\pi inm/N} \quad (6.62)$$

is N-periodic in both n and m, since

$$e^{i(n+N)t_m} = e^{2\pi i(n+N)m/N} = e^{2\pi inm/N}, \;\; e^{2\pi in(m+N)/N} = e^{2\pi inm/N}. \quad (6.63)$$

Let us also denote $f_m = f_{2\pi}(t_m)$. By (6.58), we have, using Poisson summation once more,

$$f_m = \sum_{n\epsilon\mathbb{Z}} a_n e^{int_m} = \sum_{k=0}^{N-1} \left[\sum_{n\epsilon\mathbb{Z}} a_{k+nN} \right] e^{ikt_m} \equiv \sum_{k=0}^{N-1} c_k e^{ikt_m} \quad (6.64)$$

with inverse

$$c_k = \frac{1}{N} \sum_{n=0}^{N-1} f_n e^{-ikt_n}. \quad (6.65)$$

In particular, c_0 is the mean value of the f_n. A uniformly sampled periodic function given by N data points hereby assumes a *Discrete Fourier Transform* (6.64) and (6.65), suitable for direct numerical implementation.

The DFT pair (6.64) and (6.65) is defined by a matrix transform involving $O(N^2)$ multiplications. The *Cooly-Tukey* [2] algorithm gives a recursive factorization over DFT's of smaller order, reducing the cost to $O(N \log N)$. For large N, the result gives an essential reduction in computational effort, allowing efficient *Fast Fourier Transforms* (FFTs). Further developments aim at FFTs for $N = p^m$, where p is prime not necessarily equal to 2, and optimal numerical implementation on central processing units (CPUs) and graphics processing units (GPUs). The reducted computational cost to $\log N$ per data point also ensures a dramatically improved numerical accuracy.

Since e^{ikt_n} in (6.64) has period N in k, we have

$$f_m = c_0 + \sum_{k=1}^{N/2-1} \left(c_k e^{ikt_m} + c_{N-k} e^{-ikt_m} \right) + c_{N/2}(-1)^m, \tag{6.66}$$

where the last term appears when N is even. Note that when f_m is real, then $c_{N-k} = \bar{c}_k$ and $c_{N/2}$ is real.

The expansion (6.66) shows that the maximal frequency of oscillation in f_m is across two samples, i.e., at $1/2$ the sampling frequency. We thus have the *Nyquist criterion*: to capture signals with frequencies within a bandwidth B, we must sample with a frequency of (at least) $2B$. Accordingly, we have the Nyquist-Shannon sampling theorem: a signal with finite bandwidth B is completely determined by sampling at a frequency $2B$ or higher.

In applying DFT to a periodic function with period T,

$$f(t) = \sum_{n=0}^{N-1} c_k e^{i\omega_n t}, \quad \omega_n = \frac{2\pi}{T} n, \tag{6.67}$$

we have *Parseval's theorem* as the discrete counterpart to (6.45),

$$\frac{1}{N} \sum_{k=0}^{N-1} |f_k|^2 = \sum_{n=0}^{N-1} |c_n|^2, \tag{6.68}$$

where $f_k = f(t_k)$, $t_k = kT/N$. As before, it describes conservation of "energy," whether expressed in the time by $f(t)$ or in the frequency domain by the c_k.

Example 6.6. Common numerical implementations of FFT have the factor $1/N$ moved to (6.64) from (6.65). In this event, c_0 equals N times the mean value of the function at hand. As a result Parseval's theorem (6.68) takes the form

$$\sum_{n=0}^{N-1} |f(t_n)|^2 = \frac{1}{N}\sum_{n=0}^{N-1} |c_n|^2 \tag{6.69}$$

for a discrete time series $f_n = f(t_n)$ $(n = 0, 1, \ldots N-1)$. If the function is real with zero mean value, the standard deviation σ of the samples $f(t_n)$ hereby satisfy

$$\sigma = \frac{\sqrt{2}}{N}\sqrt{\sum_{n=1}^{N/2} |c_n|^2}. \tag{6.70}$$

6.5 Fourier Series

By the above, a 2π-periodic function $g(\theta)$ has a Fourier series expansion

$$g(\theta) = \sum_{\mathbb{Z}} c_k e^{ik\theta}, \quad c_k = \frac{1}{2\pi}\int_0^{2\pi} g(x)e^{-in\theta}d\theta. \tag{6.71}$$

For a real-valued function, the Fourier coefficients satisfy $c_{-k} = \bar{c}_k$, and $g(\theta)$ is the real part of the complex analytic function $h(z) = c_0 + 2\sum_{\mathbb{N}} c_n z^n$ on S^1. If the radius of convergence of $h(z)$ is greater than 1, analyticity of $h(z)$ on S^1 assures that the c_n have exponential convergence, making (6.71) an efficient expansion, suitably computed by DFT or FFT.

If the function is merely continuous, the decay is typically algebraic, i.e., the Fourier coefficients $|c_k|$ decay as a power law in k. Specifically, the decay is $|k|^{-n+1}$ if the function is n times differentiable, the case $n = 0$ being illustrated by the block function (6.36).

Regardless of smoothness, the Fourier transform (6.71) *always* produces decay of the c_k to zero in the limit as k approaches infinity, whenever $g(x)$ is Riemann integrable, known as the *Riemann-Lebesgue theorem.* This result follows directly from the definition of the Riemann integral, in terms of the convergent limits of majoring and minoring sums [3].

Given k, consider a uniform partition $\theta_n = 2\pi n/k$ of $[0, 2\pi]$, whereby

$$c_k = \frac{1}{2\pi}\sum_{n=0}^{k-1}\int_{\theta_n}^{\theta_{n+1}} g(x)e^{-ik\theta}d\theta = \frac{1}{k}\int_0^{2\pi} g\left(\theta_n + \frac{y}{k}\right)e^{iy}dy. \tag{6.72}$$

"Subtracting and adding," we obtain

$$c_k = \frac{1}{k}\int_0^{2\pi}\left[g\left(\theta_n + \frac{y}{k}\right) - g(\theta_n)\right]e^{iy}dy + \int_0^{2\pi} g(\theta_n)e^{iy}dy, \tag{6.73}$$

where the second integral on the right hand side vanishes by 2π-periodicity of e^{iy}. Taking absolute values, we arrive at

$$|c_k| \le \frac{1}{k}\int_0^{2\pi}\left|g\left(\theta_n + \frac{y}{k}\right) - g(\theta_n)\right| dy. \tag{6.74}$$

With the majoring and minoring Riemann sums

$$S_k = \frac{2\pi}{k}\sum_{n=0}^{k-1}\sup_{[\theta_n,\theta_{n+1}]} g(\theta),\quad s_k = \frac{2\pi}{k}\sum_{n=0}^{k-1}\inf_{[\theta_n,\theta_{n+1}]} g(\theta), \tag{6.75}$$

it follows that

$$|c_k| \le \frac{1}{2\pi}\left(S_k - s_k\right) \to 0 \tag{6.76}$$

by the assumption of Riemann integrability.

In numerical implementations of (6.71) by DFT or FFT, it is pertinent to seek formulations allowing efficient representations by Fourier series. According to the above, this requires formulations in terms of smooth functions whenever possible.

When $f(x)$ is a function on $D = [-1, 1]$ which is not periodic, we may resort to expansions into non-periodic basis functions such as aforementioned Legendre polynomials. Alternatively, we can consider the following.

6.6 Chebyshev Polynomials

Consider $f(x)$ on $D = [-1, 1]$ with $x = \cos\theta$. Then $g(\theta) = f(x)$ is 2π-periodic on $[0, 2\pi]$ with Fourier expension (6.71),

$$c_n = \frac{2}{\pi}\int_0^{\pi} g(\theta)\sin n\theta\, d\theta = \frac{2}{\pi}\int_{-1}^{1} f(x)\sin n\theta\, dx. \tag{6.77}$$

This approach introduces $e^{in\theta}$ uniformly over S^1.

The *Chebyshev polynomials* as a function of $x = \cos\theta$ are defined by the real parts

$$T_n(x) = \cos n\theta = \mathrm{Re}\, e^{in\theta} = \mathrm{Re}\,(\cos\theta + i\,\sin\theta)^n\,, \tag{6.78}$$

e.g.,

$$\begin{array}{l} T_0(x) = 1,\ \ T_1(x) = \cos x,\ \ T_2(x) = 2x^2 - 1, \\ T_3(x) = 4x^3 - 3x,\ \ T_4(x) = 8x^4 - 8x^2 + 1, \ldots \end{array} \tag{6.79}$$

By (6.78), these polynomials satisfy the *nesting* property $T_n(T_m(x)) = T_{n+m}(x)$ and the recursion relation

$$T_{n+1}(x) = 2xT_n(x) - T_{n-1}(x) \tag{6.80}$$

by $\cos(n\pm 1)\theta = \cos\theta\cos n\theta \mp \sin\theta\sin n\theta$. By orthogonality of the functions $\cos\, n\theta$ over $[0, 2\pi]$, the Chebyshev polynomials are orthogonal over $[-1, 1]$ with weight $1/\sin\theta$,

$$\langle T_n, T_m\rangle = \frac{2}{\pi}\int_0^{\pi} T_n(x)T_m(x)d\theta = \int_{-1}^{1}\frac{T_n(x)T_m(x)}{\sqrt{1-x^2}}dx = \delta_{nm}(1+\delta_{n0}). \tag{6.81}$$

In the Euclidean inner product (6.3),

$$\int_{-1}^{1} T_n(x)T_m(x)dx = \left\{\begin{array}{ll} \frac{m^2+n^2-1}{[(m+n)^2-1][(m-n)^2-1]} & (n+m\ \text{even}) \\ 0 & (n+m\ \text{odd}), \end{array}\right. \tag{6.82}$$

whereby

$$\int_{-1}^{1} T_n(x)T_n(x)dx = 1 + \frac{3}{4n^2-1}. \tag{6.83}$$

Orthogonality (6.81) invites us to consider the *Chebyshev expansion*

$$f(x) = \frac{1}{2}C_0 + \sum_{\mathbb{N}} C_nT_n(x),\ \ C_n = \frac{2}{\pi}\int_{-1}^{1} f(x)T_n(x)\frac{dx}{\sqrt{1-x^2}}dx \tag{6.84}$$

in terms of the Chebyshev coefficients C_n, rather than going back to the Fourier expansion (6.71) with

$$c_n = \frac{2}{\pi} \int_{-1}^{1} f(x) \frac{x T_n(x) - T_{n+1}(x)}{\sqrt{1-x^2}} dx. \tag{6.85}$$

Functions on $[-1, 1]$ thus have various orthogonal expansions associated with different weights, here illustrated by the Legendre or Chebyshev series.

6.7 Weierstrass Approximation Theorem

The relation between Fourier and Chebyshev series (6.71) and (6.84), respectively, can be highlighted further as follows.

We recall that a function $f(x)$ on $[-1, 1]$ has an associated 2π-periodic function $u\left(e^{i\theta}\right) = f(x)$, $x = \cos\theta$, that represents the boundary value of the real part of a complex valued function $g(z)$ on the unit disk $D: |z| \leq 1$ by Cauchy's integral formula. This is made explicit by the Poisson kernel (Sect. 2.7); see Fig. 6.7.

Since $g(z)$is analytic in D with radius of convergence $R \geq 1$, its truncated Taylor series $g_N(z) = c_0 + c_1 z + c_2 z^2 + \cdots c_N z^N$ convergence uninformly on circles S' in D. If $f(x)$ is continuous, $g(z)$ is continuous on $|z| \leq 1$; $g(z)$ on S' approximates $g(z)$ on S^1, when S' is sufficiently close to S^1.

By the de Moivre's theorem, $\text{Re}\, z^n = r^n e^{in\theta}$ is a polynomial in $x' = rx$ of degree n, as in (6.78). It follows that the polynomial $\text{Re}\, g_N(z)$ in x approximates $f(x)$ uniformly by sufficiently large N.

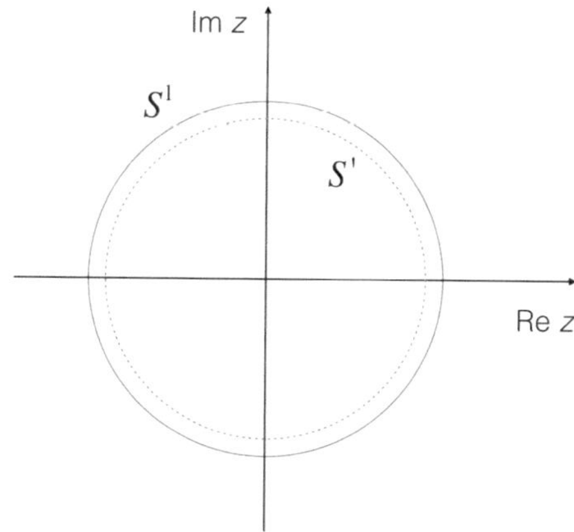

Fig. 6.7 Data $f(x)$ on $[-1, 1]$ define a complex function $g(z)$ on the unit disk with real part $f(x)$ on S^1 by the Poisson integral of Sect. 2.7. When $f(x)$ is continuous, $g(z)$ on S' in D approximates $g(z)$ on S^1 when S' is sufficiently close, while the truncated Taylor series $g_N(z)$ approximates $g(z)$ on S' for sufficiently large N. The real part of $g_N(z)$ hereby approximates $f(x)$ as a polynomial in $x = \cos\theta$ by de Moivre's theorem

6.8 Detector Signals in the Presence of Noise

Among the emerging new observatories, the advanced gravitational wave detectors LIGO-Virgo in the US and Europe and KAGRA in Japan (Fig. 6.8) pose novel problems in data-analysis: the search for minute gravitational wave signals at dimensionless strain-amplitudes of about $h \simeq 10^{-23} - 10^{-21}$ by laser interferometry. The power of this new window of observation is dramatically demonstrated by the recent detection of the gravitational wave burst GWB150914, produced by the merger of a black hole-black hole binary [4]. This first detection presents us with an entirely novel discovery with unanticipated astrophysical implications on the relatively high mass and remarkably slow spin of the black holes in this merger event [5]. A long-standing promise of *multimessenger astronomy* combining gravitational and electromagnetic observations is now becoming a concrete possibility.

In the real world, signals are accompanied by noise and the gravitational wave output of a laser-interferometric detector is no exception. Noise may originate from the source, environment and detector. For a laser-interferometric gravitational wave detector (Fig. 6.8), low frequency noise up to a few hundred Hz is typically of seismic, atmospheric and anthropogenic origin, the latter, e.g., from nearby trains and airplanes. Above a few hundred Hz, noise is dominated by shot noise due to a finite number of photons in the interferometer. Interferometer noise hereby tends to be non-Gaussian and effectively Gaussian at low, respectively, high frequencies (Fig. 6.9).

In general, the output $h(t)$ of a gravitational wave detector consists of a signal with additive noise $w(t)$,

$$h(t) = s(t) + w(t), \quad s(t) = \left(\frac{10\ \mathrm{Mpc}}{D}\right) S(t), \tag{6.86}$$

where D denotes the distance to an astrophysical source and $S(t)$ denotes the gravitational wave signal normalized to a source distance, say, 10 Mpc.

A basic tool to search for $s(t)$ in the noisy output $h(t)$ is by either cross-correlation between two (or more) detectors and matched filtering using model templates. The first may be used to search for a stochastic background in gravitational waves, e.g., of primordial origin or from an astrophysical population of sources. In searches for periodic signals, the latter is efficiently pursued by Fourier analysis. For non-periodic

Fig. 6.8 Areal views of the gravitational wave detector LIGO at Hanford (*left*, a second detector is at Livingston), Virgo at Cascina (*middle*), and KAGRA at the Kamioka Observatory (*right*)

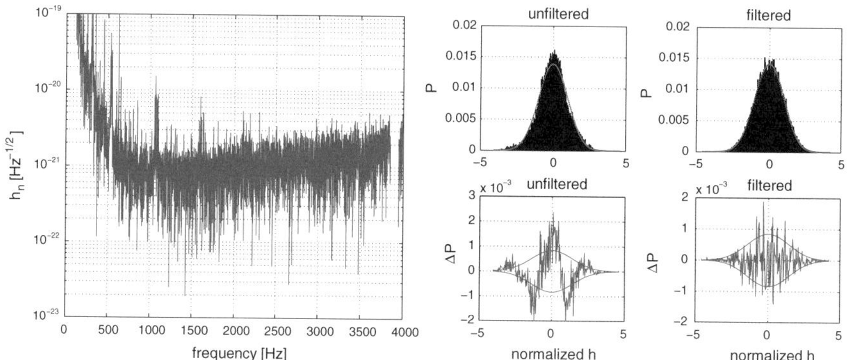

Fig. 6.9 The spectrum of dimensionless strain amplitude noise in the TAMA 300 m gravitational wave detector experiment at NAOJ (*left*), that served as the prototype for KAGRA. The noise is a super-position of non-Gaussian noise at low frequencies below about 575 Hz and effectively Gaussian noise above 575 Hz (*right*). (Reprinted from van Putten, M.H.P.M., Kanda, N., Tagoshi, H., Tatsumi, D., Masa-Katsu, F., & Della Valle, M., 2011, Phys. Rev. D, 83, 044046.)

signals, model templates are needed that must densely cover the expected parameter range of the candidate astrophysical source at hand.

6.8.1 Convolution and Cross-Correlation

The *convolution integral*

$$R(t) = \int_{-\infty}^{\infty} f(s)g(t-s)ds. \tag{6.87}$$

of two functions $f(t)$ and $g(t)$ describes the response of a linear system by the superposition of impulse excitations with infinitesimal strength $f(s)ds$. The same integral can be used to find the shift along the abscissa for an optimal match between two functions. Consider for instance the Gaussian distributions

$$f(t) = \frac{1}{\sigma\sqrt{2\pi}}e^{-\frac{t^2}{2\sigma^2}}, \quad g(t) = \frac{1}{\mu\sqrt{2\pi}}e^{-\frac{(t-a)^2}{2\mu^2}} \tag{6.88}$$

of random variables with standard deviations σ and μ, the latter with a mean a. In this event, (6.87) gives

$$R(t) = \frac{1}{\sqrt{2\pi(\sigma^2+\mu^2)}}e^{-\frac{(t-a)^2}{2(\sigma^2+\mu^2)}}. \tag{6.89}$$

Its maximum identifies $\Delta t = a$ where $f(t)$ and $g(t)$ overlap, corresponding to the maximum of $R(t)$. An efficient numerical implementation of (6.87) is by the Fourier transform, since

$$\tilde{R}(\omega) = \tilde{f}(\omega)\tilde{g}(\omega), \tag{6.90}$$

by interchanging integration with respect to s and t and factoring the two integrals:

$$\tilde{R}(\omega) = \int_{-\infty}^{\infty}\int_{-\infty}^{\infty} f(s)e^{-i\omega s}g(t-s)e^{-i\omega(t-s)}d(t-s)ds. \tag{6.91}$$

The same result obtains from the *cross-correlation* of $f(t)$ and $g(t)$,

$$Q(t) = \int_{-\infty}^{\infty} f(s)g(t+s)ds, \tag{6.92}$$

similarly evaluated in the Fourier domain,

$$\tilde{Q}(\omega) = \tilde{f}(\omega)\overline{\tilde{g}(\omega)}. \tag{6.93}$$

For discrete vectors **a** and **b**, e.g., obtained from $f(t_i)$ and $g(t_i)$ sampled at discrete times t_i, we have

$$\cos\phi = \frac{\mathbf{a}\cdot\mathbf{b}}{|\mathbf{a}||\mathbf{b}|} = \frac{\sum a_i b_i}{\sqrt{\sum a_i^2}\sqrt{\sum b_i^2}}. \tag{6.94}$$

The Pearson coefficient of fluctuations in **a** and **b** about their mean is defined by

$$\cos\psi = \frac{\mathbf{A}\cdot\mathbf{B}}{|\mathbf{A}||\mathbf{B}|},\quad A_i = a_i - \bar{a},\quad B_i = b_i - \bar{b}, \tag{6.95}$$

where $\bar{a} = N^{-1}\sum a_i$ and $\bar{b} = N^{-1}\sum b_i$. Correspondingly, we consider

$$\rho(t) = \frac{\int_{-\infty}^{\infty} f(s)g(t+s)ds}{\sqrt{\int_{-\infty}^{\infty} f^2(s)ds}\sqrt{\int_{-\infty}^{\infty} g^2(s)ds}}. \tag{6.96}$$

On a finite domain $[a, b]$, the mean of $f(t)$ and $g(t)$ need not be zero. In this event, we use

$$\rho(t) = \frac{\int_a^b F(s)G(t+s)ds}{\sqrt{\int_a^b F^2(s)ds}\sqrt{\int_a^b G^2(s)ds}} \tag{6.97}$$

with $F(s)$ and $G(t+s)$ using $F(t) = f(t) - A$, $A = (b-a)^{-1}\int_a^b f(s)ds$ and $G(t+s) = g(t+s) - B(t)$, $B(t) = (b-a)\int_a^b g(t+s)ds$.

Due to the running normalizations, (6.97) is nonlinear and a direct evaluation can be computationally expensive. However, if $f(t)$ and $g(t)$ have relatively constant variance, the computational effort practically reduces to the linear problem (6.92) by (6.93).

6.9 Signal Detection by FFT Using *Maxima*

To illustrate the above, consider searching for a periodic signal in a noisy detector output (6.86). To this end, we shall use the *Maxima* Fast Fourier Transform (FFT) implementation of DFT. For a brief instruction on using the *Maxima*'s FFT, type

```
? fft;
```

Here, fft N is a power of 2, i.e., $N = 2^m$ for some $m \geq 0$. To illustrate its implementation, consider

$$f(t) = \cos(pt) = \frac{1}{2}e^{ipt} + \frac{1}{2}e^{-ipt} \tag{6.98}$$

on a domain $D = [0, 2\pi]$ using $N = 8$ and $p = 1, 2$. The real and imaginary parts of the Fourier coefficients produced by DFT are as follows:

```
load(fft); N:8$ p:1$fpprintprec:4$ f(t):=ev(cos(p*t),numer)$
tn:array(any,N);fm:array(any,N)$
for k:0 thru N-1 do (tn[k]:ev(2*%pi/N*k,numer),fm[k]:f(tn[k]))$
for k:0 thru N-1 do display(tn[k],fm[k]);
fA:makelist(fm[k],k,0,N-1)$ c:fft(fA);
cR:makelist(realpart(c[k]),k,1,N);;
cI:makelist(imagpart(c[k]),k,1,N);
```

The output shows that all c_k are zero except for $c_1 = c_N = \frac{1}{2}$:

$$c_k : \left(0, \frac{1}{2}, 0, 0, 0, 0, 0, \frac{1}{2}\right). \tag{6.99}$$

Re-running the above for $p = 2$ similarly gives $c_2 = c_{N-1} = \frac{1}{2}$:

$$c_k : \left(0, 0, \frac{1}{2}, 0, 0, 0, 0, \frac{1}{2}, 0\right). \tag{6.100}$$

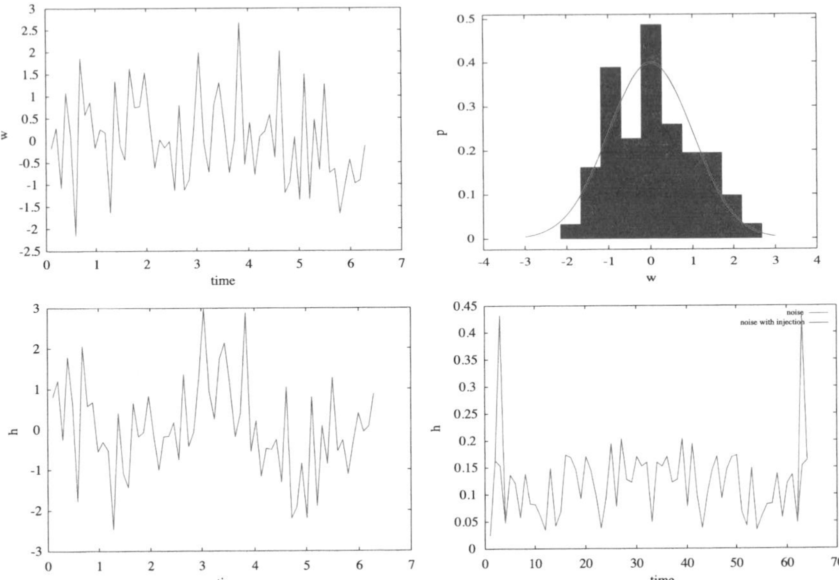

Fig. 6.10 (*Top*) Shown is Gaussian noise $w(t)$ sampled over $N = 64$ data points. The associated histogram is converted to probability density for comparison with the theoretical PDF of a Gaussian with $\sigma = 1$ and mean zero. (*Bottom*) A signal injection of $\cos(2t)$ into the noise, followed by FFT. FFT identifies the presence of the signal by large amplitude of the third and two-but-last Fourier coefficients

Next, we inject a signal $f(t)$ into Gaussian noise $w(t)$. This can be simulated using *Maxima*'s **random_normal** function:

```
load(distrib)$load(descriptive)$ m:0$sigma:1$N:64$
w:random_normal(m,sigma,N)$ xy:makelist([k,w[k]],k,1,N)$
wxplot2d([discrete,xy],[xlabel,"time"],[ylabel,"w"])$
```

A histogram of a noisy time series thus produced (Fig. 6.10) can be compared with the theoretical probability density function (PDF):

```
E:explicit(pdf_normal(x,m,sigma),x,m-3*sigma, m+3*sigma);
H:histogram_description(w,frequency=density,xlabel="w",
ylabel="p")$
draw2d(H,xrange=[-4,4],E,terminal=aquaterm)$
```

Injection of a periodic signal into detector noise (6.86) now consists of the sum of $f(t)$ and random values $w(t)$. Note that the array index of w produced by **random_normal** runs through $[1, N]$. Changing to $p = 2$ in (6.98), we write:

```
p:2$;f(t):=ev(cos(p*t),numer)$ h:array(any,N)$
for k:1 thru N do h[k]:ev(f(2*%pi/N*k),numer)+w[k]$
XY:makelist([k,h[k]],k,1,N)$
wxplot2d([discrete,XY],[xlabel,"time"],[ylabel,"h"])$
```

Now take this time-series to the Fourier domain:

```
load(fft)$ hA:makelist(h[k],k,1,N)$ c:fft(hA)$
uv:makelist([k,abs(c[k])],k,1,N)$ wxplot2d([discrete,uv])$
```

The presence of our signal (6.98) is evident from the two peaks in the third and two-but-last coefficients $|c_k|$ (see (6.100)).

Running various values of signal strength, we see that the *signal-to-noise ratio* (SNR) in the Fourier spectrum of a signal $f(t) = A\cos(pt)$ in the presence of Gaussian noise with standard deviation σ satisfies SNR = $(A/\sigma)\sqrt{N/2}$ in a time series sampled by N data points.

6.10 GPU-Butterfly Filter in $(f, \dot{f})$

The central engine of core-collapse supernovae may produce an output also in gravitational waves by virtue of conceivably non-axisymmetric mass-motion during and following angular momentum conserving collapse to a compact object, i.e., a neutron star or black hole. This poses an interesting challenge to search for signals that are potentially phase-coherent on intermediate time scales, but less likely so for the full duration of the burst. This prospect applies in particular to mass-motion in turbulent accretion disks close to or about the Inner Most Stable Circular Orbit (ISCO) of rotating black holes. From non-axisymmetric accretion flows down to the ISCO, angular momentum-rich gravitational radiation may feature slowly varying signals in the form of *chirps* with positive or negative time rate-of-change of frequency, i.e., ascending or descending chirps [6].

Searches for chirps may be pursued by applying a *butterfly filter* in $(f, \dot{f})$ space, that provides a conduit for

$$\left|\dot{f}\right| > \delta(f) > 0, \tag{6.101}$$

while suppressing signals with constant frequency. Figures 6.11 and 6.12 show the a sample of LIGO S6 data (downsampled to 4096 Hz) and the output of a butterfly filter. The output is an essentially featureless spectrogram devoid of any lines, as a suitable starting point to searches for burst sources by further correlations between two or more detectors.

A butterfly filter covers a two-dimensional parameter range in $(f, \dot{f})$. Calculating correlations (6.93) over a dense two-dimensional bank of chirp templates requires FFT on a high performance computing platform. *Graphics Processor Units* (GPU's)

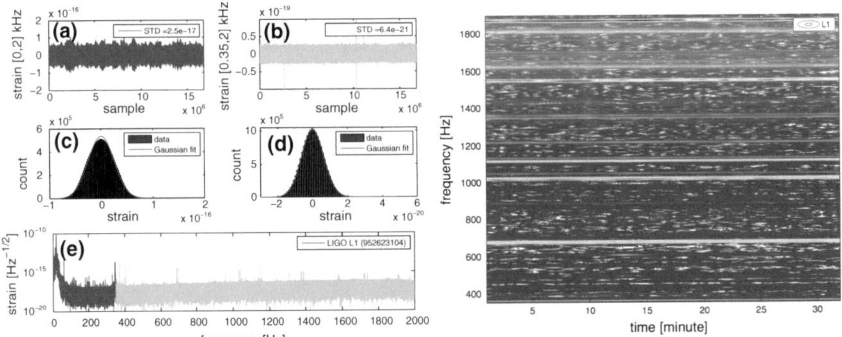

Fig. 6.11 (*Left panels*) High pass filter (350–2000 Hz) applied to LIGO S6 L1 data (0–4000Hz) in (**a**) shows high frequency shot-noise dominated output in (**b**), here applied to 64 LIGO frames of 64 s (4096^2 samples). Shot-noise is close to Gaussian (**d**), more so than the original broadband data (**c**). (*Right panel*) Fourier-based spectrogram of $N = 2^{18}$ Fourier transforms over the first 32 s, marked by the presence of various lines. (Reprinted from van Putten, M.H.P.M., 2016, ApJ, 819, 169.)

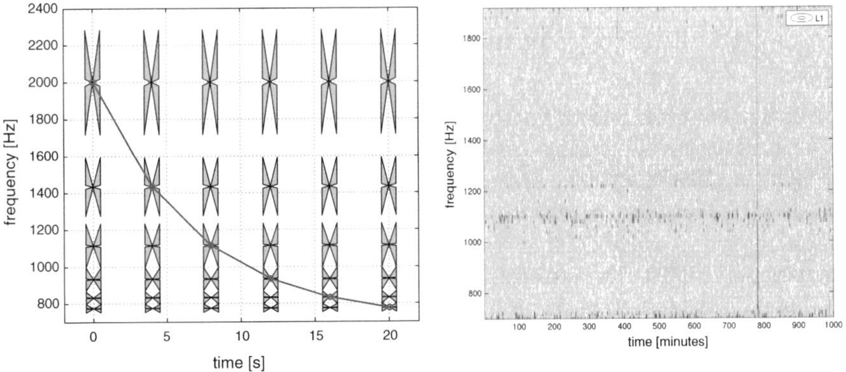

Fig. 6.12 (*Left*) A butterfly filter in the $(f, \dot{f})$ plane, to capture trajectories of ascending or descending chirps of intermediate duration. The vertex and opening angle of each butterfly represents a central frequency f and, respectively, a bandpass of slopes df/dt. (*Right*) Applied to LIGO S6 data of the L1 detector obtains an essentially featureless spectrogram with noticeable absence of any lines. A slight non-uniformity in the template bank is apparent in a weak noisy feature about 1100 Hz. (Reprinted from van Putten, M.H.P.M., 2016, ApJ, 819, 169.)

offer a cost-effective approach to *heterogeneous computing*, serviced by CPUs. A CPU-GPU computation is commonly realized using CUDA[5] or OpenCL.[6]

A successful implementation on a GPU is *compute limited* by avoiding limitations in bandwidth of the *Peripheral Component Interconnect* (PCI), and by off-loading

[5]https://developer.nvidia.com/about-cuda of NVIDIA Inc., founded in 1993.

[6]https://www.khronos.org/registry/cl/specs/opencl-2.0.pdf.

Table 6.1 Performance overview of correlations (6.103) by clFFT on a GPU. Included is allocated buffer memory (m[GB]), batch size M, and the number of correlations per second ($\dot{n}$)

$\log_2 N$	10	11	12	13	14	15	16	17	18	19	20
Gflop	580	587	324	227	264	269	268	221	200	128	118
m[GB]	1.01	1.02	1.03	1.06	1.56	1.56	1.56	1.56	1.53	1.25	1.26
$\log_2 M$	17	16	15	14	13	12	11	10	9	8	7
$\log_{10} \dot{n}$	7.1	6.7	6.1	5.6	5.4	5.0	4.7	4.3	3.9	3.2	2.9

some of the computations before and after FFT to the GPU.[7] Figure 6.14 and Table 6.1 show the performance in *Giga floating point operations per second* (Gflop s^{-1}) of correlations (6.92) and (6.93) over N data points in single precision (SP) complex-to-complex (C2C) FFT, by clfft[8] of AMD[9] under OpenCL. Here, the performance is calculated based on $5N \log_2 N$ floating point operations per FFT.

A bank of M data $\{\mathbf{z}_{(i)}\}_{i=1}^{M}$ of size N and a bank of M chirp templates $\{\mathbf{b}_{(j)}\}_{j=1}^{M}$ define $M \times M$ correlations

$$\mathbf{R}_{(i)(j)} = \mathbf{z}_{(i)} * \mathbf{b}_{(j)} \quad (i, j = 1, 2, \ldots M), \tag{6.102}$$

where $*$ refers to the correlation integral (6.92). Applied to functions sampled by N points, i.e., $\mathbf{f} = \{f_k\}_{k=0}^{N-1}$ and $\mathbf{g} = \{g_k\}_{k=0}^{N-1}$, (6.92) is

$$\rho_l = \sum_{k=0}^{N-1} f_k g_{k+l} \tag{6.103}$$

with cyclic boundary conditions on the indices ($k + l \to k + l - N$ for $k + l \geq N$).

Minimal relative overhead in FFT on a GPU obtains in *batch mode*, by *enqueuing* a *plan* for concurrent computation of M FFT's of size N, applied to a concatenated array of length NM ($8NM$ bytes in SP C2C). Furthermore, (6.93) can be handled by a *precallback* function, and the inverse FFT output can be filtered down to candidate events with signal-to-noise ratios exceeding a threshold, handled by a *postcallback* function (Fig. 6.13). At a sufficiently high threshold, *most* of the output is hereby ignored, avoiding the need for extensive reading of data buffers by the host CPU.

[7] We are distinguishing here between overall GPU compute performance subject to the limited bandwidth to the CPU, and internal GPU compute performance subject to the limited bandwidth to its Global Memory. In the application of vendor optimized GPU kernels, the user's challenge is optimization on the first by, e.g., suitable precision and data-format, batch mode and choice of pre- and post-callback functions.

[8] https://github.com/clMathLibraries/clFFT.

[9] Advanced Micro Devices Inc., founded by W.J. Sanders III in 1969.

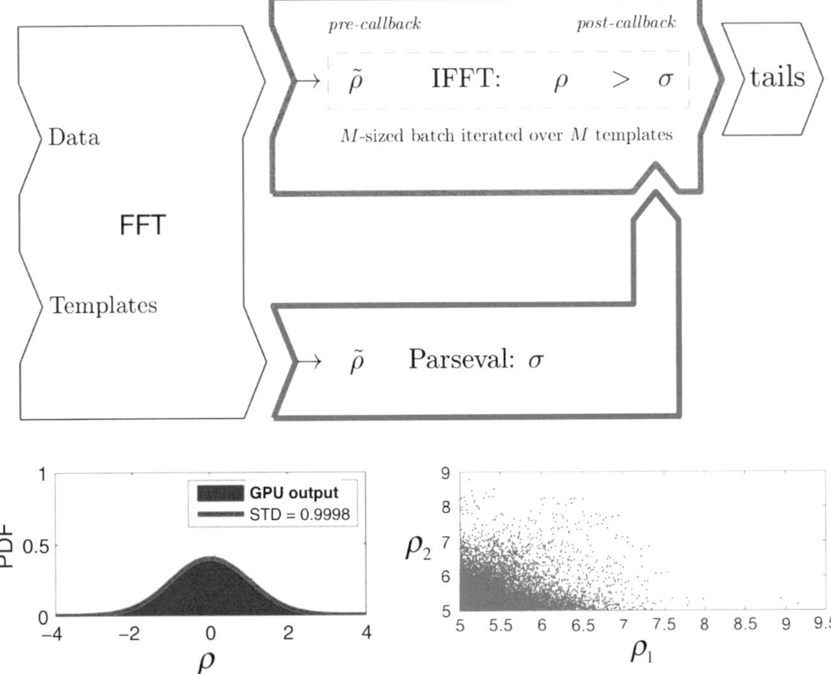

Fig. 6.13 (*Top*) GPU-CPU butterfly algorithm based on correlations (6.92) and (6.93) by the FFT with output reduced to tails of candidate signals with signal-to-noise ratios above a threshold σ. The latter obtains from a predictor step based on Parseval's Theorem. Thin and thick lines refer to, respectively, CPU and GPU computations. GPU computations use clFFT optimized in batch mode with pre- and post-callback functions for, respectively, multiplication of data and templates in the Fourier domain and reduction of data to tails. (*Bottom*) Shown is the essentially Gaussian GPU output of one channel normalized to unit variance based on the predicted STD σ by Parseval's Theorem, and the tails of two channels exceeding a cut-off $\kappa\sigma$, realizing a reduction in output by a factor about $\text{erfc}(\kappa/\sqrt{2})$

When sufficiently rare, events from a batch of M inverse FFT's of size N that meet the threshold can leave their footprint onto a single output array of size N with no over-writing. (Here, sufficiently rare results in a low density of footprints in the output array.) This statistical array projection effectively circumvents the limitations of bandwidth in reading the (inverse) FFT output by the host CPU. A consistent threshold setting for each of the M near-Gaussian output correlations can be realized by computing their standard deviations in advance.

In the presence of essentially Gaussian noise, correlations (6.102) of templates that do not match an underlying signal are essentially Gaussian. In matched filtering, therefore, the right hand side (6.70) obtained by (6.93) define predictions of σ in (6.102), *prior* to inverse FFT on the GPU. This procedure is opportune when post-callback functions are restricted to local operations. When the bank of filters is large, also (6.70) must be off loaded to the GPU.

There is further a modest gain in performance with *out of place* (inverse) FFT, in which input and output data are stored in separate allocations of *Global Memory* (rather than *in place*), further with *interleaved* data storage, in which real and imaginary parts of an array of complex numbers are stored consecutively in memory address space.

Figure 6.14 shows a typical performance of a few hundred Gflop s^{-1}. It defines an improvement over a CPU by at least an order of magnitude. This result is due to the large number of compute cores (*stream processing units* on AMD GPUs) in a high-end GPU (2048 in the AMD D700) and its fast on-chip memory.

Figure 6.14 shows a break in performance around $N = 2^{12}$ with a footprint of 32768 bytes for a complex array of size N. In OpenCL, memory is hierarchical with fast *Local Memory* supporting *Compute Units* comprising multiple *WorkItems*, each provided with even faster *Private Memory*. On the D700, Local Memory is 32768 bytes. FFT's on arrays of larger memory size inevitably invoke calls to relatively slower global memory, whence the break in Fig. 6.12.

Lastly, the above can be applied to the two H1 (Hanford) and L1 (Livingston) detector channels in one stroke, by merging their respective output $\mathbf{h}_H$ and $\mathbf{h}_L$ into complex data

$$\mathbf{z} = (\mathbf{h}_H, \mathbf{h}_L). \tag{6.104}$$

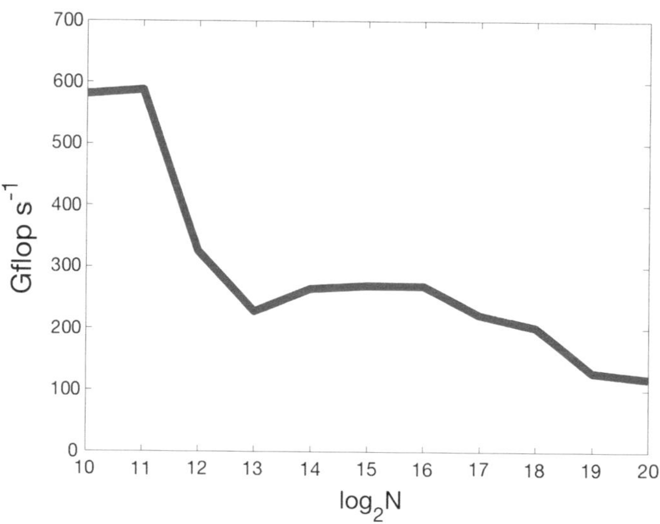

Fig. 6.14 GPU performance on correlations (6.92) and (6.93) by the Fast Fourier Transform using clFFT in single-precision, complex-to-complex and out of place. In batch mode with memory allocations of about 1 GByte and by use of pre- and post-callback functions, these results essentially represent the maximal performance of clFFT on the GPU. The calculations are performed in OpenCL on an AMD D700. The performance is better than 80% normalized to clFFT alone

Table 6.2 Overview spectral methods

1. Spectral representations of functions are efficient with orthonormal basis functions, where orthogonality is relative to a choice of domain. Notable examples are Legendre, Fourier and Chebyshev series. For smooth functions, Legendre series may be twice as efficient as Taylor series.

2. The Legendre and Chebyshev series are suitable for non-periodic problems on a finite interval. The Fourier series expansion is suitable for transients or (almost) periodic functions.

3. In a Fourier transform, the highest frequency components identified in the original signal is at one-half the sampling frequency (Nyquist criterion).

4. A smooth function tends to have a limited bandwidth in frequency space. A non-smooth function tends to have a broadband spectrum. A shallow asymptotic decay of the Fourier spectrum indicates a non-smooth function in the time-domain.

5. The Fourier spectrum of a transient function on $\mathbb{R}$ can be approximated by that of its periodic extension (6.60), provided aliasing errors are kept small by suitable choice of period T.

6. The complete spectrum of a periodic function sampled at N data points obtains by the Discrete Fourier Transform (DFT) of order N.

7. Efficient representation in Fourier series obtains for smooth functions. Fast computation is realized by FFT on CPU-GPU platforms, enabling butterfly filtering in $(f, \dot{f})$ space.

The real and imaginary parts of the output (6.102) then define the correlations for each detector.

Table 6.2 summarizes this chapter.

6.11 Exercises

6.1. The $P_n(x)$ are *even* and *odd* polynomials in x according to whether the degree n is even or odd. Determine by hand the coefficients a_i $(i = 0, 1, 2, 3)$ of

$$P_3(x) = a_3x^3 + a_2x^2 + a_1x + a_0 \tag{6.105}$$

based on orthogonality to $P_m(x)$ for $m = 0, 1, 2$.

6.2. For (6.4), consider the N-th order Legendre and Taylor series expansions $SN(x)$ and, respectively $QN(x)$. Plot the differences

$$\Delta_{P,N}(x) = SN(x) - f(x), \quad \Delta_{T,N}(x) = QN(x) - f(x) \tag{6.106}$$

for various choices of $N \leq 10$ and observe the errors in magnitude and how they are distributed over $D = [-1, 1]$. In particular, compare the maximum (the L_∞ norm) and mean squared errors (the L_2 norm)

$$E_1 = \max_{x \in D} |\Delta(x)|, \quad E_2 = \sqrt{\int_{-1}^{1} \Delta(x)^2 dx}. \tag{6.107}$$

Which expansion shows a practically uniform distribution of errors over D? By considering various N, e.g., $N = 2, 4, \ldots, N$, plot both E_1 and E_2 as a function of N. How do the expansions compare in regards to convergence?

6.3. Our discussion can be put in a more general context by considering a function domain to be some arc γ in the complex plane and an inner product given by the bilinear form

$$\langle f, g \rangle = \int_\gamma f(z)\bar{g}(z) ds, \tag{6.108}$$

where s denotes the arclength along γ. For instance, let γ_1 denote the interval $[-1, 1]$ with $s = x$ and γ_2 the unit circle $z = e^{i\phi}$ with $s = \varphi$ $(0 \leq \varphi \leq 2\pi)$ as in (6.3) and, respectively, (6.2). On γ_1, mutual orthogonality is not, respectively, *is* satisfied for the basis functions $\{z^n\}_{n=0}^{\infty}$ and $\{P_n(x)\}_{n=0}^{\infty}$. What is the general result for $f(z) = z^n$ and $g(z) = z^m$ $(n, m \geq 0)$ on γ_2?

6.4. Prove (6.47) by contour integration.

6.5. Prove (6.65) and Parseval's theorem (6.68).

6.6. For (6.4), the integral

$$\int_{-1}^{1} f(x)dx = \sqrt{2} \tanh^{-1} \frac{1}{\sqrt{2}} \tag{6.109}$$

may be evaluated by the integral of the truncated Taylor series,

$$A_n = 2a_0 + \frac{2}{3}a_2 + \frac{2}{5}a_4 + \cdots + \frac{2}{n+1}a_n. \tag{6.110}$$

The same obtains from b_0 in the Legendre expansion (6.8), which may be computed by Gauss-Legendre quadrature,

$$B_n = \sum_{i=1}^{n} w_i f(x_i). \tag{6.111}$$

Compare the rates of convergence of A_n and B_n and interpret the results.

6.7 Consider the function $f(x) = |x|$ on [−1,1]. Sketch the graph of $f(x)$ in the (x, y) plane. Does a Taylor series expansion of $f(x)$ exist at $x = 0$? Obtain and plot the Legendre expansion $S_1(x)$ of $f(x)$ of order $N = 16$; similarly for the Fourier expansion $S_2(x)$. Compare the discrepancies $D_i(x) = S_i(x) - f(x)$ $(i = 1, 2)$. Is there a preference for Legendre or Fourier expansion in this case? Study the decay of the Legendre and Fourier coefficients in the limit of large N and explain your observations.

6.8. The Legendre polynomials $\{P_n(x)\}_{n=0}^{\infty}$ are an orthogonal series of functions, satisfying (6.9). Consider the first three Legendre polynomials (cf. (6.7))

$$P_0(x) = 1, \quad P_1(x) = x, \quad P_2(x) = \frac{1}{2}\left(3x^2 - k\right). \tag{6.112}$$

By explicit calculation
(i) show $P_0(x)$ and $P_1(x)$ are orthogonal.
(ii) determine k in $P_2(x)$ in (6.112).
(iii) calculate the first two Legendre coefficients a_0 and a_1 in the Legendre expansion of $f(x) = e^x$ over the interval $D = [-1, 1]$, i.e.,

$$f(x) = \sum_{n=0}^{\infty} a_n P_n(x) = a_0 + a_1 P_1(x) + a_2 P_2(x) + \cdots. \tag{6.113}$$

6.9. (a) In Example 6.3, if $f(x) = \tan\left(\frac{1}{2}\pi x\right)$, does the Legendre expansion exist? (b) For $f(x) = \cos \pi x$ on $[-1, 1]$, determine the first three Taylor coefficients in the expansion

$$f(x) = \sum_{k=0}^{2} c_k x^2 + C_2(x) \tag{6.114}$$

and the first three Legendre coefficients in the expansion

$$f(x) = \sum_{k=0}^{2} a_k P_k(x) + R_2(x). \tag{6.115}$$

Compare the remainders $C_2(x)$ and $R_2(x)$.

6.10. For the Gaussian, $f(t) = e^{-(t/\tau)^2}$, derive the frequency spectrum by the Fourier integral (6.24). For a given $T \gg \tau$, obtain the approximate frequency of $f_T(t)$ in (6.60) by DFT and estimate the aliasing error.

6.11. Show that the cosines of x^n and x^m in (6.17) in the Euclidean inner product (6.3) decay as

$$\sim 1 - \frac{k}{2n+1} \quad (0 \le k \ll n); \quad \sim \sqrt{\frac{2n+1}{2k}} \quad (k \gg n), \tag{6.116}$$

where $m = n + k$. Next, consider the linear map

$$x^n = \sum_{m=0}^{n} Z_{nm} P_m(x). \tag{6.117}$$

For each N, the matrix $Z = [Z_{nm}]$ $(0 \le n, m \le N)$ represents the expansion of $\{x^n\}_{n=0}^N$ in $\{P_n\}_{n=0}^N$. The leading block

$$Z \simeq 10^{-3} \begin{pmatrix} 1000 & 0 & 0 & 0 & 0 & 0 & 0 & 0 & 0 & 0 \\ 0 & 1000 & 0 & 0 & 0 & 0 & 0 & 0 & 0 & 0 \\ 1667 & 0 & 667 & 0 & 0 & 0 & 0 & 0 & 0 & 0 \\ 0 & 1400 & 0 & 400 & 0 & 0 & 0 & 0 & 0 & 0 \\ 1800 & 0 & 1029 & 0 & 229 & 0 & 0 & 0 & 0 & 0 \\ 0 & 1571 & 0 & 698 & 0 & 127 & 0 & 0 & 0 & 0 \\ 1857 & 0 & 1238 & 0 & 450 & 0 & 69 & 0 & 0 & 0 \\ 0 & 1667 & 0 & 909 & 0 & 280 & 0 & 37 & 0 & 0 \\ 1889 & 0 & 1374 & 0 & 634 & 0 & 169 & 0 & 20 & 0 \\ 0 & 1727 & 0 & 1063 & 0 & 425 & 0 & 100 & 0 & 11 \end{pmatrix}, \tag{6.118}$$

shows that x^n mostly projects onto $P_m(x)$ $(m < n)$ with a minor projection onto $P_n(x)$: x^n is only slightly tilted out of the plane spanned by the polynomials of degree $n - 1$, expressed by the diagonal elements

$$Z_{nn} = \frac{< x^n, P_n >}{< P_n, P_n >} = \frac{2n+1}{2} \int_{-1}^{1} x^n P_n(x) dx \ll 1 \tag{6.119}$$

in terms of (6.3). Though each such block of Z is non-singular, show that it becomes ill-posed when N is large. [*Hint*. Show that the eigenvalues Z_{nn} of Z are the reciprocals of (5.99).]

6.12. Derive (6.82) from, e.g.,

$$\int_{-1}^{1} T_n(x)T_m(x)dx = \int_0^{\pi} \cos n\theta \cos m\theta \sin\theta d\theta, \tag{6.120}$$

and show that the cosines of $T_n(x)$ and $T_m(x)$ in (6.82) and (6.83) in (6.3) decay as

$$\sim \frac{1}{n^2}\ (0 \le k \ll n); \quad \sim \frac{1}{k^2}\ (k \gg n), \tag{6.121}$$

where $m = n + k$. Next, consider the basis transformation

$$T_n(x) = \sum_{m=0}^{n} Z_{nm} P_m(x) \tag{6.122}$$

with 5×5 diagonal blocks

$$[Z_{nm}] \simeq 10^{-3} \begin{pmatrix} 1000 & 0 & 0 & 0 & 0 \\ 0 & 1000 & 0 & 0 & 0 \\ -333 & 0 & 1333 & 0 & 0 \\ 0 & -600 & 0 & 1600 & 0 \\ -67 & 0 & -762 & 0 & 1829 \end{pmatrix} \tag{6.123}$$

$(0 \le n, m \le 4)$ and

$$[Z_{nm}] \simeq 10^{-3} \begin{pmatrix} 7213 & 0 & 0 & 0 & 0 \\ 0 & 7267 & 0 & 0 & 0 \\ -3633 & 0 & 7321 & 0 & 0 \\ 0 & -3660 & 0 & 7375 & 0 \\ -915 & 0 & -3687 & 0 & 7428 \end{pmatrix} \tag{6.124}$$

$(66 \le n, m \le 70)$. The eigenvalues of Z_{nm} are given by the diagonal elements

$$Z_{nn} = \frac{< T_n, P_n >}{< P_n, P_n >} = \frac{2n+1}{2} \int_{-1}^{1} T_n(x)P_n(x)dx. \tag{6.125}$$

Establish well-posedness by evaluation of Z_{nn} for large n. [*Hint.* Consider $T_n(x) = p_{n-1}(x) + A_n x^n$, where $p_n(x)$ is a polynomial of degree $n-1$, so that $< T_n, P_n >= A_n < x^n, P_n >$, and use **6.13**.]

6.13. Based on **6.14**, how are efficiencies in Chebyshev and Legendre expansions expected to compare when applied to smooth functions on $[-1, 1]$?

6.14. SNR is defined by he square root of the ratio of energy in the signal to the energy in the background noise. For a periodic signal $f(t) = A\cos(pt)$, show that SNR $\propto N^{1/2}$ in a Fourier-based detection algorithm.

6.15. Consider injection of a pulse

$$f(t) = Ae^{-(t/2)^2} \tag{6.126}$$

into noisy data and recover the signal by (a) cross-correlation and (b) Fourier analysis. Determine for both (a) and (b) the minimum SNR for signal identification.

6.16. Write a program for direct numerical evaluation of DFT on $[0, 2\pi]$ in single and double precision. For $N = 256, 1024, 4096$, consider the numerical error in the output in single precision, by comparing with the exact solutions for the coefficients in response to harmonic data of the form $\cos px$ for various p or by comparing with results obtained in double precision. (a) Evaluate numerical errors in $\{c_k\}_{k=0}^{N/2}$ as a function of k. (b) Evaluate (in double precision) the numerical discrepancies in Parseval's theorem as a function of N. (c) Doing the same using FFT in single precision, note the improvement in time *and* accuracy over DFT.

6.17. Write a program to evaluate (6.103) by DFT or FFT on a CPU. Benchmark its performance according to Table 6.1.

6.18. Prove the statement following (6.104) and give the Fourier transforms on $[0, 2\pi]$ of

$$Z = (\cos px, 0),\ \ Z = (\sin px, 0),\ \ Z = (0, \cos px),\ \ Z = (0, \sin px) \tag{6.127}$$

with $p = 0, 1, 2, \ldots$.

6.19. Let $\{x_k\}_{k=1}^{N}$ be a sequence of random numbers of size N. Consider calculating the standard deviation σ of a subsequence $\{y_{k_m}\}_{m=1}^{M}$ satisfying, e.g., $y_{k_m} > a$ $(1 \leq M \leq N)$, where $a > 0$ is some threshold. Moden *Central Processing Units* (CPUs) obtain fast numerical implementations by vectorization based on constant array sizes. In calculating σ, perform benchmarks on implementations that do and do not explicitly extract sub-arrays of size M. What is the result as a function of M/N as N becomes large?

6.20. In Example 6.5, show that

$$\alpha = \frac{a}{\lambda} \tag{6.128}$$

in (6.31), where a denotes the acceleration of the mirror in Fig. 6.5 and λ is the wave length in the uniform distribution of phase over Y. What physical constants are needed to make α dimensionless? With the dispersion relation $\omega = ck$ of photons in vacuum, $2\pi\alpha = \hbar\omega/k_BT$, the result gives the *Unruh temperature* [7]

$$k_BT = \frac{a\hbar}{2\pi c} \tag{6.129}$$

of essentially thermal radiation (Appendix C), where k_B is the Boltzmann constant, $\hbar$ is Planck's constant and c is the velocity of light.

References

1. Moses, J., 2012, J. Symb. Comput. 47, 123; doi:10.1016/j.jsc.2010.08.018.
2. J.W. Cooley and J. Tukey, 1965, most commonly applied to $N = 2^m$ $(m \in \mathbb{N})$.
3. van Putten, M.H.P.M., 1998, SIAM Rev., 40, 33.
4. Abott, B.P., et al., 2016, *Phys. Rev. Lett.*, 116, 061102.
5. van Putten, M.H.P.M., & Della Valle, M., 2017, MNRAS, 464, 3219.
6. Levinson, A., van Putten, M.H.P.M., & Pick, G., 2015, ApJ, 812, 124.
7. Unruh, W., 1976, *Phys. Rev. D.*, 14, 870.

Chapter 7
Root Finding

Root finding is one of the oldest numerical problems that goes back to times well before Newton. It illustrates various numerical issues such as initial guesses, convergence rates and stability in various numerical formulations.

7.1 Solving for $\sqrt{2}$ and π

An illustrative problem is the computation of $\sqrt{2}$. The existence of $\sqrt{2}$ as a root of $f(x) = x^2 - 2$ follows from continuity of $f(x)$ on $[-1, 3]$ and the fact that $f(1) = -1 < 0$, $f(3) = 7 > 0$. Consider an initial guess x and compute a correction ξ according to

$$(x + \xi)^2 = 2: \; x^2 + 2x\xi + \xi^2 = 2. \tag{7.1}$$

An *approximate* correction $\epsilon \simeq \xi$ obtains by linearization, dropping the quadratic term ξ^2 in case of $\xi < 1$, i.e.,

$$x^2 + 2x\epsilon = 2: \; \epsilon = \frac{2 - x^2}{2x}. \tag{7.2}$$

A guess $x = 1.5$ hereby obtains $\epsilon = -1/12$ and hence the approximate root

$$x' = x + \epsilon = \frac{17}{12} = 1.4167, \tag{7.3}$$

which differs from $\sqrt{2} = 1.4142\ldots$ by 2.5×10^{-3}. We are at liberty to repeat this procedure with $x' = 17/12$, obtaining $x'' = 577/408 = 1.414215686\ldots$, which

M.H.P.M. van Putten, *Introduction to Methods of Approximation in Physics and Astronomy*, Undergraduate Lecture Notes in Physics,
DOI 10.1007/978-981-10-2932-5_7

differs from $\sqrt{2}$ by 2.1×10^{-6}. The next approximation shows a residual discrepancy of 1.6×10^{-12}, suggesting a pattern of *quadratic convergence* in which the error in each approximation is the error squared in the previous step.

For a more systematic treatment of the above, we devise an iterative method for the equivalent problem of finding a root

$$f(x) = 0 \tag{7.4}$$

of

$$f(x) = x^2 - 2 \tag{7.5}$$

by repeated use of an associated linear problem, illustrated in Fig. 7.1. Taking advantage of the smoothness of $f(x)$ in (7.5), consider the Taylor series expansion

$$f(x) = f(x_0) + f'(x_0)(x - x_0) + \frac{1}{2} f''(\xi)(x - x_0)^2, \tag{7.6}$$

where ξ is between x_0 and x. If x_0 is our first guess for $\sqrt{2}$, then the leading two terms suggest to consider x satisfying

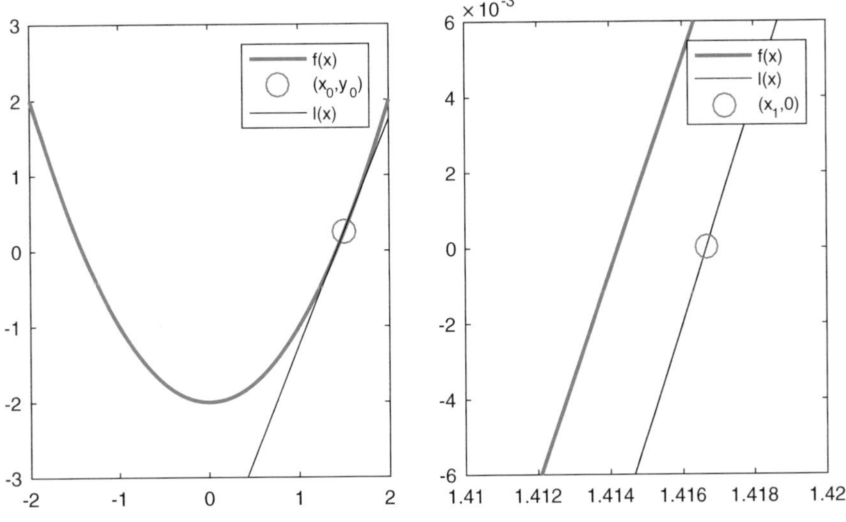

Fig. 7.1 Root finding by Newton's method applied to $f(x) = x^2 - 2$ approaches $a = \sqrt{2}$ by successively solving linear problems defined by the tangent approximation, here shown for $l(x) = 0$, $l(x) = f(x_0) + (x - x_0) f'(x_0)$, giving $x_1 = 17/12$ for $x_0 = 3/2$. At finite curvature ($f''(a) \neq 0$), the method shows quadratic convergence

Table 7.1 Newton's method applied to $f(x) = x^2 - 2$

n	x_n	ϵ_n
0	2	0.6
1	1.5	0.0858
2	1.41666	2.5×10^{-3}
3	1.414215	2×10^{-6}
4	1.41421356237	1.6×10^{-12}

$$f(x_0) + f'(x_0)(x - x_0) = 0: \quad x = x_0 - \frac{f(x_0)}{f'(x_0)}. \tag{7.7}$$

As per above, we repeat this procedure, taking x as our next starting point. The result is a sequence $\{x_n\}_{n=0}^N$, generated by

$$x_{n+1} = x_n - \frac{f(x_n)}{f'(x_n)}, \quad n = 0, 2, \ldots, N - 1. \tag{7.8}$$

Numerical evaluation (Table 7.1) shows the resulting approximations and accompanying errors $\epsilon_n = x_n - \sqrt{2}$

Example 7.1. The scheme (7.8) is very general. It suggests application also to problems in $\mathbb{C}$. Consider computing $\pi = 3.14159\ldots$ as the root of

$$f(z) = e^{iz} + 1, \tag{7.9}$$

that defines Euler's identity

$$e^{i\pi} + 1 = 0. \tag{7.10}$$

According to (7.8), we consider the iterations

$$z_{n+1} = z_n - \frac{e^{iz} + 1}{ie^{iz}}. \tag{7.11}$$

Numerical evaluation gives the approximations and errors $\epsilon_n = z_n - \pi$ shown in Table 7.2.

In each iteration, the number of significant digits *doubles*. Why does this *Newton's method* (7.8) work so well?

Table 7.2 Newton's method applied to $f(z) = e^{iz} + 1$

n	z_n	ϵ_n
0	1	2.14
1	$2.909 + 0.58i$	0.6
2	$3.32 - 0.16i$	2.4×10^{-1}
3	$3.169 + 0.0015i$	2.8×10^{-2}
4	$3.14155 + 0.00038i$	3.8×10^{-4}
5	$3.141592668 - 7 \times 10^{-8}i$	7×10^{-8}
6	$3.141592653589793 - 2.48 \times 10^{-15}i$	2.7×10^{-15}
7	$3.141592653589793\ldots - 2.5 \times 10^{-30}i$	3.7×10^{-30}
8	$3.141592653589793\ldots$	6.7×10^{-60}

7.2 Convergence in Newton's Method

In light of the iterative scheme (7.8), consider a Taylor series expansion about the root. With $\epsilon = x - a$, we have

$$\begin{aligned} f(x) &= f(a) + f'(a)(x-a) + \tfrac{1}{2}f''(a)(x-a)^2 + O\left((x-a)^3\right) \\ &= \epsilon f'(a) + \tfrac{1}{2}\epsilon^2 f''(a) + O\left(\epsilon^3\right), \end{aligned} \tag{7.12}$$

$$f'(x) = f'(a) + f''(a)(x-a) + O\left((x-a)^2\right) = f'(a) + \epsilon f''(a) + O\left(\epsilon^2\right),$$

so that

$$\frac{f(x)}{f'(x)} = \frac{\epsilon f'(a) + \frac{1}{2}\epsilon^2 f''(a) + O\left(\epsilon^3\right)}{f'(a) + \epsilon f''(a) + O\left(\epsilon^2\right)} = \epsilon \frac{1 + \frac{1}{2}\epsilon f''(a)/f'(a) + O\left(\epsilon^2\right)}{1 + \epsilon f''(a)/f'(a) + O\left(\epsilon^2\right)}. \tag{7.13}$$

The result can be expanded further, using $1/(1+z) = 1 - z + z^2 - z^3 + \ldots$ ($|z| < 1$), whereby

$$\frac{f(x)}{f'(x)} = \epsilon \left[1 - \frac{1}{2}\epsilon f''(a)/f'(a) + O\left(\epsilon^2\right)\right]. \tag{7.14}$$

Applied to (7.8), we find

$$\epsilon_{n+1} = \epsilon_n - \frac{f(x_n)}{f'(x_n)} = \frac{1}{2}\frac{f''(a)}{f'(a)}\epsilon_n^2 + O(\epsilon^3). \tag{7.15}$$

Whenever $f'(a)$, $f''(a) \neq 0$, (7.15) shows that the error in each iteration is the previous error squared, whereby the number of significant digits doubles in each step as shown in Table 7.1.

In fact, (7.15) explicitly shows a dependence of the rate of convergence on the *curvature* of the graph of $f(x)$ at the root $x = a$: convergence is faster when the graph is more linear at $x = a$. This opens a suggestion for a possibly *cubic* convergence, whenever

$$f''(a) = 0, \quad f''(x) = f^{(3)}(a)\epsilon + O(\epsilon^2) \tag{7.16}$$

whereby the first term on the right hand-side in (7.15) gains an additional power in ϵ_n.

7.3 Contraction Mapping

Newton's method (7.8) is a special case of a contraction mapping,

$$x_{n+1} = G(x_n) \tag{7.17}$$

in seeking *fixed points*

$$a = G(a). \tag{7.18}$$

Here, convergence of $x_{n+1} = G(x_n)$ derives from

$$\begin{aligned} G(x_n) &= G(a) + G'(a)(x_n - a) + O((x_n - a)^2) \\ &= a + G'(a)(x_n - a) + O((x_n - a)^2), \end{aligned} \tag{7.19}$$

whereby

$$\epsilon_{n+1} = x_{n+1} - a = G'(a)\epsilon_n + O(\epsilon_n^2). \tag{7.20}$$

It follows that

$$|\epsilon_{n+1}| \simeq \left|G'(a)\right| |\epsilon_n| \simeq L^n |\epsilon_0|, \quad L = \left|G'(a)\right|. \tag{7.21}$$

Specializing, Newton's method (7.8) obtains with $G(x) = x - f(x)/f'(x)$, satisfying

$$G'(x) = \frac{f(x) f''(x)}{(f'(x))^2} \tag{7.22}$$

as illustrated in Fig. 7.2. In this event, $L \propto f(x) \propto \epsilon$ in approaching the root $f(x) = 0$; and since $L \propto f''(x)$, one more power of convergence is attained if $f(x)$ is linear at the zero $x = a$.

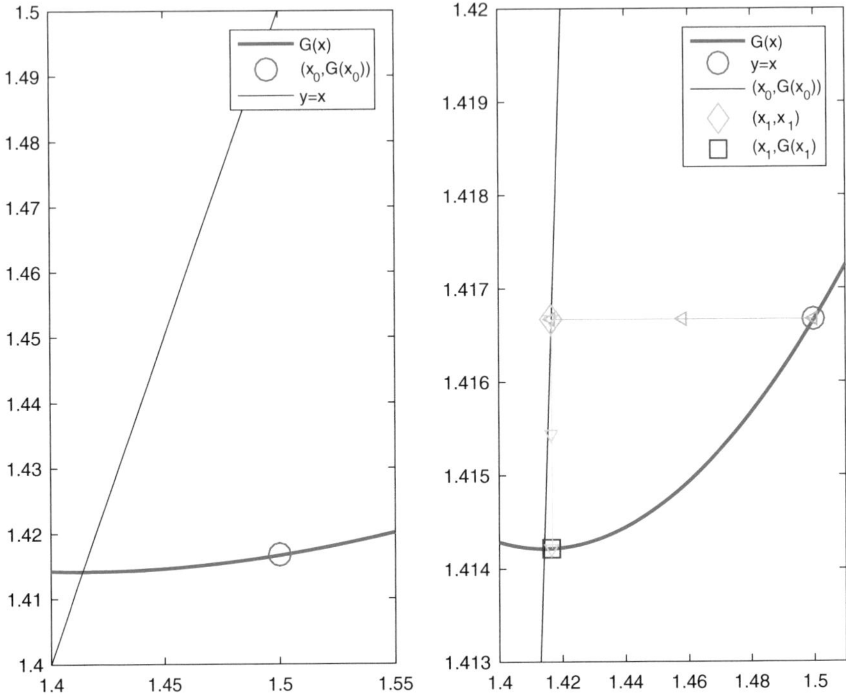

Fig. 7.2 Root finding by a contraction mapping $x_{n+1} = G(x_n)$ typically satisfies algebraic convergence $\epsilon_n = O(L^n)$, $L = \left|G'(a)\right|$ at the fixed point $a = G(a)$. When $L = 0$ as shown for $G(x) = x - f(x)/f'(x)$, $f(x) = x^2 - 2$, the method specializes to Newton's method with quadratic convergence

Example 7.2. Let us turn once more to calculating π starting with $f(x) = \sin x$. Formally, Newton's method gives

$$G_0(x) = x - \frac{\sin x}{\cos x}. \tag{7.23}$$

A slight simplification obtains by neglecting second order deviations from 1 in $\cos x$ away from $x = \pi$, using, instead,

$$G(x) = x + \sin x. \tag{7.24}$$

At this point, we can ignore its derivation and apply (7.17); see Table 7.3.

Table 7.3 Cubic convergence in the contraction mapping $G(x) = x + \sin x$

n	x_n	ϵ_n
0	1	−2.14
1	1.8414	−1.3
2	2.805	−0.3
3	3.1352	-6×10^{-3}
4	3.14159261159 ...	-4×10^{-8}
5	3.141592653589793 ...	-1×10^{-23}
6	3.141592653589793 ...	-3×10^{-70}

Convergence in Newton's method for calculating a square root $a = \sqrt{c}$ by solving $f(x) = 0$, $f(x) = x^2 - c$, can be accelerated to quartic order upon considering the root of $h(x) = g(x)f(x)$ with pre-factor

$$g(x) = 1 - \frac{1}{4c}f(x) + \frac{1}{8c^2}f(x)^2. \tag{7.25}$$

The coefficients in $g(x)$ are chosen so as to satisfy the local flatness conditions

$$G''(a) = G'''(a) = 0 \tag{7.26}$$

by $h''(a) = h'''(a) = 0$. The contraction mapping $G(x) = x - h(x)/h'(x)$ hereby obtains quadratic convergence. Slightly simplified expressions satisfying (7.26) obtain with (Fig. 7.3)

$$G(x) = x - \gamma(x)\frac{f(x)}{f'(x)}, \tag{7.27}$$

$$\gamma(x) = 1 + \frac{1}{4c}f(x) - \frac{1}{8c^2}f(x)^2. \tag{7.28}$$

Example 7.3. Explicitly, the case for $c = 2$ gives

$$G(x) = \frac{40 + 60x^2 - 10x^4 + x^6}{64x}. \tag{7.29}$$

Iterations starting at $x_0 = 1$ give the following as shown in Table 7.4.

Thus, x_{20} approaches $\sqrt{2}$ to within 10^{12} significant digits. The square root of c can thus be evaluated efficiently to arbitrary precision in a moderate number of iterations.

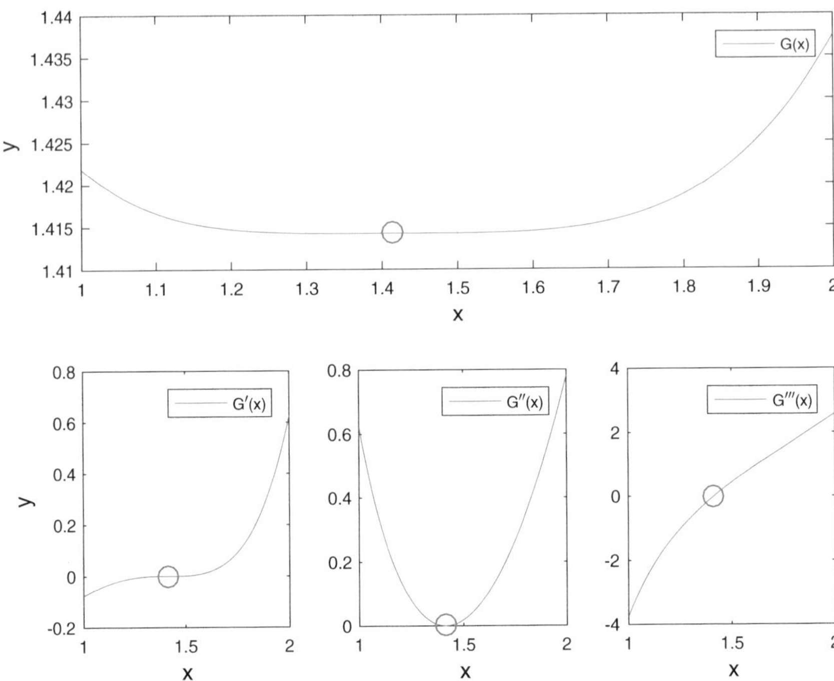

Fig. 7.3 The contraction map $G(x)$ of (7.29) has vanishing derivatives to third order, producing quartic convergence in approaching the fixed point $a = G(a)$ (*circles*)

Table 7.4 Quartic convergence in a modified Newton's method for $f(x) = x^2 - 2$

n	x_n	$\lvert x_n - \sqrt{2}\rvert$	$\log(\epsilon_n)/\log(\epsilon_{n-1})$
0	1	0.41	
1	1.421875000000000	0.00766	1.4930
2	1.414213563132792 ...	7.6×10^{-10}	4.3104
3	1.414213562373095 ...	7.3×10^{-38}	4.0719
4	1.414213562373095 ...	6.4×10^{-149}	4.0177
5	1.414213562373095 ...	6.4×10^{-597}	4.0044
6	1.414213562373095 ...	6.4×10^{-2390}	4.0011

7.4 Newton's Method in Two Dimensions

Consider Newton's method to calculate a root of

$$\mathbf{F}(x, y) = \begin{pmatrix} f_1(x, y) \\ f_2(x, y) \end{pmatrix} = \begin{pmatrix} x^2 + y^2 - 1 \\ y - x^2 \end{pmatrix}. \tag{7.30}$$

As before, a Taylor series expansion

$$\mathbf{F}(x, y) = \mathbf{F}(x_0, y_0) + DF_{(x_0,y_0)} \begin{pmatrix} x - x_0 \\ y - y_0 \end{pmatrix} + \text{h.o.t.} \tag{7.31}$$

where h.o.t refers to higher order terms, obtains a linearized problem: an initial guess (x_0, y_0) introduces an estimate (x_1, y_1)

$$\mathbf{F}(x_0, y_0) + DF_{(x_0,y_0)} \begin{pmatrix} x_1 - x_0 \\ y_1 - y_0 \end{pmatrix} = \mathbf{0}, \tag{7.32}$$

where DF is the Fréchet derivative

$$DF = \begin{pmatrix} \partial f_1/\partial x & \partial f_1/\partial y \\ \partial f_2/\partial x & \partial f_2/\partial y \end{pmatrix} = \begin{pmatrix} 2x & 2y \\ -2x & 1 \end{pmatrix}. \tag{7.33}$$

When determinant of DF is nonzero,

$$|DF| = \frac{\partial f_1}{\partial x}\frac{\partial f_2}{\partial y} - \frac{\partial f_2}{\partial x}\frac{\partial f_1}{\partial y} \neq 0, \tag{7.34}$$

the inverse DF^{-1} exists,

$$DF^{-1} = \frac{1}{2x(1+2y)} \begin{pmatrix} 1 & -2y \\ 2x & 2x \end{pmatrix}. \tag{7.35}$$

For Newton's method to apply, iteractions (x_n, y_n) must stay away from $x = 0$ and $y = -\frac{1}{2}$. In this event, the solution to (7.32) is

$$\begin{pmatrix} x_1 - x_0 \\ y_1 - y_0 \end{pmatrix} = DF^{-1}_{(x_0,y_0)} \mathbf{F}(x_0, y_0). \tag{7.36}$$

As before, we consider (7.36) to define Newton iterations

$$\begin{pmatrix} x_{n+1} \\ y_{n+1} \end{pmatrix} = \begin{pmatrix} x_n \\ y_n \end{pmatrix} - DF^{-1}_{(x_n,y_n)} \mathbf{F}(x_n, y_n). \tag{7.37}$$

Numerically, we find the approximations and errors $\epsilon_n = z_n - \pi$ shown in Table 7.5. The results of this table show convergence to one of the two roots,

$$(x, y) = (\pm\sqrt{u}, u) = (0.7862, 0.6180), \tag{7.38}$$

where $u = -\frac{1}{2} + \frac{1}{2}\sqrt{5}$. Figure 7.4 shows convergence for two different initial guesses.

Table 7.5 Newton's method applied to $\mathbf{F}(x, y)$ in (7.30)

n	x_n	y_n	ϵ_n	x_n	y_n	ϵ_n
0	0.5	0		1.5	0	
1	0.8917	1		1.08	1	
2	.	.		0.8494	0.6667	
3	.	.		.	.	
4	0.7862	0.6180		0.7862	0.6180	
8	0.7862	0.6180		0.7862	0.6180	

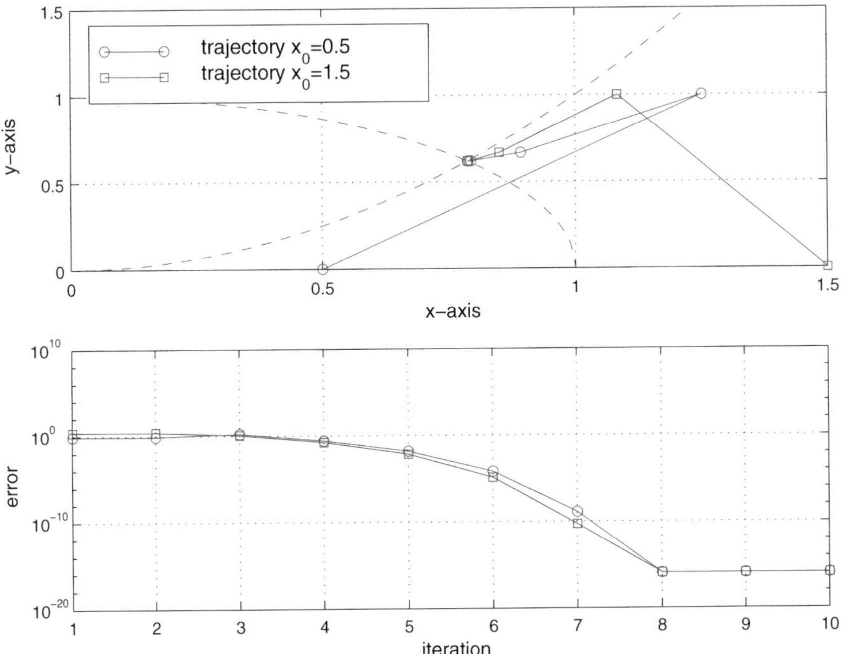

Fig. 7.4 Shown are the discrete trajectories of the iterations by Newton's root finding method applied to (7.30)

7.5 Basins of Attraction

Newton's method in two dimensions is naturally encountered in solving roots of functions of a complex variable by a contraction mapping. Extending the domain of initial guesses z_0 to the problem of root finding of (7.27) to $\mathbb{C}$, we are led to consider regions in the complex plane about each root, where z_0 gives convergence of

$$z_{n+1} = G(z_n). \tag{7.39}$$

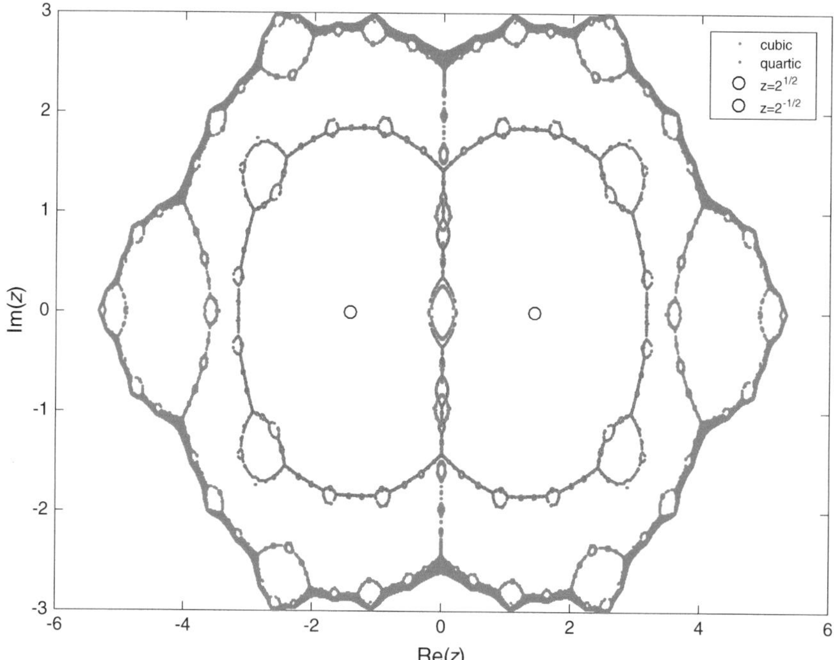

Fig. 7.5 Shown are basins of attraction about the roots of $f(z) = z^2 - 2 = 0$ (*circles* at $\pm\sqrt{2}$) by the contraction map $G(z) = z - \gamma(z)f(z)/f'(z)$ with cubic convergence ($\gamma(z) = 1 + f(z)/8$, outer contour) and quartic convergence within (7.28, inner contour)

Quite typically, the *basin of attraction* about each fixed point $a = G(a)$ is bounded (though it may be semi-infinite). The boundary can be determined numerically, by identifying z_0 for which convergence fails. As the numerical results show, these boundaries take complex shapes with self-similar structures, commonly referred to as *fractals* [1], that appear with remarkable universality in Nature (e.g., Fig. 7.7). Fractals derive their complicated appearance from the property that, viewed as a point set in the plane, a fractal path between two points has infinite length yet zero surface area inferred from, loosely speaking, the divergence in length estimated from a finite cover using boxes of arbitrarily small size. A fractal assumes with dimension $1 < d < 2$. Arising from detailed structure to arbitrarily small scales, they possess no tangents and hence are not differentiable.

Example 7.4. Figure 7.5 shows the result for (7.17) with cubic and quartic convergence. Noticeable is the relatively smaller size of the basins of the latter about each of the two roots $\pm\sqrt{2}$. Figure 7.6 shows the same for $f(z) = z^3 - 1$. The boundary of a basin of attraction is found to have a fractal dimension. Numerically, convergence in a basin is decided by z_N for some finite maximum N in the number of iterations, leading to a finite numerical resolution of the boundaries shown.

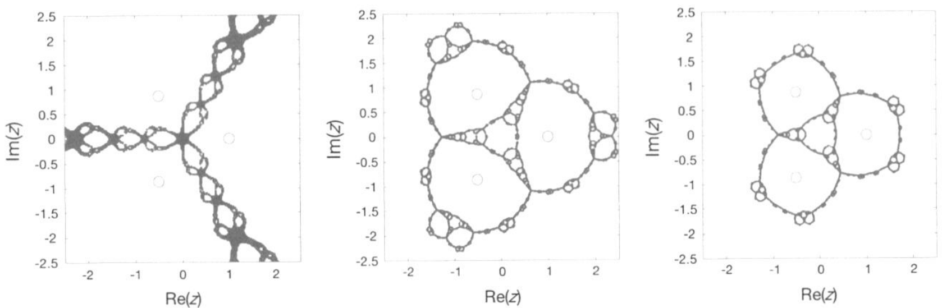

Fig. 7.6 The basins of attraction about the roots of $f(z) = z^3 - 1 = 0$ (*circles* on S^1) by Newton's method with quadratic convergence (*left*), a contraction mapping with cubic (*middle*) and quartic convergence (*right*). The basins of attraction about each of the three roots diminishes in size with order of convergence

7.6 Root Finding in Higher Dimensions

Application of Newton's method in higher dimensions is sometimes aided by *numerical continuation*, e.g., Example 11.1 [2]. It can take one of the following two forms.

- Eliminate one the variables, analytically or numerically, leaving a system of reduced rank for Newton's method to solve. In the present example, we may considering $y = y(x)$ as a solution to the implicit equation

$$f_1(x, y) = 0 \tag{7.40}$$

  and subsequently solve the one-dimensional root finding problem

$$f_2(x, y(x)) = 0. \tag{7.41}$$

- By a *homotopy* $\mathbf{F}_{\cdot}(x, y)$ with a continuation parameter $0 \leq \lambda \leq 1$ such that $\mathbf{F}_1 = \mathbf{F}$ denotes our original problem (7.30), whereas $\mathbf{F}_\lambda$ for $0 \leq \lambda < 1$ is an approximation such that near or at $\lambda = 0$ a suitable choice of starting values $(x_0(\lambda), y_0(\lambda))$ are well understood.

Both these avenues are potentially effective to solving large or complex nonlinear problems.

Table 7.6 summarizes this discussion.

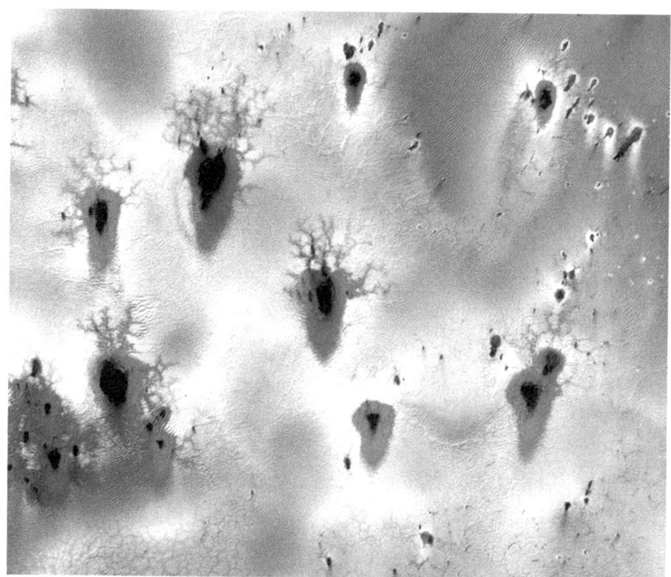

Fig. 7.7 Fractal pattern formation in a 1 km^2 areal view of frozen structures of CO_2 on the Southern hemisphere of Mars. (*Source* NASA HiRISE camera, Mars Reconnaissance Orbiter.)

Table 7.6 Overview root finding

1. Newton's method efficiently converges to a root $x = a$ of $f(x)$ if $f'(a) \neq 0$ with quadratic convergence if $f''(a) \neq 0$. The initial guess must be in the basin of attraction of a. It applies to one or more dimensions.
2. A contraction map $G(x)$ defines roots in terms of a fixed point $a = G(a)$. Convergence is typically algebraic (of order L^n), provided that $L = |G'(a)| < 1$. If $L = 0$, convergence is quadratic or of higher order, determined by the first non-zero derivative of $G(x)$ at a.
3. The basin of convergence of a is typically of complex shape, that can be determined numerically. It diminishes in size with the order of convergence of the root fining method, whereby higher order methods require more specific initial guesses.
4. The problem of finding suitable initial guesses when the basin of attraction is small, e.g., in higher dimensions, can sometimes be circumvented by application of numerical continuation.

7.7 Exercises

7.1. Consider Euler's identity $e^{i\pi} + 1 = 0$. Use Euler's identity to formulate an implicit equation for e given π and use Maxima *Newton* root finding routine to solve for e. Switch the role of e and π, and solve for π in the same fashion. (In fact, Newton allows solving for these two equations at the same time.)

7.2. Apply Newton's method to solve for the *double root* $x = 1$ of $f(x) = (x-1)^2$. In this event, convergence reduces to a power law. Explain why quadratic convergence fails in this limiting case.

7.3. Consider the roots $f(x) = 0$ of the function

$$f(x) = x^2 - a. \tag{7.42}$$

For $a > 0$, $x_\pm = \pm\sqrt{a}$ are two distinct solutions, whereas $a = 0$ has one double root $x = 0$. By Newton iteration, solve for $f(x) = 0$ according to

$$x_{n+1} = x_n - \frac{f(x_n)}{f'(x_n)}, \quad \epsilon_n = x_n - \sqrt{a}. \tag{7.43}$$

Make a graph of the logarithm of the errors

$$E_n = -\log|\epsilon_n| \tag{7.44}$$

for $a = 1$, $a = 1/2$, $a = 1/4$ and $a = 0$ and interpret the results.

7.4. Define a root finding scheme by Newton iteration for

$$\text{(a)}\ x^2 - 4x = 1, \quad \text{(b)}\ \frac{1}{1+x^2} = \frac{1}{3}. \tag{7.45}$$

For (a) and (b) in problem 1, chose a starting value and compute the first few iterations, sufficiently so to identify convergence. Can you estimate the final error?

7.5. Consider the contraction map for roots of a smooth function $f(x)$ with $\gamma = 1 + af + bf^2$ in (7.28). Obtain the general expressions

$$a = \frac{f''}{2f'^2}, \quad b = \frac{f'''f' - f''^2}{2f'^4} \tag{7.46}$$

that produce quartic convergence when evaluated at the root. If these expressions for a and b explicitly involve the unknown root, how can they be suitably approximated?

7.6. Following the two-dimensional root finding problem $\mathbf{F}(x, y) = \mathbf{0}$ for

$$\mathbf{F}(x, y) = \begin{pmatrix} f_1(x, y) \\ f_2(x, y) \end{pmatrix} = \begin{pmatrix} xy \\ x^2 + y^2 - 1 \end{pmatrix} \tag{7.47}$$

- Determine the roots (x, y) by inspection.
- Compute the Fréchet derivative $\mathbf{A} = D\mathbf{F}(x, y)$ and its determinant.
- Compute the inverse $\mathbf{A}^{-1}$ and determine the locus where its determinant vanishes.
- Following an initial guess $(x_0, y_0) = (1, 1/2)$, calculate (x_1, y_1) by Newton's method and sketch this iteration in the (x, y) plane.
- Compute subsequent iterations and observe the rate of convergence.

7.7. An abstract algorithm to compute basins of attraction on a square $[-a, a] \times [-a, a]$ such as (Fig. 7.6) is:

(a) Set initial values on a Cartesian grid
$$z_{kl} = x_k + iy_l, x_k = -a + kh, \ y_l = -a + lh \ (k, l = 1, 2, \ldots M)$$

(b) Iterate N times
$$z_{kln} = G(z_{kln}), (n = 1, 2, \ldots N, k, l = 1, 2, \ldots M)$$

(c) Identify the set of points that divergence
$$Z = \left\{z_{klN} | \ \left|z_{klN}^3 - 1\right| > \epsilon, \ k, l = 1, 2, \ldots M\right\},$$

where $h = 2a/M$ and $0 < \epsilon \ll 1$ serves as a convergence threshold. On modern CPUs, efficient computation depends crucially on vectorization. Write each of (a–c) in terms of operations on arrays and ensure that (b) takes most of the CPU time. If the grid size M^2 is large, a further limitation is memory size. How is the above formulated to allow efficient scaling to any M?

References

1. Mandelbrot, M., *Fractals: Form, Chance and Dimension* (Freeman and Co, 1977).
2. Keller, H.B., 1987, *Numerical Methods in Bifurcation Problems* (Berlin: Springer Verlag/Tata Institute for Fundamental Research).

Chapter 8
Finite Differencing: Differentiation and Integration

A number of problems of physics and astronomy appear as Ordinary Differential Equations (ODEs). A practical approach to their solution starts with an approximation to the derivative operator. Broadly, this can be approached by *finite differencing* or by exact differentiation following a spectral representation. Here, we focus on the first, motivated by an elementary consideration of the problem of estimating velocities and accelerations from particle trajectories. The methods discussed here will be applied to numerical solutions of some illustrative ODEs.

8.1 Vector Fields

Vector fields assign vectors to all points in some region of space, by allowing vector coefficients to be functions of the coordinates, e.g.,

$$\mathbf{a}(x, y, z) = a_1(x, y, z)\mathbf{i} + a_2(x, y, z)\mathbf{j} + a_3(x, y, z)\mathbf{k}. \tag{8.1}$$

It becomes of interest to differentiate vector fields with respect to one of the coordinates, e.g.,

$$\partial_x \mathbf{a} = (\partial_x a_1)\mathbf{i} + (\partial_x a_2)\mathbf{j} + (\partial_x a_3)\mathbf{k}, \tag{8.2}$$

using the fact that the ONB $\{\mathbf{i}, \mathbf{j}, \mathbf{k}\}$ is constant. Similarly, we introduce integration of vector fields, integrating each component using the rules of calculus.

Consider $\mathbf{r} = x(t)\mathbf{i} + y(t)\mathbf{j} + z(t)\mathbf{k}$ at the time-dependent position vector of a moving object. Differentiation obtains the velocity

$$\mathbf{v}(t) = \frac{d\mathbf{r}(t)}{dt} = \dot{x}(t)\mathbf{i} + \dot{y}(t)\mathbf{j} + \dot{z}(t)\mathbf{k}. \tag{8.3}$$

M.H.P.M. van Putten, *Introduction to Methods of Approximation in Physics and Astronomy*, Undergraduate Lecture Notes in Physics,
DOI 10.1007/978-981-10-2932-5_8

Example 8.1. Recall the proverbial apple falling from a tree, e.g., by a gust of wind with possibly nonzero initial velocity. With gravitational acceleration $\mathbf{g} = -g\mathbf{k}$ along the vertical direction $\mathbf{k}$ and conservation of momentum in the horizontal direction, we have Newton's equation $\mathbf{a}(t) \equiv \ddot{\mathbf{r}}(t) = -g\mathbf{k}$ with

$$[\dot{\mathbf{r}}(t)]_x \equiv \dot{\mathbf{r}} \cdot \mathbf{i} = V, \ \mathbf{r}(0) = h\mathbf{k}. \tag{8.4}$$

Integrating, we obtain $\mathbf{v}(t) = \dot{\mathbf{r}}(t) = V\mathbf{i} - gt\mathbf{k}$, and hence a position

$$\mathbf{r}(t) = Vt\mathbf{i} + \left(h - \frac{1}{2}gt^2\right)\mathbf{k}. \tag{8.5}$$

With a change of variables $x = Vt$, the latter can also be expressed as

$$\mathbf{r}(x) = x\mathbf{i} + \left(h - \frac{g}{V^2}x^2\right)\mathbf{k}, \tag{8.6}$$

to bring about the parabolic shape of the trajectory.

When problems have *symmetry*, it is often advantageous to choose a corresponding coordinate system.

Example 8.2. Following Sect. 3.2.2, consider further a particle in a circular motion, e.g., an orbit with radius R is described by a position vector with length R and poloidal angle φ, i.e.,

$$\mathbf{r}(\varphi) = x(\varphi)\mathbf{i} + y(\varphi)\mathbf{j} = (R\cos\varphi)\mathbf{i} + (R\sin\varphi)\mathbf{j} = R\mathbf{i}_r \tag{8.7}$$

expressed in terms of a basis vector along the radial direction,

$$\mathbf{i}_r = \cos\varphi\mathbf{i} + \sin\varphi\mathbf{j} = \begin{pmatrix} \cos\varphi \\ \sin\varphi \end{pmatrix}. \tag{8.8}$$

Differentiation of (8.7) with respect to φ, $d\mathbf{r}/d\varphi = R\,\mathbf{i}_\varphi$, gives a second basis vector

$$\mathbf{i}_\varphi = -\sin\varphi\mathbf{i} + \cos\varphi\mathbf{k} = \begin{pmatrix} -\sin\varphi \\ \cos\varphi \end{pmatrix}. \tag{8.9}$$

The pair $\{\mathbf{i}_r, \mathbf{i}_\varphi\}$ forms an orthonormal basis, that rotates along with the particle position. If $\varphi = \varphi(t)$, then

$$\mathbf{v}(t) \equiv \dot{\mathbf{r}}(t) = \frac{d\varphi(t)}{dt}\frac{d\mathbf{r}}{d\varphi} = R\,\omega(t)\mathbf{i}_\varphi \tag{8.10}$$

in terms of the instantaneous *angular velocity* $\omega(t) = d\varphi(t)/dt$. If the angular velocity is constant, then it equals the mean, orbital angular velocity $\Omega \equiv 2\pi/P$ for an orbital period P.

Vector fields (8.1) can be added following the rules of pointwise addition of vectors.

Example 8.3. Consider the magnetic field along a straight wire carrying a current I. In cgs units, the magnetic field satisfies

$$\mathbf{B}(\sigma) = \frac{2I}{r}\mathbf{i}_\theta \tag{8.11}$$

according to Ampere's law of current induced circulation $2\pi\sigma B = 4\pi I$, where σ denotes the radial distance to the wire in a cylindrical coordinate system (σ, θ, z). Consider submerging this wire in an external uniform magnetic field $\mathbf{B}_0 = B_0\mathbf{i}$ in the plane of $\mathbf{B}$, expressed in a Cartesian basis $(\mathbf{i}, \mathbf{j}, \mathbf{k})$ in which

$$\mathbf{i}_\theta = \mathbf{i}\cos\theta + \mathbf{j}\sin\theta. \tag{8.12}$$

Figure 8.1 shows the total magnetic field $\mathbf{B} + \mathbf{B}_0$ obtained by linear superposition. The resulting magnetic field is asymmetric. The bending of magnetic field lines to the left of the wire are associated with enhanced magnetic pressure $B^2/(8\pi)$, relative to the right. This pressure difference is associated with a Lorentz force on the wire.

8.1.1 Estimating Velocity

Velocities can be estimated from snapshots of positions obtained by sampling (Fig. 8.2). In this process, we may use a model such the particle trajectory (8.5) or (8.6) and perform fit to data obtained at selected points of measurement, and proceed with a best-fit model trajectory for further analysis. For instance, velocities may be inferred based on a given value of the Earth's acceleration $\mathbf{g} = -g\mathbf{k}$, following an estimate of the velocity at one or more intermediate instances in time.

Often, there is a need for *model independent* analysis. We may wish to measure all the relevant physical parameters, perhaps even to determine Earth's gravitational

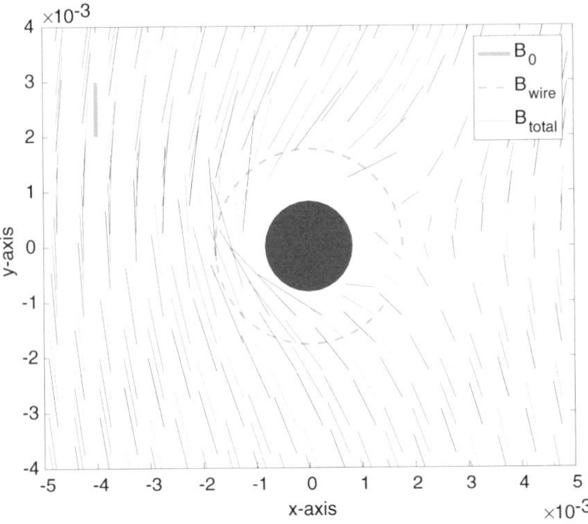

Fig. 8.1 The total magnetic field obtained by linear superposition of that produced by a current (orthogonal to the plane) in an external magnetic field (tangent to the plane) is asymmetric. The resulting magnetic pressure difference is associated with a Lorentz force on the wire

Fig. 8.2 Two subsequent snapshots of traffic flow in Hanoi separated by a sub-second time-interval, overlayed with a uniform grid in the field of view. (*Source* http://www.YouTube.com.)

acceleration, e.g., from a free fall or the period of a pendulum. With no model, we resort to approximate derivatives based on discrete sets of data. This is the art of *finite differencing*, illustrated below on functions of one variable. These ideas are readily applied to the coefficients of vector fields.

The derivative of a function $f(t)$ at a time t is defined as the limit

$$f'(t) \equiv \frac{df(t)}{dt} = \lim_{t' \to t} \frac{f(t') - f(t)}{t' - t}. \tag{8.13}$$

It means that

$$f'(t) = \lim_{n\to\infty} \frac{f(t+h_n) - f(t)}{h_n} \tag{8.14}$$

exists for any sequence of arbitrarily small deviations $\{h_n\}_{n=0}^{\infty}$, $h_n \to 0$ as $n \to \infty$. Let

$$D_+ f(t) = \lim_{h\to 0^+} \frac{f(t+h) - f(t)}{h}, \quad D_- f(t) = \lim_{h\to 0^-} \frac{f(t) - f(t-h)}{h} \tag{8.15}$$

denote the right and left sided derivatives. If both exist and $D_+ f(t) = D_- f(t)$, then $f(t)$ is differentiable at t in the sense of (8.13). In the real world, functions are often differentiable everywhere except for a finite number of points, e.g., at kinks or discontinuities, and functions may be differentiable up to a finite number of times.[1]

Based on the above, we define *one-sided* and *two-sided* finite differences

$$\delta_h^+ f(t) = \frac{f(t+h) - f(t)}{h}, \quad \delta_h^- f(t) = \frac{f(t) - f(t-h)}{h}, \tag{8.16}$$

$$\Delta_h f(t) = \frac{1}{2}\left(\delta_h^+ + \delta_h^-\right) f(t) = \frac{f(t+h) - f(t-h)}{2h}. \tag{8.17}$$

Note that the first in (8.16) approximates the right sided derivative $D^+ f(t)$ (if it exists). How good are the approximations (8.16–8.17)?

Consider the following three points along the graph of $f(t)$ (Fig. 8.3)

$$A = (t-h, f(t-h)),\ B = (t, f(t)),\ C = (t+h, f(t+h)). \tag{8.18}$$

A, B and C introduce three approximate tangents according to (8.16–8.17). If $f(t)$ is smooth and h is small, then $\Delta_h f(t)$ is expected to do a better job than the $\delta_h^{\pm} f(t)$, as Δ_h represents their mean.

To estimate the degree of finite difference approximations to (8.13), consider a smooth function $f(t)$ and its Taylor series expansion about t,[2]

$$f(t+h) = f(t) + hf'(t) + \frac{1}{2}h^2 f''(t) + \frac{1}{6}h^3 f^{(3)}(t) + O(h^4). \tag{8.19}$$

Substitution into the expressions (8.16–8.17) gives

---

[1]There exists a nowhere differentiable continuous function, constructed by Karl Weierstrass (1872). For each real t, there exist sequences $\{t_n\}$ and $\{t_n'\}$ converging to t such that his function satisfies the inequality $\liminf (f(t_n) - f(t))/(t_n - t) > \limsup f(t_n') - f(t))/(t_n' - t)$. See further Sect. 7.5.

[2]A function $f(t)$ is analytic at t if $f(t)$ has a convergent Taylor series at t to all orders (Chap. 2).

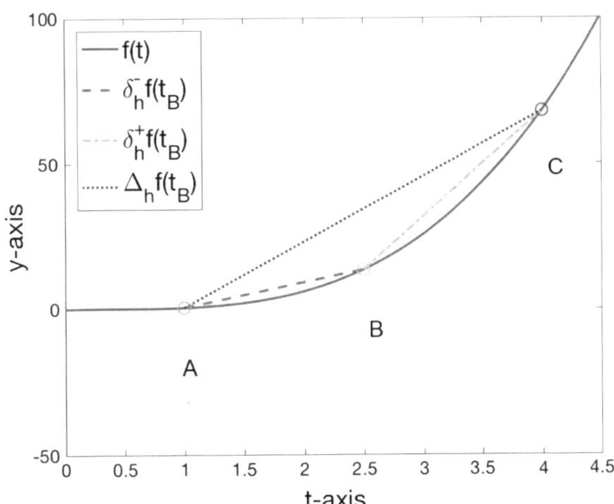

Fig. 8.3 Illustration of the one-sided and symmetric finite difference approximations $\delta_h^\pm$ and, respectively, Δ_h to the slope at B of a function with finite curvature

$$
\begin{aligned}
\delta_h^+ f(t) &= \tfrac{1}{h}\left([f(t) + hf'(t) + O(h^2)] - f(t)\right) \\
\Delta_h f(t) &= \tfrac{1}{2h}\left([f(t) + hf'(t) + \tfrac{1}{2}h^2 f''(t) + O(h^3)]\right. \\
&\qquad \left. -[f(t) - hf'(t) + \tfrac{1}{2}h^2 f''(t) + O(h^3)]\right).
\end{aligned}
\tag{8.20}
$$

After some cancellations, we have

$$
\delta_h^+ f(t) = f'(t) + O(h), \quad \Delta_h f(t) = f'(t) + O(h^2). \tag{8.21}
$$

Convergence to $f'(t)$ is linear and quadratic in h by δ_h^+ and, respectively, $\Delta_h f(t)$ as h becomes small. *If* $f(t)$ has a second derivative at t, then it may be advantageous to use $\Delta_h f(t)$, giving a good approximation at relatively modest step size h.

Example 8.4. The finite differences $\delta_h f(t)$ and $\Delta_h f(t)$ are exact up to linear, respectively, quadratic polynomials:

$$
f(t) = a + bt: \quad \delta_h f(t) = b, \tag{8.22}
$$

and

$$
f(t) = a + bt + ct^2: \quad \Delta_h f(t) = b. \tag{8.23}
$$

This follows readily by the definitions of δ_h and Δ_h.

Example 8.5. Continuing with our proverbial apple trajectory (8.5) or (8.6), consider two snapshots of the apple's position at instances t and $t+h$. A constant horizontal velocity is adequately measured by either δ_h or Δ_h. In estimating the velocity downwards from two snapshots, we note the identity

$$\delta_h^+ f(t) = \Delta_{\frac{h}{2}} f\left(t + \frac{1}{2}h\right). \tag{8.24}$$

For a parabolic trajectory $f(t) = a + bt + ct^2$,

$$\delta_h^+ f(t) = \frac{[a + b(t+h) + c(t+h)^2] - [a + bt + ct^2]}{h} = b + 2ct + ch. \tag{8.25}$$

The error in estimating $f'(t)$ is ch. By (8.24) and (8.23), δ_h *is* exact for the velocity at intermediate instances $t + \frac{1}{2}h$ between consecutive snapshots.

Evidently, some care in choosing h is needed. For the periodic motion of a circular orbit with period P, $\mathbf{r}(t) = \mathbf{r}(t+P)$, an estimate of the instantaneous velocity by finite differencing of the position vector is $\mathbf{v}(t) \equiv d\mathbf{r}(t)/dt \simeq \Delta_h \mathbf{r} = [\mathbf{r}(t+h) - \mathbf{r}(t-h)]/2h$. Evidently, $h \ll P$, since $h = P/2$ in would produce zero.

8.1.2 *Estimating Acceleration*

Finite differencing (8.17) is readily generalized to acceleration. Consider

$$\delta_h^2 f(t) = \frac{f(t+h) - 2f(t) + f(t-h)}{h^2} \tag{8.26}$$

to curvature in the graph of $f(t)$. The slope at midpoints $t \pm h/2$ obtains to second order accuracy

$$\Delta_h f(t - h/2) = \delta_h^- f(t), \quad \Delta_h f(t + h/2) = \delta_h^+ f(t) \tag{8.27}$$

and (8.26) follows from

$$h^{-1}\left[\Delta_h f(t+h/2) - \Delta_h f(t-h/2)\right] = h^{-1}\left[\delta_h^+ - \delta_h^-\right] f(t). \tag{8.28}$$

How good is (8.26)? Again, the answer may be obtained by inspection of the graph of $f(t)$. However, changes in slopes refer to curvature in the graph, and they are generally difficult to determine by geometrical means or inspection. If $f(t)$ is smooth and permits a Taylor series expansion to second order, then by (8.19)

$$\delta_h^2 f(t) = h^{-2}\left[\left\{f(t) + hf'(t) + \frac{h^2}{2}f^{(2)}(t) + \frac{h^3}{6}f^{(3)}(t) + O(h^4)\right\} - \{\}\right], \quad (8.29)$$

where {} refers to a repeat expression with h replaced by $-h$. After cancellations, we are left with

$$\delta_h^2 f(t) = f''(t) + O(h^2), \quad (8.30)$$

showing second order accuracy. Consider once more the linear and quadratic polynomials (8.22–8.23). By direct evaluation, we find

$$\delta_h f(t) = 2c. \quad (8.31)$$

So Δ_h^2 will be just fine for the measurement of the gravitational acceleration of the Earth in between successive snapshots taken at equidistant times $t_i = hi$ ($i = 0, 1, 2, \ldots$) independent of h.

8.2 Gradient Operator

The *nabla* operator ∇ is a three-dimensional vector operation. Extending (8.2), it is defined in Cartesian coordinates by

$$\nabla = \mathbf{i}\partial_x + \mathbf{j}\partial_y + \mathbf{k}\partial_z. \quad (8.32)$$

Applied to a scalar $f = f(x, y, z)$, such as pressure or temperature, it produces the vector

$$\nabla f = \frac{\partial f}{\partial x}\mathbf{i} + \frac{\partial f}{\partial y}\mathbf{j} + \frac{\partial f}{\partial z}\mathbf{k} = \begin{pmatrix} f_x \\ f_y \\ f_z \end{pmatrix}, \quad (8.33)$$

where subscripts refer to partial differentiation with respect to coordinates.

Example 8.6. Consider a contour plot of a two-dimensional pressure distribution $P(x, y)$ in the atmosphere shown in Fig. 8.4. Isobars are curves of constant pressure. Along tangents $d\boldsymbol{\tau} = dx\boldsymbol{i} + dy\boldsymbol{j}$ to these isobars

$$0 = dP = P_x dx + P_y dy, \quad (8.34)$$

that is, ∇P defined by (8.33) is orthogonal to $d\boldsymbol{\tau}$. Along the isobar of 1022 mbar shown, where is ∇P the largest, i.e., where is the maximum of

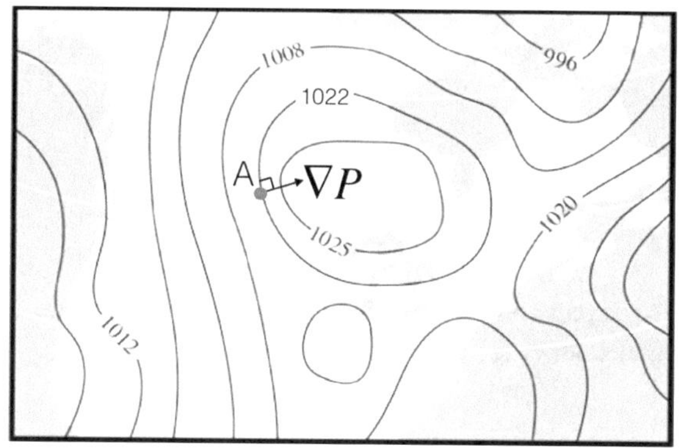

Fig. 8.4 Isobars are curves of constant atmospheric pressure P, here shown in mbar ranging from 1008 (low) to 1025 (high). These isobars are implicitly defined by the equation $P = P_0$. Along each isobar, ∇P is a vector orthogonal to it

$$|\nabla P| = \sqrt{P_x^2 + P_y^2}\ ? \tag{8.35}$$

In a finite difference approximation, we estimate

$$\nabla P = \left(\frac{P_2 - P_1}{\delta} + O(\delta)\right)\mathbf{n} \tag{8.36}$$

along an isobar $P = P_1$ with unit normal $\mathbf{n}$ pointing to a neighboring isobar $P = P_2$, where δ is the distance between them. Considering $P_1 = 1022$ mb and $P_2 = 1025$ as shown, we have

$$|\nabla P| = \frac{3\,\text{mb}}{\delta} + O(\delta), \tag{8.37}$$

maximum over $P = P_1$ occurs at A by inspection.

8.3 Integration by Finite Summation

The converse of differentiation by finite differencing is integration by finite summation. Consider, for instance, the sum

$$s_1 = \sum_{i=0}^{n} v(t_i)\Delta t, \quad v(t_i) = \delta_h^+ x(t_i), \; t_i = hi, \; h = \Delta t. \tag{8.38}$$

Expanding this sum following the definition of δ_h, we have

$$s_1 = \sum_{i=0}^{n-1} \frac{x(t_i + h) - x(t_i)}{h} \Delta t = \sum_{i=0}^{n-1} (x_i - x_{i-1}), \tag{8.39}$$

that is,

$$s_1 = -x_0 + x_1 + (x_2 - x_1) + \cdots + (x_n - x_{n-1}). \tag{8.40}$$

The expansion on the right is a *telescoping* sequence, in which all intermediate terms cancel, leaving

$$s_1 = x_1 - x_0. \tag{8.41}$$

Clearly, in estimating the velocity by first-order finite differencing δ_h gives the exact result. How about Δ_h? A similar calculation applied to

$$s_2 = \sum_{i=1}^{n-1} \Delta_h x(t_i) \Delta t \tag{8.42}$$

shows

$$s_2 = \frac{1}{2} \left[(x_2 - x_0) + (x_3 - x_1) + (x_4 - x_2) + \cdots + (x_n - x_{n-2}) \right]. \tag{8.43}$$

After cancellations, we find

$$s_2 = \frac{1}{2} \left[-x_0 - x_1 + x_n + x_{n-1} \right] = x_n - x_0 - \frac{1}{2} \left[(x_1 - x_0) + (x_n - x_{n-1}) \right]. \tag{8.44}$$

As an estimate, $s_2 = x_n - x_0 + O(h)$ is therefore first order accurate due to boundary terms.

In the limit as $h = \Delta t$ drops to zero, s_2 agrees with s_1, and we write the result as a Riemann integral

$$s = \int_0^T v(t) dt, \tag{8.45}$$

where $T = nh$ is kept constant in the process of taking $h \to 0$.

The above serves to integrate a consistency in approximation of integration and differentiation by the first-order Riemann integral and δ_h^+.

8.4 Numerical Integration of ODE's

In the geometric relationship between tangent vectors and trajectories, the latter is the integral of the first. Mathematically, this is described by a first-order initial value problem, defining trajectories by their tangents supplemented with an initial direction at time zero. The general form is a first order ODE is

$$\dot{y} = f(t, y), \quad y(t_0) = y_0, \tag{8.46}$$

where $f(t, y)$ smooth in that it has a Taylor series expansion about the domain of time t and y of interest.

The following three integration schemes are illustrative, based on piece-wise linear approximations to the solution using integration along tangent lines. Their applicability relies heavily on the assumed degree of smoothness of the solutions, commonly considered by Taylor series expansions. (That is, a certain degree of differentiability of the solutions.) We mention

- *Forward Euler*, based on finite differencing in time to the right,

$$\dot{y}|_{t_n} \simeq \frac{y(t_n + h) - y(t_n)}{h} : \quad y_{n+1} = y_n + hf(t_n, y_n), \tag{8.47}$$

  where we use the notation $t_n = nh$. Forward Euler creates a piece-wise linear approximation based on extrapolation to t_{n+1} along the tangent at time t_n;
- *Backward Euler*, based on finite differencing in time the time to the left,

$$\dot{y}|_{t_n} \simeq \frac{y(t_n) - y(t_n - h)}{h} : \quad y_{n+1} = y_n + hf(t_{n+1}, y_{n+1}). \tag{8.48}$$

  Whenever f depends on y, the iteration scheme is implicit. It may, in each time step, require numerical root finding to solve for the y_{n+1}. Backward Euler creates a piece-wise linear approximation based on extrapolation to t_n along the tangent at time t_n;
- *Midpoint rule*, representing a democratic compromise between the forward and backward Euler,

$$\dot{y}|_{t_{n+1}} \simeq \frac{y(t + h) - y(t)}{h} : \quad y_{n+1} = y_n + hf(t_{n+1/2}, y_{n+1/2}), \tag{8.49}$$

  where $y_{n+1/2} = y_n + \frac{1}{2}hf(t_n, y_n)$ is an intermediate forward Euler step.

It can be shown that forward and backward Euler are first order and the midpoint rule is second order over a given time interval.

Figure 8.5 illustrates the above on the ODE

$$f'(t) = \sin t, \quad f'(t) = e^t, \tag{8.50}$$

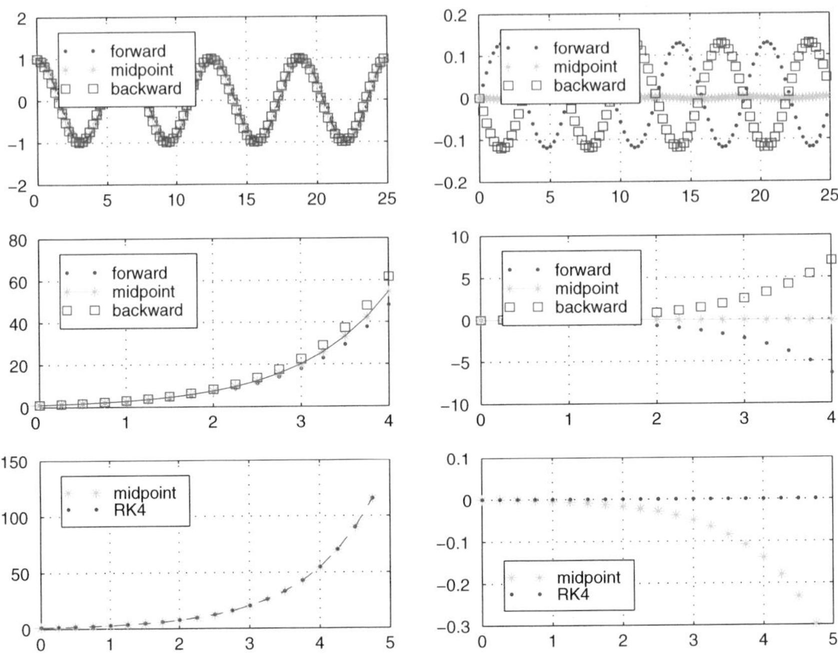

Fig. 8.5 Shown are the numerical solution of the ODE's $f'(t) = \sin t$ and $f'(t) = e^t$ (*left panels*) with associated errors (*right panels*)

explicitly showing the different behaviors of the two different Euler methods and the midpoint rule.

The *Runge-Kutta method* aims at fourth-order accuracy. It again uses integration along tangents, now using four slopes, one evaluated at t_n, two times at the mid point $t_{n+1/2}$ and one at t_{n+1}, namely

$$\begin{aligned} k_1 &= f(t_n, y_n), \\ k_{1+i} &= f(t_{n+1/2}, y_n + \tfrac{1}{2}hk_i) \quad (i = 1, 2) \\ k_4 &= f(t_{n+1}, y_n + hk_3), \end{aligned} \tag{8.51}$$

and

$$y_{n+1} = y_n + \frac{h}{6}\left(k_1 + 2k_2 + 2k_3 + k_4\right), \tag{8.52}$$

where $t_n = nh$ as before. It can be shown that the truncation error $\delta_n = y_n - y(t_n)$ is fourth order in the time step h, provided that $f(t, y)$ is sufficiently smooth.

If $f(t, y) = f(t)$ in (8.46) does not depend on y, then the Runge-Kutta method reduces to (Fig. 8.6)

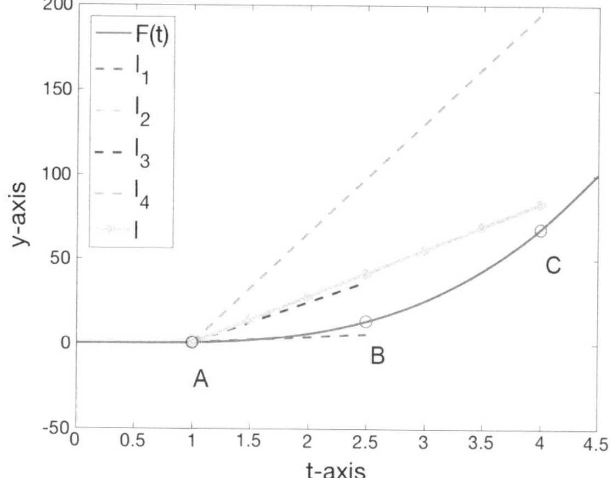

Fig. 8.6 Illustration of the Runge-Kutta method applied to the ODE $df(t)/dt = F(t)$ by an explicit time step to t_C from t_A along $l:\ y(t) = f(t_A)+k_4(t-t_A)$, based on a weighted average k_4 of slopes at times t_A, t_B and t_C

$$\begin{aligned} k_1 &= f(t_n), \\ k_2 &= f(t_{n+1/2}), \quad k_3 = k_2 \\ k_4 &= f(t_{n+1}, y_n + hk_2), \end{aligned} \tag{8.53}$$

and

$$y_{n+1} = y_n + \frac{h}{6}(k_1 + 4k_2 + k_4). \tag{8.54}$$

8.5 Examples of Ordinary Differential Equations

For illustrative purposes, we here discuss three nonlinear problems in ordinary differential equations solved by numerical integration.

8.5.1 Osmosis

Following Sect. 4.2, consider osmosis through the semi-permeable membrane of an egg (*hypertonic*) in a glass of water (*hypotonic*), after dissolving its hard outer shell. The egg takes in water to dilute concentrations of various solutes within. It hereby increases its total mass $m(t)$ with time. According to van't Hoff, $dm(t)/dt$ is proportional to the excess concentration inside over that outside,

$$[c] = c - c_0, \quad c = \frac{N}{V}, \tag{8.55}$$

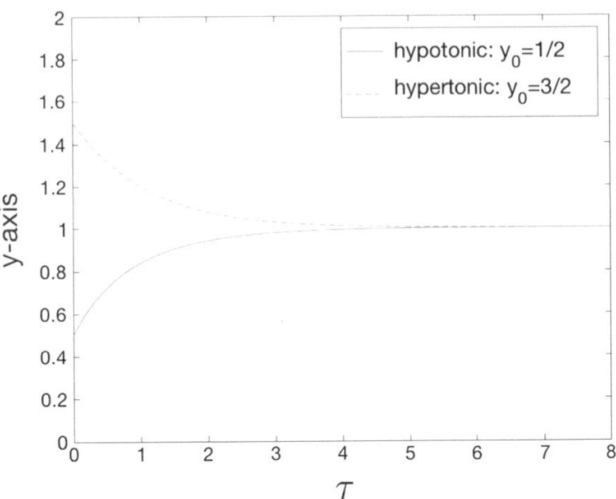

Fig. 8.7 Numerical integration by a Runge-Kutta method of (8.57) modeling osmosis by an egg put in a hypotonic and hypertonic solution with initial data $y_0 = 1/2$ and, respectively, $y_0 = 3/2$. Equilibrium attains at the stationary point $y = 1$

and the product of temperature T, permeability μ and area A of the membrane. Here, c is expressed in terms of the total number of solute molecules N in the volume $V = m/\rho$ of the egg with mass-density $\rho \simeq \rho_0$ effectively equal to that of water (ρ_0). As a result,

$$\frac{dm(t)}{dt} = A\,\mu k_B T \left[\frac{\rho_0 N}{c_0 m(t)} - 1\right] c_0, \tag{8.56}$$

where $A = 4\pi r^2 = (36\pi)^{\frac{1}{3}}(N/c_0)^{2/3}y^{2/3}$ is the surface area of the membrane, $k_B T$ denotes the Boltzmann constant and $y = mc_0/(N\rho_0)$ is the dimensionless mass of the egg. Then

$$\frac{dy(\tau)}{d\tau} = \left[y^{-\frac{1}{3}} - y^{\frac{2}{3}}\right], \tag{8.57}$$

where τ is dimensionless time, scaled by $c_0^{4/3}(36\pi N)^{\frac{1}{3}}\mu k_B T$, that depends only on temperature. (We shall ignore variations μ as the egg expands.) This nonlinear equation (8.57) shows a stationary point $y = 1$, corresponding to vanishing concentration differences. The initial condition $c - c_0- > 0$ defines $y(0) < 1$, giving $dy(\tau)/d\tau > 0$. A detailed picture of the solutions obtains by numerical integration shown in Fig. 8.7, here by application of a Runge-Kutta integration.

8.5.2 *Migration Time of the Moon*

The Earth-Moon system is a binary with most of the angular momentum in the Moon's orbital motion. Table 8.1 labels the pertinent quantities.

The Moon is receding at a rate of $v = 3.8$ cm yr^{-1} at a mean orbital distance of $r = 3.8 \times 10^{10}$ cm. This puts the Moon back to Earth on a characteristic time scale

$$t_m = \frac{r}{v} \simeq 10\,\text{Gyr}, \tag{8.58}$$

consistent with the Earth-moon age of about 4.5 Gyr. There has been a gradual expansion of the orbit by a gain in orbital angular momentum $J = mj$, $j = \sqrt{GM(1-e^2)a}$, powered by a *tidal torque* from the Earth's rotational motion. Here, m denotes the Moon mass, $m/M_\oplus \simeq 1/81$, $M_\oplus = 5.97 \times 10^{27}$ is the Earth mass, a is the Moon's orbital semi-major axis with ellipticity e, G is Newton's constant. Presently, e is a mere 5%. In what follows, we shall assume circular orbits with $a = r$.

To streamline our discussion, we consider the total energy and total angular momentum

$$H_0 \simeq E + H_☾\,, \quad J_0 \simeq J_\oplus + J_☾ \tag{8.59}$$

in terms of the Earth's energy E in angular momentum $J_\oplus$ and the Moon's orbital energy $H_☾$ and angular momentum $J_☾$. In the approximation of circular orbit, the Earth-Moon system is *virialized*, $2E_k + U_N = 0$, where E_k is the total kinetic energy in orbital motion of the Earth and Moon about the center of mass (CM) and U_N denotes their Newtonian binding energy.

Example 8.7. The equation $M_\oplus r_1 = mr_2$ expresses the distances r_1 and r_2 of the Earth and, respectively, Moon to the CM. As a result, $r_1 = mr/(m+M_\oplus) \simeq rm/M_\oplus \simeq r/81$, where $r = r_1 + r_2$ denotes the Earth-Moon separation. With $r \simeq 380{,}000$ km and $R_\oplus \simeq 6000$ km, it follows that $r_1 \simeq 4700$ km $< R_\oplus$, i.e., the CM falls well within the Earth's radius of the Earth. As a result, (a) the Earth's total energy is E essentially in angular momentum $I\Omega_\oplus$, i.e., $E = (1/2)I\Omega_\oplus^2$ in terms of the moment of inertia $I \simeq (2/5)M_\oplus R_\oplus^2$, $R_\oplus = 6 \times 10^8$; (b) $H_☾ \simeq -(1/2)U_N$, and hence $H_☾ \simeq -GMm/2a \simeq GMm/(2r)$ when specialized to a circular orbit satisfying $a = r$;

Ignoring spin, we have $H_☾ \simeq -GMm/2r$ in the approximation of a circular orbit. The total energy and angular momentum in the Earth-Moon system, therefore, satisfies

$$H_0 \simeq E - \frac{GMm}{2r}, \quad J_0 \simeq J_\oplus + mj, \tag{8.60}$$

where $j = \sqrt{GM_\oplus r}$. Importantly, J_0 is conserved whereas H_0 is not, in the process of tidal dissipation into heat.

In the above, kinetic energy in the Moon's radial migration is completely ignored. Though intuitively obvious, the following is makes this precise.

Example 8.8. In radial migration, the kinetic energy $e_k = \frac{1}{2}m\,(dr/dt)^2 = 5.35 \times 10^11\,\mathrm{erg\,s^{-1}} = 530\,\mathrm{kJ}$ is no more than a one daily exercise, hereby satisfying $e/H_{☾} = 1.4 \times 10^{-24}$.

Earth's spin-down by tidal interaction conserves J_0 in (8.60). Tidally locked to its orbit, the Moon's angular velocity is so low that its spin angular momentum does not partake in this process. By now, most of the Earth's spin angular momentum has been transferred to the Earth-Moon orbital motion,

$$\frac{I\Omega}{mj} = \frac{2}{5}\left(\frac{M}{m}\right)\left(\frac{R}{r}\right)^2\left(\frac{\Omega}{\omega}\right) \simeq \frac{1}{4}. \tag{8.61}$$

By the first law of thermodynamics, the Earth's rotational energy $\frac{1}{2}I\Omega^2$ is gradually transferred to the Moon, $H_{☾} \simeq -GMm/(2a)$, and heat, by tidal dissipation at

$$W = (\Omega - \omega)\,\frac{dJ_m}{dt} \simeq 3\,\mathrm{TW}, \tag{8.62}$$

where 1 TW = 10^{12} W, or about 6 mW m^{-2}. Since presently $\Omega/\omega \simeq 30$, the Earth's rotational energy is mostly dissipated into heat (below), producing expansion of the moon's orbit at a mere three percent efficiency. Even so, W is negligible compared to the Earth's luminosity of infrared radiation $L = 4\pi R_{\oplus}^2\sigma T^4 \simeq 1.6 \times 10^{24}\,\mathrm{erg\,s^{-1}}$. This result (8.62) follows directly from $\dot{H}_{☾} = -H_{☾}\,\dot{r}/r$, $\dot{J}_{☾} = J_{☾}\,\dot{r}/2r$, $\dot{r}/r = 1.205 \times 10^{-7}\,\mathrm{s^{-1}}$. With $H_{☾} = 3.86 \times 10^{35}$ erg, $J_{☾} = 2.87 \times 10^{41}$ erg s, we have

$$\dot{H}_{☾} = 1.22 \times 10^{18}\,\mathrm{erg\,s^{-1}} = 0.12\,\mathrm{TW},\ \dot{J}_{☾} = 4.55 \times 10^{23}\,\mathrm{erg}. \tag{8.63}$$

where 1TW = 10^{12} denotes one terawatt.

Transfer of angular momentum in the Earth's spin to the orbit of the moon is principally a dissipative process, proportional to the phase-lag $\Delta\varphi$ between the Earth's tidal bulge and direction to the Moon. Gain in specific angular momentum of the Moon hereby scales as

$$\frac{dj}{dt} \propto \frac{h}{r^3}\sin\left(2\Delta\varphi\right), \tag{8.64}$$

where $h \propto r^{-3}$ denotes the Earth's tidal deformation amplitude and r^{-3} is the gravitational coupling of the Earth's quadrupole moment back to the moon. It moves over the Earth's surface with tidal frequency

$$\omega' = 2(\Omega - \omega). \tag{8.65}$$

Conserving J_0, we have $\dot{J}_\oplus = -\dot{J}_☾ = 4.55 \times 10^{23}$ erg. Given $E = J_\oplus^2/(2I)$, we have

$$-\dot{E} = -\frac{J_\oplus}{I}\dot{J}_\oplus = -\Omega_\oplus \dot{J}_\oplus = D + \dot{H}_☾ \tag{8.66}$$

where D denotes the tidal dissipation rate. Given $\Omega_\oplus \dot{J}_\oplus = 3.3 \times 10^{19}$ erg s^{-1} = 3.3 TW, *most of the Earth's spin energy is dissipated* at a rate $D = -\dot{E} - \dot{H}_☾$ = 3.2 TW or about 6.6 mW m^{-2}—much too small to raise the surface temperature of the Earth.

For the Earth, tidal dissipation is mostly in the oceans, more so than in the Earth's mantle by a factor of a few. The rigidity μ of the latter has a *Maxwell time* $\tau_M = \nu/(\mu/\rho)$, where ν (cm^2 s^{-1}) is the kinematic viscosity, $[\mu/\rho]$ = cm^2 s^{-2} and ρ is the mass density. The dimensionless tidal frequency $\tau_M(\Omega - \omega)$ is sufficiently high [1] to give a near-elastic deformation of the mantle with relatively small dissipation compared to (8.62) [2].

Tidal dissipation in ocean currents is presently near-resonance with the Atlantic at tidal frequency (8.65) at high Q-factor. A starting point for modeling (8.64), then, is the forced pendulum equation,

$$\ddot{h}(t) + 2i\epsilon\omega_0\dot{h}(t) + \omega_0^2 h(t) = \omega_0^2 A \sin(\omega' t), \tag{8.67}$$

where ω_0 denotes its eigenfrequency, ϵ is the damping coefficient and $A \propto r^{-3}$ denotes the tidal interaction strength. Figure 8.8 illustrates that (8.67) features distinct limits to forcing to the left or right of ω_0 with phase-lag $\Delta\varphi$ (and hence dissipation) proportional to $\epsilon(\omega/\omega_0)$ and $\epsilon(\omega_0/\omega)$ at low, respectively, high frequency. The amplitude approaches a constant or decays with $(\omega_0/\omega)^2$ in the same two limits.

Let $\tau = (t_0 - t)/t_m$ denote the look back time satisfying $\tau = 0$ at present and $\tau = 0.45$ a the time of birth of the Moon, scaled by the migration time t_m. Normalizing (8.64) to $j = 1$ and $dj/d\tau = 1$ at present, we have

$$\frac{dj}{d\tau} = -\frac{p}{Zj^{12}} \tag{8.68}$$

where $p = \sin(2\Delta\varphi)/\sin(2\Delta\varphi_0)$ and $Z = z/z_0$ are given by $\tan\Delta\varphi = V/|U|$ and $z = \sqrt{U^2 + V^2}$, $U = F^2 - \omega_0^2$, $V = 2\epsilon\omega_0 F$, normalized by the present values $p_0 = p(1)$ and $z_0 = z(1)$ at $j = 1$, with forcing at the normalized tidal frequency

$$F(j) = 5.016\left(1 - 0.7938j - 0.006833j^{-3}\right), \tag{8.69}$$

satisfying $F(j) = 1$ at present $j = 1$ conform (8.61). Importantly, (8.69) defines two stationary points in (8.68)

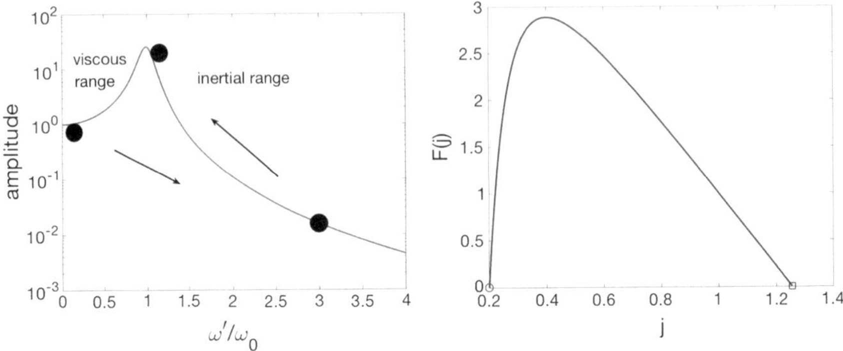

Fig. 8.8 (*Left*) Trajectory of the Earth-moon system shown in the amplitude-frequency diagram of a forced pendulum with constant Q-factor and resonance frequency ω_0, the latter close to today's tidal frequency. Starting from an unstable geostationary orbit with vanishing tidal angular frequency ω', passage through resonance reaches a state of maximal tidal frequency in the inertial range $\omega'/\omega_0 \gg 1$. As the Earth continues to spin down, the trajectory approaches resonance from the right, representative for the current state of the Earth-moon system. (*Right*) Shown is the dimensionless tidal angular frequency $F(j)$ as a function of the dimensionless specific angular momentum j of the moon's orbit, normalized to $j = 1$ at present, satisfying $F(1) = 1$. It shows an extremum of about 3 when the moon was at 16% of its present distance and two stationary point $F(j) = 0$, corresponding to synchronous orbits satisfying $\Omega = \omega$

Table 8.1 Relevant quantities to the Earth-Moon system at present

Object	Mass	Radius	ω	E_{rot}	J_{spin}	J_{orb}
Earth	$M_\oplus$	$R_\oplus$	$\Omega_\oplus$	E	$J_\oplus$	~ 0
Moon	m	$R_☾$	$\Omega_☾$	~ 0	~ 0	$J_☾$

$$F(j) = 0: \ j_0 = 0.2011, \ j_1 = 1.2554 \tag{8.70}$$

in the past and, respectively, future. At these points, $\Omega = \omega$ suppresses any tidal torque. The first corresponds to a position when the Moon was about 28 times closer to the Earth at $r_0 \simeq 2.2 R_\oplus$; the second will occur at about 100 $R_\oplus$. Our model, therefore, describes migration across this range.

In the formulation (8.68), we can explore various channels for dissipation, by specifying scalings for damping as a function of tidal amplitude and frequency (Table 8.1).

Insisting on a migration time equal to the age of the Moon, Table 8.2 lists the present-day damping coefficient ϵ_0 in various scalings of ϵ with respect to tidal amplitude and frequency. The results obtain by Runge-Kutta integration of (8.68) (Fig. 8.9), and indicate that most of the migration time of the moon is associated with the last 40% of its current distance following a sharp initial transition. This early start is the eviction from the instable synchronous orbit at the first root $F(j_0) = 0$. Observational data point to a present-day Q-factor $Q_0 = 1/2\epsilon_0$ satisfying $5 < Q < 30$ $(0.017 < \epsilon_0 < 0.100)$. According to Table 8.2, this suggests a preferred scaling with both tidal amplitude and frequency.

Table 8.2 Present-day damping coefficient ϵ_0 and Q-factor in models of scaling with respect to tidal amplitude and frequency for consistency with the moon's age of 4.52 Gyr, assuming near-resonance today ($\omega_0'/\omega_0 = 1.11$)

Scaling	ϵ	ϵ_0	Q
Stokes' limit	ϵ_0	0.1365	3.7
Tidal frequency	$\epsilon_0 F$	0.0860	5.8
Tidal amplitude	$\min\{1, \epsilon_0 h\}$	0.0527	9.5
Tidal amplitude-frequency	$\min\{1, \epsilon_0 hF\}$	0.0366	13.7

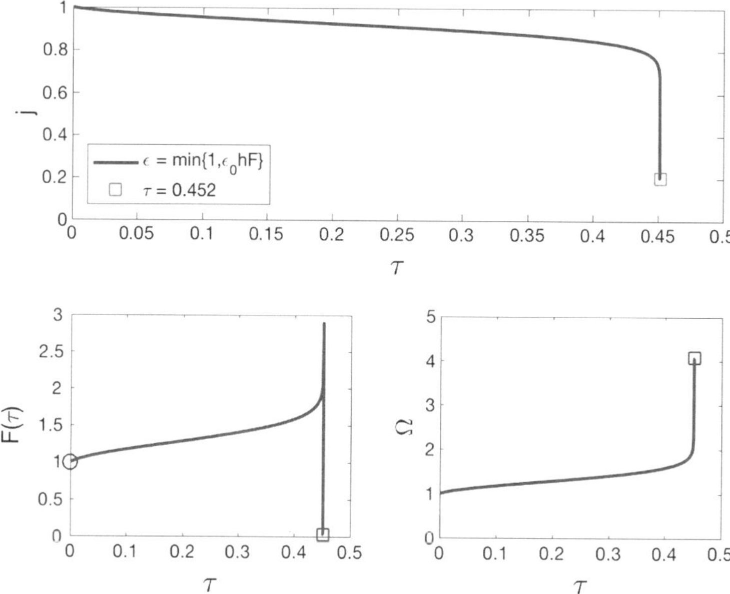

Fig. 8.9 Integration of (8.68) on the Moon's migration with damping $\epsilon = \min\{1, \epsilon_0 hF\}$ ($\epsilon_0 = 0.0366$) as a function of dimensionless look back time τ from $j = 1$ at $\tau = 0$ to $\tau = 0.452$ at $j = 0.2011$. Note the sharp transition at $\tau = 0.452$, representing rapid eviction of the Moon's from its initial position

Integration back to the onset of migration at the first zero $j = 0.2011$ of $F(j)$ shows that the Earth's initial angular velocity Ω_0 was quite high. (It then contained all of the angular momentum.) Let $\Omega_b = \sqrt{GM_\oplus/R_\oplus^3}$ denote the Earth's angular velocity at break-up. Then by (8.61), $\Omega_0/\Omega_b \simeq 5\Omega/\Omega_b \simeq 30\%$ prior to the formation of the Moon. Such high spin rate (normalized to break-up) is essentially that of Jupiter today ($\Omega/\Omega_b = 28\%$). The associated Coriolis forces give rise to Jupiter's extreme weather patterns. By virtue of the moon as a deposit for J_0, Earth attained slow spin allowing for our presently clement global climate [8].

Table 8.3 Cosmological evolution according to the scale factor in three-flat Friedman-Robertson-Walker line-elements (8.71) in units with velocity of light equal to 1

Phase	$a(t)$	H	R_H
de Sitter	$e^{H_0 t}$	H_0	H_0^{-1}
Radiation dominated	$a_0\left(\frac{t}{t_0}\right)^{\frac{1}{2}}$	$\frac{1}{2}t^{-1}$	$2t$
Matter dominated	$a_0\left(\frac{t}{t_0}\right)^{\frac{2}{3}}$	$\frac{2}{3}t^{-1}$	$\frac{3}{2}t$

All of the above was quite different in the past, inferred from the scaling of torque $\dot{J}_\oplus \propto r^{-6}$. When the Moon was at 50% (10%) its current distance, the tidal torque was 64 (one million) times stronger. While we live in a quiet time now, times were quite violent for some time following the Moon's birth.

8.5.3 Cosmological Expansion

On the largest scales, the Universe is well described by a homogeneous isotropic universe described by the Friedmann-Robertson-Walker line-element

$$ds^2 = -dt^2 + a(t)^2(dx^2 + dy^2 + dz^2). \tag{8.71}$$

Choices of $a(t)$ describe various possible phases in the evolution of the universe (Table 8.3).

Observations of Type Ia supernovae [3] show a finite Hubble parameter and a negative deceleration parameter

$$H = \frac{\dot{a}}{a} \simeq 70\,\mathrm{km\,Mpc^{-1}\,s^{-1}}, \quad q = -\frac{\ddot{a}a}{\dot{a}^2} < 0. \tag{8.72}$$

These data reveal a Universe in a state of *accelerated expansion*.

We also detect a *Cosmic Microwave Background* (CMB) in black body radiation at a temperature of about 2.7 K [4] in radio emission with peak brightness frequency $\nu = 160$ GHz. This forms a relic of the surface of last scattering at an age of about 379 kyr at recombination to neutral matter (mostly hydrogen and helium) at a temperature of $T_{SLS} \simeq 3000$ K. At this point in time, matter and radiation decoupled. Subject to gravity, matter aggregated to form stars and galaxies and radiation formed the present CMB following adiabatic expansion by a factor of $T_{CMB}/T_{SLS} \simeq 1/1100$. During most of this epoch, the universe was essentially matter dominated. Gravitational self-binding suggests, therefore, a gradual deceleration in cosmological expansion, i.e., $q > 0$. (Whence the name "deceleration parameter.") However, observed is accelerated expansion!

The content of the universe can be normalized to the critical closure density

$$\rho_c = \frac{3H^2}{8\pi G} \tag{8.73}$$

delineating re-collapse in the distant future from eternal expansion in a three-flat universe with $\Lambda = 0$. Gravitational attraction by matter at a density $\rho = \rho_c$ would bring H to zero at late times. Let $\Omega_M = \rho_M/\rho_c$ defines the fractional content of mass in the Universe. A further content of energy unseen, *dark energy*, is Ω_Λ. For a three-flat universe, conservation of total energy[3] implies

$$\Omega_M + \Omega_\Lambda = 1. \tag{8.74}$$

The Einstein equations provide a suitable starting point for a dynamical equation of motion for $a(t)$. Reduced to (8.71) above, we have

$$q = \frac{1}{2}\Omega_M - \Omega_\Lambda. \tag{8.75}$$

Since $\Omega_M > 0$, the detection (8.72) shows $\Omega_\Lambda > 0$, leaving us with one the greatest mysteries in the Universe, to explain the nature of dark energy.

To explore (8.75), consider a constant Λ. This framework is commonly referred to as ΛCDM, where CDM refers to cold dark matter,

$$\rho = \rho_0 \left(\frac{a_0}{a}\right)^3, \tag{8.76}$$

mostly also unseen since the observed galaxies satisfy $\Omega_b \simeq 4.5\%$. In this event, (8.75–8.76) define the Friedman equations

$$\left(\frac{H}{H_0}\right)^2 = \omega_M \left(\frac{a}{a_0}\right)^{-3} + \omega_\Lambda, \quad \frac{\ddot{a}}{a} = -\frac{1}{2}\omega_M \left(\frac{a}{a_0}\right)^{-3} + \frac{1}{3}\Lambda, \tag{8.77}$$

where ω_M and ω_Λ are the values of Ω_M and, respectively, Ω_Λ today with Hubble parameter H_0. These two equations are consistent. With $\omega_\Lambda = \Lambda/3$, the second equation is the time derivative of the first.

With the initial conditions $a(0) = a_0$ and $H(0) = H_0$, either one of the two Friedman equations (8.77) is readily integrated numerically backwards in time, e.g., by Runge-Kutta. Note that the second equation can be cast into a first-order system $b(t) = \dot{a}(t)$ and $\dot{b}(t)/a(t)$ defined by the right hand side of (8.77). Figure 8.10 shows the results as a function of redshift

$$\frac{a}{a_0} = \frac{1}{1+z}. \tag{8.78}$$

Figure 8.10 shows results for both ΛCDM and a candidate dynamical dark energy

$$\Lambda = (1-q)H^2 \tag{8.79}$$

---

[3]This is a consequence of the Bianchi identity—the boundary of a boundary is the empty set—in the Einstein equations.

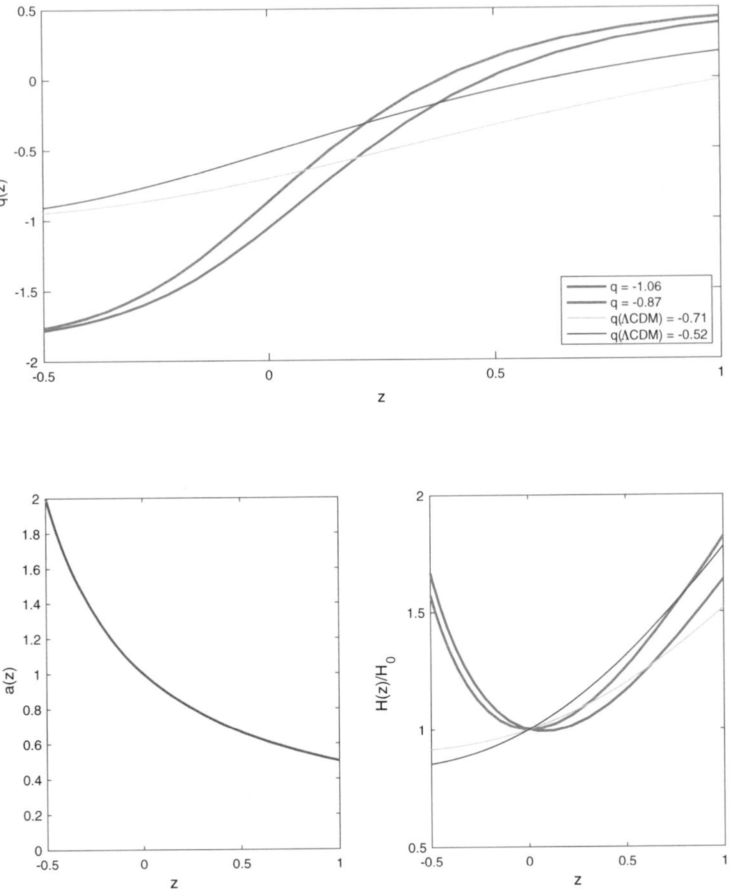

Fig. 8.10 Numerical integration by a Runge-Kutta method of (8.77) modeling the expansion of the Universe with constant Λ (ΛCDM, *thin lines*) and a dynamical Λ (*thick lines*), shown as a function of redshift in a window around $z = 0$. The two curves in each case refer to different values of q today. These two alternatives produce different rates of transition to the present $q(0) < 0$, which is relatively slow in ΛCDM

representing low energy vacuum fluctuations associated with the cosmological horizon [5]. While a detailed derivation of (8.79) falls outside the scope of the present discussion, a hint derives from dimensional analysis based on c and G alone. The quantities

$$L_0 = \frac{c^5}{G}, \quad p_0 = -\frac{L_0}{cA_H} \tag{8.80}$$

define a unit of liminosity and a unit of pressure on the cosmological horizon with radius $R_H = c/H$ and area $A_H = 4\pi R_H^2$. In the de Sitter limit of an empty universe, all quantities are Lorentz invariant, whereby p_0 has an associated energy density

$\rho_0 = -p_0$. It follows that

$$\Omega_\Lambda = \frac{\rho_0}{\rho_c} = \frac{2}{3}, \tag{8.81}$$

corresponding to the limit $q = -1$ in (8.79) and close to today's state of the Universe.[4]

In geometrical units, (8.81) is seen to derive from $[\sqrt{\Lambda}] = \mathrm{cm}^{-1}$ as a surface energy density, rather than $[\Lambda] = \mathrm{cm}^{-2}$ as a volume energy density. The first points to holographic origin of our accelerated expansion in the cosmological horizon. As a surface density, $[\Lambda]$ defines a transition radius $r_t = \sqrt{R_g R_H}$ around a mass M with gravitational radius $R_g = GM/c^2$. Applies to a galaxy of mass $M = M_{11} 10^{11} M_\odot$, it defines a characteristic radius

Table 8.4 Overview of finite differencing and its applications

1. Vector fields can be differentiated and integrated by applying the rules of calculus to their components in a given coordinate system. Differentiation can be approximated by finite differencing based on one- or two-sided difference operators $\delta_h^\pm$ and Δ_h. For smooth functions, they are first and second order accurate, respectively. For linear or quadratic polynomials, $\delta_h^\pm$ and Δ_h, respectively, are exact. Integration by finite summation is a converse of finite differencing. It recovers exact results - the Riemann integral - in the limit as the step size goes to zero.

2. Euler's method provides numerical integration of ordinary differential equations based on successive linear extrapolations, forwards or backwards in time.

3. Runge-Kutta integration provides fourth-order accuracy, based on averaging linear extrapolations at intermediate time-steps.

4. Numerical integration provides a powerful tool to study a great diversity of dynamical systems, e.g., osmosis (8.57), the Earth-Moon migration (8.67) and cosmological evolution (8.77). The latter is model dependent on dark matter and dark energy. Dimensional analysis in geometrical units points to a holographic origin in or association with the cosmological horizon.

[4] In geometrical units, $L_0 = 1$ and $\Omega_\Lambda = 2/3$ represent scales in the hyperbolic and, respectively, elliptic part of the Einstein equations defined by c and G.

$$r_t = 4.7\,\mathrm{kpc}\, M_{11}^{1/2}, \tag{8.82}$$

which is typical for distances beyond which dark matter becomes apparent in galaxy rotation curves.

Table 8.4 summarizes this discussion.

8.6 Exercises

8.1. Consider the magnetic flux passing a sphere of radius R about a magnet at its center, aligned with the vertical axis corresponding to $\theta = 0$ in a spherical coordinate system (r, θ, φ). The magnetic field $\mathbf{B}$ at the north pole $\theta = 0$ is $\mathbf{B} = B_0\mathbf{k}$, where $\mathbf{k}$ denotes the unit normal along $\theta = 0$. We define

$$B_n = \mathbf{B} \cdot \mathbf{n} \tag{8.83}$$

to be the normal component of $\mathbf{B}$ to the sphere in terms of the unit outgoing normal $\mathbf{n}$ and consider the total magnetic flux through a polar cap

$$\Phi(\theta) = 2\pi R^2 \int_0^\theta B_n(\theta) d\theta \tag{8.84}$$

Sketch the magnetic field lines around the magnet and determine B_n at $\theta = 0, \pi/2, \pi$. What is $\Phi(0)$ and $\Phi(\pi)$? Obtain a Taylor series approximation to $\Phi(\theta)$ about $\theta = 0$. For what θ is $\Phi(\theta)$ maximal? Sketch $\Phi(\theta)$ for all θ. Compare your results with electric flux $\Psi(\theta)$ similarly defined for a central charge Q.

8.2. A superconducting quantum interference device (SQUID) is a magnetic sensor that probes magnetic flux down to the quantum level $\Phi_0 = hc/e \simeq 4 \times 10^{-7}$ G cm^2. Consider a SQUID of surface area 1 cm^2. Plot the graph of the magnetic field detected by the SQUID as a function of an external magnetic field of strength that ranges continuously from 0 to 15 G.

8.3. A magnetic dipole is located next to a straight current wire of infinite length. Initially, the magnetic moment $\boldsymbol{\mu}$ of this dipole is parallel to the current wire. Sketch the subsequent motion of the dipole moment, when left free to move. [*Hint.* A dipole with an external magnetic field has a potential energy $U = -\boldsymbol{\mu} \cdot \mathbf{B}$ and experiences a torque $\boldsymbol{\tau} = \boldsymbol{\mu} \times \mathbf{B}$.]

8.4. Consider a function $f(t)$ as a function of t on the real line and the finite difference expression

$$D_h f(t) = \frac{f(t+h) - f(t-h)}{2h} \quad (0 < h \ll 1). \tag{8.85}$$

1. If $f(t) = t^2$, sketch the graph of $f(t)$ and its derivative $f'(t) = df(t)/dt$.
2. If $f(t) = t^2$, show that $D_h f(t) = f'(t)$ and explain your result.
3. If $f(t) = e^t$, is $D_h f(t) > f'(t)$ or $D_h f(t) < f'(t)$?

8.5. Consider the initial value problem describing the electrical discharge of a capacitor over a resistor with time scale τ,

$$\dot{y}(t) = -\frac{1}{\tau} y(t), \quad y(0) = 0. \tag{8.86}$$

Choose $\tau = 1$.

- Write down the analytic solution for $t \geq 0$.
- Write down an iterative scheme for *forward Euler* using a step size $\Delta t = h$ and integrate up to $T = 10$ using $h = 1/2^n$ for various n.
- Make a graph of the logarithm of the error at $T = 10$ as a function of n. Verify that accuracy is *first order*.
- Apply *backward Euler* method and the *midpoint rule*. Verify that they produce first and second order accuracy, respectively.

8.6. Consider the initial value problem for osmosis in Sect. 8.3. Allow the permeability of the membrane to increase with area, e.g., $\mu = \mu_0 + \mu_1 A$. Modify (8.57) accordingly in consider the effect of $\mu_1 > 0$ on the swelling time by numerical integration.

8.7. Following the two-dimensional root finding problem $\mathbf{F}(x, y) = \mathbf{0}$ for

$$\mathbf{F}(x, y) = \begin{pmatrix} f_1(x, y) \\ f_2(x, y) \end{pmatrix} = \begin{pmatrix} xy \\ x^2 + y^2 - 1 \end{pmatrix} \tag{8.87}$$

- Determine the roots (x, y) by inspection.
- Compute the Fréchet derivative $\mathbf{A} = D\mathbf{F}(x, y)$ and its determinant.
- Compute the inverse $\mathbf{A}^{-1}$ and determine the locus where its determinant vanishes.
- Following an initial guess $(x_0, y_0) = (1, 1/2)$, calculate (x_1, y_1) by Newton's method and sketch this iteration in the (x, y) plane.
- Compute subsequent iterations and observe the rate of convergence.

8.8. Radial free fall of a test particle in Newtonian gravity satisfies[5]

$$\ddot{r} = -\frac{GM}{r^2}, \tag{8.88}$$

where r denotes the distance to a central mass M and G is Newton's constant. Its trajectory satisfies the Hamiltonian

$$H = \frac{1}{2}\dot{r}^2 - \frac{GM}{r}. \tag{8.89}$$

[5]Free fall in Schwarzschild spacetime is reviewed in Appendix C.

For free fall of particle initially at rest at a distance R, show $H < 0$ and derive the free fall time scale

$$t_{ff} = \frac{\pi R^{\frac{3}{2}}}{2\sqrt{2GM}}. \tag{8.90}$$

The problem of free fall is singular as time approaches t_{ff}. For numerical integration, we can regularize (8.88) as

$$\ddot{r} = -\frac{GM}{r^2 + \epsilon^2} \tag{8.91}$$

with ϵ small. Show that (8.91) derives from

$$F = \mathrm{Re}\left\{\frac{1}{(r + i\epsilon)^2}\right\}. \tag{8.92}$$

Since (8.91) has a Hamiltonian, initial data with $H < 0$ produce bound orbits that extend for infinite time. Argue that these orbits are periodic. Putting $GM = 1$ and $R = 2$, use numerical integration to calculate t_{ff} from the minima in $r(t)$.

8.9. In Fig. 8.2, assume that the two snapshots are separated by 0.5 s. Consider the car center left. Estimate its projected west-bound grid velocity by finite differencing. Next, assume a true motion north-west and a grid spacing of about 2 m at the location of the car (along the horizontal). Estimate the true velocity of the car.

8.10. For a quadratic expression $u = x^2$, use the midpoint rule for an exact result $\Delta u = A_n h$ for $\Delta u = u_{n+1} - u_n$ on a uniform grid $x_n = nh$.

8.11. Apply $\delta_h^\pm$ and Δ_h to

$$(i)\ f(x) = 2x;\quad (ii)\ f(x) = x^2;\quad (iii)\ f(x) = e^{2x} \tag{8.93}$$

and identify the order of accuracy by explicit error estimates.

8.12. Based on the midpoint rule (8.42), consider the *trapezoidal rule* of integration

$$h\sum_{k=0}^{k-1} \frac{f(x_k) + f(x_{k+1})}{2} = \frac{h}{2}\left(f(x_0) + f(x_n)\right) + h\sum_{n=1}^{n-1} f(x_k), \tag{8.94}$$

where $x_k = kh$. Apply it to integrate the functions in (8.93) and obtain error estimates.

8.13. Consider the explicit time-stepping scheme defined by the approximation $\delta_\eta^+ u = \delta_h^+ u$ for the first order wave equation $u_t = u_x$ in (t, x), where $\eta = \Delta t$ and

$h = \Delta x$. Derive the Courant-Friedrichs-Lewy (CFL) condition $C = \Delta t/\Delta x < 1$ necessary for stability. [*Hint.* Consider $u(t,x) = e^{i(\omega t + kx)}$ with real k. Stability requires Im $\omega > 0$].

8.14. Apply $\delta_h^{\pm}$ and Δ_h to

$$(i)\ f(x) = 2x;\quad (ii)\ f(x) = x^2;\quad (iii)\ f(x) = e^{2x} \tag{8.95}$$

and identify the order of accuracy by explicit error estimates.

8.15. In the limit of a self-gravitating fluid of mass M and radius R (with zero elasticity), show that the zero-frequency tidal deformation amplitude h_0 by a perturber of mass m at a distance r is of order

$$h_0 = \left(\frac{R}{r}\right)^3 \left(\frac{m}{M}\right) R \simeq 30\,\text{cm}, \tag{8.96}$$

where result on the right hand-side applies to the Earth-moon system.

8.16. Tidal dissipation in the oceans is complicated by the diversity of linear and non-linear dissipation processes. The Ursell number [6] expresses the ratio of nonlinear to dispersive interaction in ocean waves. Based on scaling, argue that for shallow water waves with propagation speed $c_s = \sqrt{gd}$ in oceans of depth d, where $g = 9.8\,\text{ms}^{-2}$ is Earth's gravitational acceleration, the Ursel number satisfies

$$\text{Ur} = \frac{hg}{\nu^2 d^2} \simeq 300 \times \left(\frac{h}{0.3\,\text{m}}\right)\left(\frac{d}{4000\,\text{m}}\right)^{-2}\left(\frac{P}{12\,\text{h}}\right)^2 \gg 1 \tag{8.97}$$

8.17. For a rotating black hole of mass M described by the Kerr metric, we can parametrize the angular momentum as $J = M^2 \sin\lambda$ [7], so that $\Omega \cdot \mathbf{J} = M \sin^2(\lambda/2)$ and $E_{rot} = 2M \sin^2(\lambda/4)$. Derive (3.33).

8.18. Consider a neutron star of mass $M = pM_\odot$ with uniform mass density ρ and radius $R = q10$ km.

1. Calculate the moment of inertia from the volume integral

$$I = \int_V \rho r^2 d^3x \tag{8.98}$$

   over the volume $V = (4\pi/3)R^3$ of the star and express the result in terms of the scale factors p and q.
2. Held by self-gravity, calculate the maximal angular velocity Ω before break-up.

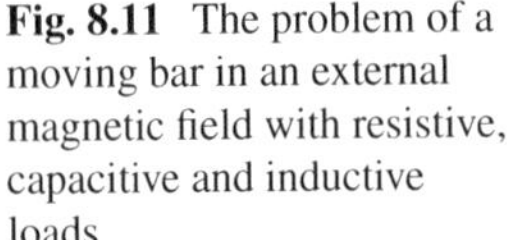

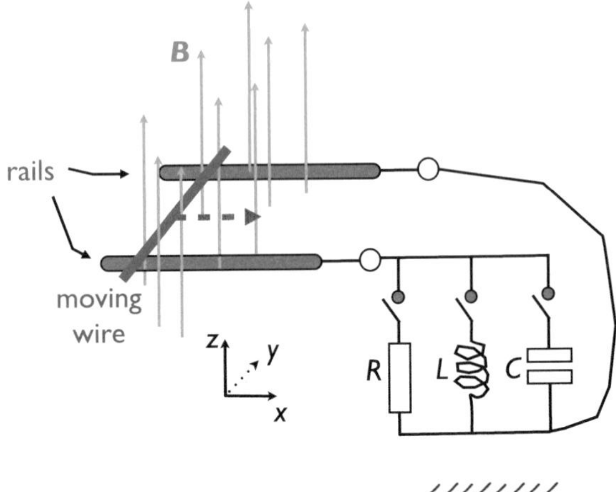

Fig. 8.11 The problem of a moving bar in an external magnetic field with resistive, capacitive and inductive loads

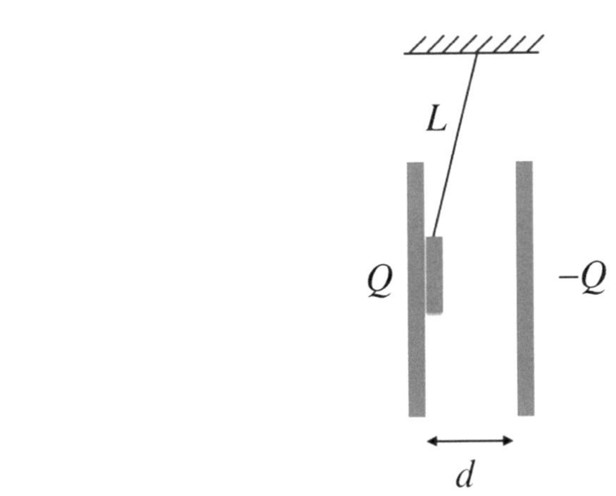

Fig. 8.12 The problem of an oscillating coin in a parallel plate capacitor

3. What is the maximal rotational energy E_{rot} of the neutron star? [*Hint:* The rotational energy satisfies $E_{rot} = (1/2)I\Omega^2$.]
4. What is the maximal ratio E_{rot}/E_0, expressed relative to the rest mass energy $E_0 = pM_\odot c^2$. Here, c denotes the velocity of light.

8.19. In electrodynamics, the motion of a conductor in an external magnetic field can as a dynamo or an actuator by, respectively, Faraday's law of induction and Lorentz forces. Consider a bar of mass m in an external magnetic field, sliding over two electrically conducting rails connected to different kinds of loads, namely, a resistor R, a capacitor C and an inductor L. If the bar is set into motion by a constant force and left alone after some time, (a) derive the ordinary differential equation for the motion for each of these three different loads, and (b) indicate the associated Poynting flux $\mathbf{S} = c\mathbf{E} \times \mathbf{B}/(4\pi)$ about the bar and load (Fig. 8.11).

8.20. A coin in the form of a thin disk of radius R and mass m is suspended by a long wire of length L in a gap of size d of a parallel plate capacitor (Fig. 8.12). The plates have initially a charge Q. As a result, the coin will electrostatically bounce off the plates, transferring a discrete current between them. (a) Show that the frequency at which the coin bounces is a linear function of the capacitor voltage. What is the current mediated by the coin and how does Q evolve over time? [*Hint.* Assume $R \ll d \ll L$ and work in the small angle approximation.] (b) If the capacitor is held horizontally, so that the coin moves up and down in Earth's gravitational field, does the frequency change?

8.21. Consider the motion of an electron with charge $q = -e$ and mass m moving freely in a uniform magnetic of strength B. Derive the Larmor frequency $\omega = eB/mc$ of circular motion from (a) Lorentz forces in the laboratory frame of reference and (b) Faraday induction in the frame comoving with the electron, both in the nonrelativistic limit (velocities much smaller than the velocity of light).

References

1. Efroimsky, M. 2012, Cel. Mech. and Dyn. Astron, 112, 283; Efroimsky, M., 2015, AJ, 150, 98.
2. Ray, R.D., Eanes, R.J., & Chao, B.F., 1996, Nature, 381, 595.
3. Riess A. et al., 1998, ApJ, 116, 1009, Perlmutter S. et al., 1999, ApJ, 517, 565.
4. Penzias, A. A., & Wilson, R. W. 1965, ApJ, 142, L419.
5. van Putten, M.H.P.M., 2015, MNRAS, 450, L48.
6. Ursell, 1953, Proc. Camb. Philos. Soc., 49, 685.
7. van Putten, M.H.P.M., 1999, Science, 284, 115.
8. van Putten, M.H.P.M., NewA, 54, 115.

Chapter 9
Perturbation Theory, Scaling and Turbulence

Systems often contain various scales that parameterize solutions. Their limits, large or small, may define leading order behavior that can be derived directly from approximate equations, rather than limits of exact solutions to the original equations. Occasionally, multiple scales may coexist, e.g., short periods of oscillation subject to slow decay by dissipation. Furthermore, the order of a system may depend on the choice of scale. For instance, the Reynolds number determines whether fluid flow on macroscopic scales is dominated by inertia or viscosity; or whether a rheological medium is effectively fluid or elastic, in the limit of small or large dimensionless frequency $\tau_M\omega$ given by the Maxwell time times the angular frequency ω. For instance, suspending a pendulum in air or syrup produces the fundamentally different results of damped oscillations or exponential decay, effectively described by a second and, respectively, first order ordinary differential equation. In more complex fluid motion, such change of order naturally occurs in transition to free flow away from a solid boundary.

Below, we look at some illustrative examples in gravitation, seeking approximation solutions to problems with a small parameter by perturbation theory.

9.1 Roots of a Cubic Equation

The function

$$N = 1 - \frac{2M}{r} - \frac{1}{3}\Lambda r^2 \tag{9.1}$$

describes the redshift factor in Schwarzschild black hole of mass M in a de Sitter space defined by the cosmological constant $\Lambda > 0$. The zeros $N = 0$ denote event horizons, given by the zeros of

M.H.P.M. van Putten, *Introduction to Methods of Approximation in Physics and Astronomy*, Undergraduate Lecture Notes in Physics,
DOI 10.1007/978-981-10-2932-5_9

$$p(r) = r - 2M - \frac{1}{3}\Lambda r^3. \tag{9.2}$$

As a cubic polynomial, $p(r)$ attains a double zero when $p(r) = p'(r) = 0$, i.e., when

$$\Lambda M^2 = \frac{1}{9}. \tag{9.3}$$

Three distinct zeros exist whenever $\Lambda m^2 < \frac{1}{9}$, when M is inside a larger de Sitter space. When $\Lambda = 0$, the black hole event horizon is at the Schwarzschild radius $R_S = 2M$; when $M = 0$, the cosmological de Sitter horizon is at $R_{dS} = \sqrt{3/\Lambda}$. To calculate these radii when both M and Λ are nonzero, we might use exact solutions to $p = 0$. However, envisioning a small cosmological constant, it seems prudent to consider $\Lambda M^2 \ll \frac{1}{9}$ as a small parameter. We therefore seek roots of

$$u - 1 - \epsilon u^3 = 0 \tag{9.4}$$

by a *regular perturbation* of the normalized Schwarzschild radius $u = r/2M$ due to Λ as a function of $\epsilon = \frac{4}{3}\Lambda M^2$. Substitution of the expansion

$$u = 1 + \epsilon u_1 + \epsilon^2 u_2 + \cdots \tag{9.5}$$

into (9.4) gives

$$(\cdots)\,\epsilon + (\cdots)\,\epsilon^2 + \cdots = 0 \tag{9.6}$$

is to hold *for all* $0 \leq \epsilon \ll 1$. Hence, the coefficients $(\cdots)$ should all vanish. From the coefficient of ϵ, we have $u_1 = 1$, showing

$$u = 1 + \epsilon : \;\; r = 2M\,(1 + \epsilon)\,, \tag{9.7}$$

i.e., the black hole slightly expands in a de Sitter space. To consider the perturbation of the de Sitter horizon by M, we multiplying (9.4) by $\sqrt{\epsilon}$,

$$v - \sqrt{\epsilon} - v^3 = 0, \;\; v = \sqrt{\epsilon} u, \tag{9.8}$$

and purse a regular expansion n the small parameter $\delta = \sqrt{\epsilon}$,

$$u = \frac{1}{\delta}\left(1 + \delta u_1 + \delta^2 u_2 + \cdots\right). \tag{9.9}$$

Substitution into (9.8) gives

$$1 + \delta u_1 + \delta^2 u_2 + \cdots - \delta - \left(1 + 3\delta u_1 + O(\delta^2) + \cdots\right), \tag{9.10}$$

showing $u_1 = -\frac{1}{2}$ and hence

$$u = 1 - \frac{1}{2}\delta: \; r = \sqrt{\frac{3}{\Lambda}}\,(1-\delta)\,. \tag{9.11}$$

The above shows that, for $0 < \epsilon \ll 1$ (9.4) presents a *regular* perturbation equation for the event horizon of the black hole, yet a *singular* perturbation for the de Sitter horizon—the latter being non-existent in the limit $\epsilon = 0$.

9.2 Damped Pendulum

To illustrate these ideas, consider motion of a damped linearized pendulum of a mass m suspended by a rod of length l in terms of a deflection angle $\dot{x}$, described by

$$\ddot{x} + \alpha\dot{x} + \omega_0^2 x = 0, \tag{9.12}$$

where $\omega_0 = \sqrt{g/l}$ denotes the fundamental frequency in the presence of Earth's gravitational acceleration g. Here, $ml\ddot{x}$ is the inertial force, mgx the restoring force and $\alpha\dot{x}$ ($\alpha \geq 0$) is a viscous force with $[\alpha] = 1/\mathrm{s}$. As a linear equation, the general solution of (9.12)

$$x(t) \propto \mathrm{Re}\, e^{i\lambda t} \tag{9.13}$$

is determined by the eigenvalue λ, given by the complex-valued frequency

$$\lambda = \begin{cases} \sqrt{\omega_0^2 - \frac{1}{4}\alpha^2} + i\frac{1}{2}\alpha & (\alpha < 2\omega_0), \\ i\frac{1}{2}\left[\sqrt{\alpha^2 - 4\omega_0^2} + \alpha\right] & (\alpha > 2\omega_0), \end{cases} \tag{9.14}$$

allowing us to write down exact solutions, i.e., a damped oscillation or purely exponential decay

$$x(t) = \begin{cases} e^{-\frac{1}{2}\alpha t}\cos\left(\sqrt{\omega_0^2 - \frac{1}{4}\alpha^2 t}\right) & (\alpha < 2\omega_0) \\ e^{-\frac{1}{2}\left[\alpha + \sqrt{\alpha^2 - 4\omega_0^2}\right]t} & (\alpha < 2\omega_0). \end{cases} \tag{9.15}$$

However, it is often more instructive to focus on limiting cases. In case of small viscosity, using $\sqrt{1+\epsilon^2} \simeq 1 + \frac{1}{2}\epsilon^2$ ($0 \leq \epsilon \ll 1$) and setting $\omega_0 = 1$, we obtain the asymptotic expression

$$x(t) \simeq e^{-\frac{1}{2}\alpha t} \cos\left(\left[1 - \frac{1}{8}\alpha^2\right] t\right). \tag{9.16}$$

Further insight is gained by deriving (9.16) directly from (9.12), rather than the exact solution (9.15). To study the limit of small viscosity, we consider the (9.12) with $\omega_0 = 1$ and $\epsilon = \alpha$ as a small parameter,

$$\ddot{x} + \epsilon \dot{x} + x = 0 \ \ (0 \leq \epsilon \ll 1), \tag{9.17}$$

and consider the perturbation expansion

$$x(t) = x_0(t_*, \tau) + \epsilon x_1(t_*, \tau) + \epsilon^2 x_1(t_*, \tau) + \cdots \tag{9.18}$$

in terms of a fast and slow time[1]

$$t_* = \left(1 + \epsilon^2 \omega_2 + \epsilon^3 \omega_3 + \cdots\right) t, \ \ \tau = \epsilon t. \tag{9.19}$$

In the application to (9.17), we suppress the first-order term $\epsilon\omega_1$ (Exercise 9.1.) Substitution into (9.17) with

$$\begin{aligned} \tfrac{d}{dt} x_i &= x_{it_*}\left(1 + \epsilon^2 \omega_2 + \cdots\right) + \epsilon x_{i\tau} \\ \tfrac{d^2}{dt^2} x_i &= x_{it_*t_*}\left(1 + 2\epsilon^2 \omega_2 + \cdots\right) + 2\epsilon x_{it_*\tau} + \epsilon^2 x_{i\tau\tau} + O\left(\epsilon^3\right) \end{aligned} \tag{9.20}$$

and collecting terms of the same order in ϵ obtains

$$\begin{aligned} L(x_0) &= 0, \\ L(x_1) &= -x_{0t_*} - 2x_{0t_*\tau}, \\ L(x_2) &= -x_{1t_*} - 2x_{1t_*\tau} - 2\omega_2 x_{0t_*t_*} - x_{0\tau\tau} \end{aligned} \tag{9.21}$$

in the notation $L(x) = x_{t_*t_*} + x$. For $\epsilon \geq 0$, (9.17) the total energy is non-increasing, i.e.,

$$H = \frac{1}{2}\dot{x}^2 + \frac{1}{2}x^2 : \ \ \frac{d}{dt}H = -\epsilon \dot{x}^2 \leq 0, \tag{9.22}$$

and hence we seek solutions that are bounded. For this reason, the right hand side in the second and third equation of (9.21) cannot be on resonance, i.e., should not contain the fundamental harmonic $\omega_0 = 1$. Consider

$$x_0(t_*, \tau) = A(\tau) \cos t_* + B(\tau) \sin t_* \tag{9.23}$$

[1] With $\omega_0 = 1$, time refers to total phase. In the present two time-scale problem, stretched time $\left(1 + \epsilon\omega_1 + \epsilon^2\omega_2 + \epsilon^3\omega_3 + \cdots\right) t$ is here accompanied by a slow time scale $\tau = \epsilon t$.

as a solution to the first equation. Since (9.12) is autonomous, we are at liberty to put $B = 0$. The right hand side of the second equation of (9.21) is

$$-x_{0t_*} - 2x_{0t_*\tau} = \left(2A' + A\right)\sin t_*. \tag{9.24}$$

Suppressing on-resonance forcing, we insist $A + 2A' = 0$, and obtained the exponential decay $A(\tau) = A_0 e^{-\frac{1}{2}\tau}$. We next proceed with

$$x_1(t_*, \tau) = C(\tau)\cos t_* + D(\tau)\sin t_*. \tag{9.25}$$

Since (9.12) is linear, we may proceed with $C = 0$. (A nonzero C can be absorbed in $A = A_0 + \epsilon A_1 + \cdots$.) Repeating the same arguments for the right hand side of the third equation of (9.21),

$$\left(2\omega_2 A + A''\right)\cos t_* - \left(D + 2D'\right)\sin t_* \tag{9.26}$$

we obtain a fast time perturbation

$$\omega_2 = -\frac{1}{8} \tag{9.27}$$

explicit in (9.16) and once more the exponential decay $D = D_0 e^{-\frac{1}{2}\tau}$.

9.3 Orbital Motion

Since gravitational forces are conservative, orbital motion of binary stars or planets allows for a Hamiltonian $H = E_k + U$ given by the sum of kinetic energy E_k and a potential U describing gravitational binding energy. For a binary of two masses M_1 and M_2, we have

$$H = \frac{1}{2}M_1 v_1^2 + \frac{1}{2}M_2 v_2^2 - \frac{GM_1M_2}{r}, \tag{9.28}$$

where $r = |\mathbf{r_1} - \mathbf{r_2}|$ denotes the separation distance between the two masses with velocities $\mathbf{v}_{1,2} = \dot{\mathbf{r}}_{1,2}$.

It is convenient to use a polar coordinate system (r, φ) about the center of mass of the binary,

$$M_1 r_1 = M_2 r_2, \quad r = r_1 + r_2, \quad \mathbf{p}_1 + \mathbf{p}_2 = \mathbf{0}, \quad \mathbf{p}_i = M_i \mathbf{v}_i \ (i = 1, 2), \tag{9.29}$$

where the $\mathbf{p}_i$ refer to the momenta of the M_i. Thus, we may write

$$H = \frac{p_1^2}{2M_1} + \frac{p_2^2}{2M_2} - \frac{GM_1M_2}{r} = \frac{p_1^2}{2\mu} - \frac{GM_1M_2}{r}, \tag{9.30}$$

where $\mu = \frac{M_1M_2}{M_1+M_2}$ denotes the *reduced mass*. Let v denote the magnitude $|\mathbf{v}_1 - \mathbf{v}_2|$ of the velocity difference between the two stars, satisfying

$$v^2 = v_1^2 + v_2^2 - 2\mathbf{v}_1 \cdot \mathbf{v}_2 = v_1^2 + v_2^2 + 2v_1v_2 = \frac{p_1}{\mu^2} \tag{9.31}$$

in view of (9.29). With $M = M_1 + M_2$. The results is a one-body Hamiltonian

$$H = \frac{p_1^2}{2M_1} + \frac{p_2}{2M_2} - \frac{GM_1M_2}{r} = \mu\left(\frac{1}{2}v^2 - \frac{GM}{r}\right) \tag{9.32}$$

in terms of the separation between M_1 and M_2, describing equivalent motion of a test particle (of unit mass) around a centre with the total mass $M = M_1 + M_2$. Because M and v are invariant with respect to a Galilean transformation of the system, *the sign of H is a Galilean invariant.*

The Hamiltonian (9.32) can be conveniently rewritten following the Möbius transformation $u = 1/r$ with $u = u(\varphi)$, upon exploiting conservation of specific angular momentum $j = r^2\dot{\varphi}$. (This formally follows from symmetries of the Lagrangian as discussed in Chap. 10.) With $v^2 = \dot{r}^2 + r^2\dot{\varphi}^2$ and $\dot{r} = -ju'$, $u' = du(\varphi)/d\varphi$ and dropping the coefficient μ in (9.32), it follows that

$$H = \frac{1}{2}j^2\left(u'^2 + u^2\right) - GMu = \frac{1}{2}j^2u'^2 + U_N \tag{9.33}$$

with Newtonian potential

$$U_N = \frac{1}{2}\left(u - u_0\right)^2 + U_0, \quad u_0 = \frac{GM}{j^2}. \tag{9.34}$$

On the basis of (9.32 and 9.33), bound motions ($H < 0$) have $u(\varphi)$ harmonic in φ. Explicitly, differentiation of (9.32) gives

$$\frac{d^2u}{d\theta^2} + u = \frac{GM}{j^2} \tag{9.35}$$

with solutions

$$u = \frac{GM}{j^2}\left(1 + e\cos\varphi\right) \tag{9.36}$$

that are bound whenever the ellipticity $e < 1$. The corresponding trajectory in (r, φ) is an ellipse with maximum and minimum radius

$$r_{max} = \frac{j^2}{GM(1-e)} = a(1+e), \quad r_{min} = \frac{j^2}{GM(1+e)} = a(1-e), \quad (9.37)$$

expressed in terms of the major semi-axis a (with $\varphi = 0, \pi$, $r_{max} + r_{min} = 2a$) with corresponding specific angular momentum

$$j = \sqrt{GMa(1-e^2)}. \quad (9.38)$$

The orbital period P follows from the fact that specific angular momentum represents area change per unit time, $j = 2dA/dt = 2A/P$, where $A = \pi ab$ is the area enclosed by the ellipse and b is the minor semi-axis.

Note that ae is the distance of a focal point (the origin of our polar coordinate system) to the the center of the ellipse. Thus, $b = r \sin\varphi$ when $r\cos(\pi - \varphi) = ae$, i.e., $-\cos\varphi = e$, that recovers the familiar expression $b = a\sqrt{1-e^2}$. Combining, $A = \pi a^2\sqrt{1-e^2}$, whereby $P = 2A/j$ gives Kepler's law

$$\Omega^2 = \frac{GM}{a^3}, \quad \Omega = \frac{2\pi}{P}. \quad (9.39)$$

9.3.1 Perihelion Precession

Newton's theory of gravitation gives a satisfactory explanation of Keplerian orbits in the solar system, except for Mercury. A small but distinct precession in Mercury's slightly elliptical orbit is of great historical significance in identifying a non-Newtonian feature of gravitation, that is well accounted for by geodetic pression in Einstein's theory of general relativity. The equation describing this precession presents an interesting example for two-timing, further illustrating the method of the previous section.

Following the usual Möbius transformation $u = 1/r$ with $u = u(\varphi)$ on orbital motion in polar coordinates (r, φ), we obtain the Hamiltonian for test particles (mass equal to 1) around a central mass M with specific angular momentum j, given by

$$H = \frac{1}{2}j^2u'^2 + U \quad (9.40)$$

with the Newtonian and Einstein potentials (9.34) and, respectively,

$$U_E = U_N - R_g u^3. \quad (9.41)$$

Here, we use the gravitational radius

$$R_g = \frac{GM}{c^2} = 1.5 \times 10^5 \times \left(\frac{M}{2 \times 10^{33}\mathrm{g}}\right) \mathrm{cm} \quad (9.42)$$

of a mass M, where G denotes Newton's constant and c the velocity of light.

In considering orbits of small ellipticity, we focus on perturbations of u about the inverse radius $u_0 = 1/r_0$ at a minimum of U, satisfying $U'(u_0) = 0$: $u_0 \simeq M/j^2$ (exact for $U = U_N$ and approximate for $U = U_E$).

It is evident from (9.41) that orbits in general relativity are not the same as those in Newtonian gravity, due to the perturbation $R_g u^3$ in the potential U_E. Instead of exact periodic behavior in closed Keplerian orbits in Newton's theory, implied by the equation of motion

$$u'' + u = u_0, \tag{9.43}$$

we anticipate open orbits in

$$u'' + u = u_0 + 3R_g u^2. \tag{9.44}$$

For an approximately circular orbit with inverse radius $u_0 = 1/r_0$, consider the parameter $\epsilon = R_g u_0$,

$$\epsilon = \frac{R_g}{r_0} \simeq 10^{-8} \times \left(\frac{\mathrm{AU}}{r_0}\right), \tag{9.45}$$

where we scale r_0 to one astronomical unit AU $= 1.3 \times 10^{13}$ cm defined by the Earth's distance o the Sun. For planetary motions in our solar system, ϵ is a small parameter in (9.44) introduced by Einstein's theory of gravitation. With $u = \epsilon v$, (9.44) normalizes to

$$v'' + v = 1 + 3\epsilon v^2. \tag{9.46}$$

Due to (9.45), U_E in (9.41) is not symmetric about its minimum. For $\epsilon > 0$, perturbations may produce departures from the Keplerian elliptical orbits in the Newtonian theory, in terms of amplitude and period. The discrepant time scales of orbital motion and changes is period suggest using a two-timing method with (9.19), anticipating once again $\omega_1 = 0$. For (9.46), we therefore seek

$$v = 1 + a\epsilon + A(\tau)\sin\left(t_* + \varphi(\tau)\right) + \cdots. \tag{9.47}$$

With $\alpha = t_* + \varphi(\tau)$, we have

$$\begin{aligned} v' &= \epsilon A' \sin\alpha + A\cos\alpha\left[1 + \epsilon\varphi'(\tau)\right] + O(\epsilon^2), \\ v'' &= 2\epsilon A'\cos\alpha - 2\epsilon\varphi' A\sin\alpha - A\sin\alpha + O(\epsilon^2), \end{aligned} \tag{9.48}$$

so that

$$v'' + v = 1 + a\epsilon + 2\epsilon A' \cos\alpha - 2\epsilon\varphi' A \sin\alpha + O(\epsilon^2). \tag{9.49}$$

Expanding the right hand side of (9.46),

$$1 + 3\epsilon v^2 = 1 + 3\epsilon \left(1 + 2A \sin\alpha + \frac{1}{2}A^2 (1 - \cos(2\alpha))\right). \tag{9.50}$$

Substituting (9.49 and 9.50) into (9.46), we shall match to constant term and suppress secularity, the latter for bound orbits by virtue of the Hamiltonian (9.40). Thus,

$$1 + a\epsilon = 1 + 3\epsilon \left(1 + \frac{1}{2}A^2\right), \quad 2A' \cos\alpha - 2\varphi' A \sin\alpha = 6A \sin\alpha \tag{9.51}$$

showing that

$$A' = 0, \quad \varphi' = -3, \quad a = 3 + \frac{3}{2}A^2. \tag{9.52}$$

The result

$$v \simeq 1 + 3\left(1 + \frac{1}{2}A^2\right)\epsilon + A \sin\left(t\left[1 - 3\epsilon\right]\right) \tag{9.53}$$

shows a positive perturbation in u_0 (a diminishing of r_0) illustrated in Fig. 9.1, and a new period of motion

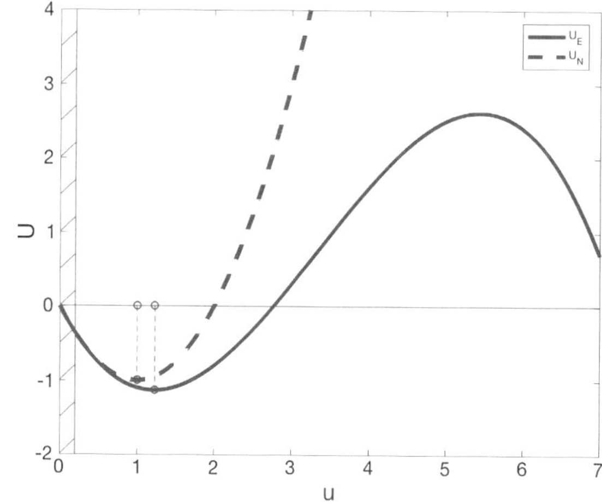

Fig. 9.1 Shown are the Newtonian and Einstein potentials U_N and, respectively, U_E in the Hamiltonian of orbital motion (9.40). In U_E, symmetry about its local minimum is lost. The shaded strip about $u = 0$ refers to the cosmological limit of large r, when the Rindler horizon of the associated accelerations drops beyond the cosmological horizon with possible perturbations of away from Newtonian inertia

$$\frac{2\pi}{1-3\epsilon} = 2\pi + 6\pi\epsilon + O(\epsilon^2) \tag{9.54}$$

with an excess $\Delta\varphi = 6\pi\epsilon$ in each period, compared to the Kepler's value 2π. This outcome is known as *perihelion* precession.

9.4 Inertial and Viscous Fluid Motion

In fluids, energy and momentum transport is most apparent in convection. More hidden from view is diffusive transport due to internal thermal motions of the fluid constituents. To leading order, such diffusion is linear in concentration gradients, in temperature, momentum or vorticity.

In a Eulerian description, convective and diffusive momentum transport is given by the Navier-Stokes equations for a velocity field, that represents local averages of particle velocities. Alternatively, we can use a Langrangian description, giving each fluid element a position vector $\boldsymbol{\xi}_i$ $(i = 1, 2, \ldots N)$, and tracking its displacement in time. A Lagrangian description is natural for, say, N-body systems, whose equations of motion derive from a Hamiltonian description. It can be advantageous when the system covers an extended region of space with a large dynamic range in particle density and pressure.

Here, we focus on a Eulerian description using a velocity field $\mathbf{u} = \mathbf{u}(x, y, z, t)$, describing the velocity of fluid elements at a coordinate position (x, y, z, t). It is a mean field description, in that $\mathbf{u}$ describes the mean velocity of all the individual fluid elements in a small box centered at (x, y, z, t). In the Eulerian description, the acceleration follows by tracking the velocity $\mathbf{u}$ across a displacement $\mathbf{u}\delta t$ along the physical trajectory of a given box of particles. Doing so over a small time interval δt and taking the limit in which δt approaches zero gives for the acceleration $\mathbf{a}$ the *Lagrangian* or *convective derivative*

$$D_t\mathbf{u} \equiv \lim_{\delta t \to 0} \frac{\mathbf{u}(\mathbf{x} + \mathbf{u}\delta t, t + \delta t) - \mathbf{u}(\mathbf{x}, t)}{\delta t} = \partial_t\mathbf{u} + (\mathbf{u} \cdot \nabla)\mathbf{u}, \tag{9.55}$$

where $\mathbf{x} = (x, y, z)$. Converting the right hand side of (9.55) to force per unit volume by multiplication with the mass density, ρ, we obtain

$$\rho D_t\mathbf{u} = -\nabla p + \cdots \tag{9.56}$$

in the presence of a pressure p. The dots refer to other terms that may come into the picture, such as $-\nabla\Phi$ from and external potential such as the gravitational binding energy to the Earth and viscous forces.

Transport of momentum by diffusion is described by a viscosity. The *dynamical viscosity* μ defines the constant of proportionality between a tangential force to a plate of area A, moving at a velocity V parallel to an adjacent plate at a separation

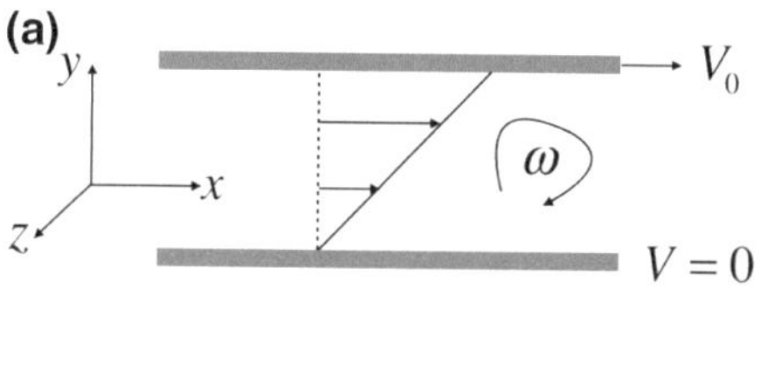

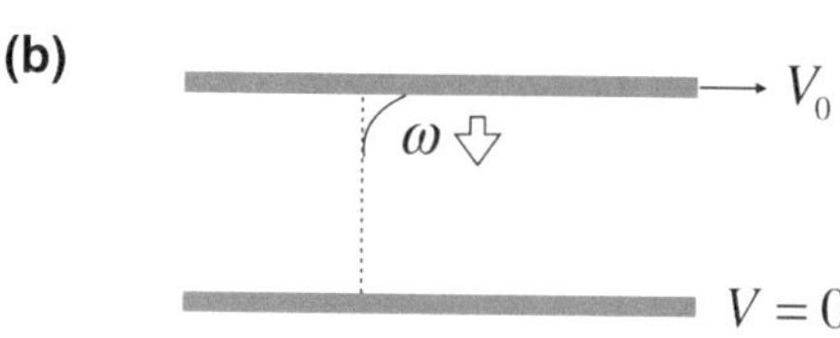

Fig. 9.2 Shown is the linear velocity profile of a fluid in a gap between two plates with relative velocity V_0 (A). This Poisseuille flow appears in a stationary state at sufficiently low Reynolds numbers. It results from diffusion of vorticity ω due to no-slip boundary conditions on the plates at large time (B)

h, according to

$$F = \mu A \frac{V}{h} \tag{9.57}$$

In a stationary state, the moving plate at height $y = h$ and the wall at height $y = 0$ experience the exact same force F in (9.57), leaving the fluid in between with a linear velocity profile $u = y/h$ $(0 \leq y \leq h)$ shown in Fig. 9.2a. A more local description of the force (9.57) is

$$F(y) = \mu A \frac{du(y)}{dy}, \quad F_2 = F(L), \quad F_1 = -F(0), \tag{9.58}$$

where the minus sign is included to preserve our definition of force, to be applied externally to keep the wall in place (e.g., by its foundations). Applied to the moving plate and the wall in steady state, we have $F_2 = F_1$.

Suppose we start with zero initial velocity and suddenly change to $V > 0$ at $t = 0^+$ (Fig. 9.2b). In this event, $F_2(0^+) > 0$, while $F_1(0^+) = 0$, i.e., the forces are initially out-of-balance. The result is a net force $F_2 - F_1 > 0$ on the fluid in between. For a fluid with density ρ, it produces a mean acceleration satisfying

$$\rho a = \mu \frac{u'(h) - u'(0)}{h} \simeq \frac{\mu}{h} \int_0^h u''(y')dy', \tag{9.59}$$

that subsequently decays to zero in approaching Fig. 9.2a over some relaxation time $\tau = V/a$. The expression (9.59) shows the appearance of a force density due to a local curvature in the velocity profile. This force may either produce time dependence or be balanced by a pressure gradient.

Example 9.1. Consider applying a uniform pressure to drive a fluid through a gap between two plates. A steady state *Poisseuille flow* appears in response to an external forcing, by applying some constant pressure p at an inlet. For a fluid between two plane parallel plates, it gives the velocity distribution

$$u = c\alpha(1-\alpha), \ \ \alpha = \frac{y}{h}, \tag{9.60}$$

where c is some constant. The quadratic relation (9.60) satisfies $\Delta u = -2ch^{-2}$, which refers to a constant pressure $p(y) = p_0$ at the inlet, assuming zero pressure at the outlet. Following (9.58), the net force $F = phW$ on the fluid suspended between the plates across some horizontal width W equals the sum of the two forces that the fluid applies to each plate,

$$phW = -\mu A u'(h) + \mu A u'(0), \ \ A = WL, \tag{9.61}$$

over some length L along the $x-$direction of motion. It follows that

$$h\frac{dp}{dx} = -h\frac{p_0}{L} = \mu h u''(y) = -2\mu c h^{-1}, \tag{9.62}$$

that is,

$$c = \frac{h^2 p_0}{2\mu L} \tag{9.63}$$

in (9.60). It has the dimension of velocity. Note the dimension of pressure over dynamical viscosity:

$$\left[\frac{p_0}{\mu}\right] = \mathrm{s}^{-1}, \tag{9.64}$$

so that (9.60) has dimensions of velocity. Indeed, the mean velocity across the flow satisfies

$$< u >= \frac{1}{h}\int_0^h u(y)dy = c\int_0^1 \alpha(1-\alpha)d\alpha = \frac{h^2 p_0}{12\mu L}. \tag{9.65}$$

A similar calculation in cylindrical geometry gives

$$\dot{m} = \rho < u > \pi r^2 = \frac{\pi p_0 r^4}{8\nu L}, \tag{9.66}$$

where $\nu = \mu/\rho$ denotes the kinematic viscosity. This result shows a dramatic dependence of the mass flow $\dot{m}$ on radius, r. A relatively minor decrease in 15% in radius causes a 50% decrease in flow rate for a given pressure. It explains why accumulation of even minor amounts of fat on the inner walls of our veins can be dangerous.

For a *Newtonian fluid*, the result of velocity curvatures in transverse the direction of propagation x superimpose. The general result for the force density due to viscous momentum transport, therefore, is

$$\mathbf{f} = \mu\Delta\mathbf{u}, \quad \Delta = \frac{\partial^2}{\partial_x^2} + \frac{\partial^2}{\partial_y^2} + \frac{\partial^2}{\partial_z^2}, \tag{9.67}$$

where $\Delta = \nabla \cdot \nabla$ is the Laplace operator. The Navier-Stokes equations now follow by including viscous momentum transport (*diffusion of momentum*) in (9.56), that is

$$\rho\left(u_t + (\mathbf{u} \cdot \nabla)\mathbf{u}\right) = -\nabla p + \mu\Delta\mathbf{u} \tag{9.68}$$

supplemented with mass continuity

$$\partial_t \rho + \nabla \cdot \rho\mathbf{u} = 0. \tag{9.69}$$

The Navier-Stokes equation (9.68) can be expressed in normalized (dimensionless) variables to explicitly bring about the role of the

$$\mathrm{Re} = \frac{U_0 L}{\nu}, \tag{9.70}$$

defined by characteristic scales of velocity and length, U_0 and, respectively, L. To this, we write

$$u = U_0\tilde{u}, \quad p = \rho_0 U_0^2 \tilde{p}, \quad \rho = \rho_0\tilde{\rho}, \quad \mathbf{x} = L\tilde{\mathbf{x}}, \quad t = \frac{U_0}{L}\tilde{t} \tag{9.71}$$

where all tilde quantities are now dimensionless, using additional scales of pressure and density, denoted by p_0 and, respectively, ρ_0. Dropping tildes, these variables satisfy the dimensionless Navier-Stokes equation

$$u_t + (\mathbf{u} \cdot \nabla)\mathbf{u} = -\frac{1}{\rho}\nabla p + \mathrm{Re}^{-1}\Delta\mathbf{u}. \tag{9.72}$$

9.4.1 Large and Small Reynolds Numbers

For large Reynolds numbers, (9.72) reduces to the inviscid *Euler equations of motion*, whereas small Reynolds numbers obtains *Stokes' equations*:

$$u_t + (\mathbf{u} \cdot \nabla)\mathbf{u} = -\frac{1}{\rho}\nabla p, \quad \nabla P = \mu \Delta u. \tag{9.73}$$

These equations may be considered for incompressible flows,

$$\nabla \cdot \mathbf{u} = 0. \tag{9.74}$$

In Euler's equation (9.73), of particular interest is the quadratic nonlinearity defined by the convective term $(\mathbf{u} \cdot \nabla)\mathbf{u}$ (Fig. 9.3). In compressible gas dynamics, including the motion of dust, it can lead to *steepening*, which is a starting point for the formation of *shocks* and associated entropy creation. Furthermore, it provides a mechanism for *period doubling* that gives rise to break-up of large to small eddy motion. In sufficiently high Reynolds number flows, it gives rise to *turbulence* by a cascade that terminates when small scale eddies dissipate into heat by viscosity.

Stokes' equation (9.73) describes balance between pressure and viscous forces. For incompressible flows (*solenoidal flows*), $\nabla \cdot \mathbf{u} = 0$, the pressure field is harmonic: $\Delta p = \mu \Delta(\nabla \cdot \mathbf{u}) = 0$. Consequently, the velocity satisfies a *biharmonic equation* and the vorticity field $\omega = \nabla \times \mathbf{u}$ is *harmonic*, i.e.:

$$\Delta\Delta\mathbf{u} = 0, \quad \Delta\omega = \mathbf{0}. \tag{9.75}$$

Exploiting incompressibility once more, we may write

$$\mathbf{u} = \nabla \times \mathbf{B} \tag{9.76}$$

for some vector field $\mathbf{B}$. Here, we may insist $\nabla \cdot \mathbf{B} = \mathbf{0}$, e.g., $\mathbf{B} = \nabla \times \mathbf{A}$ for some vector field $\mathbf{A}$.[2] The vorticity hereby expands to $\nabla \times \mathbf{u} \equiv \nabla(\nabla \cdot \mathbf{B}) - \Delta\mathbf{B} = -\Delta\mathbf{B}$. By the second equation of (9.75), it follows that $\mathbf{B}$ satisfies the biharmonic equation as well

$$\Delta\Delta\mathbf{B} = \mathbf{0}. \tag{9.77}$$

In the Navier-Stokes equation (9.68), we can alternatively focus on the effect of steepening by the convective term on the left hand side (Fig. 9.3a). In a pressureless medium such dust or traffic flow, steepening arises from particles taking over other particles (or vehicles) ahead in the presence of a negative velocity gradient along the direction of motion. (Hydronamically, the limit of dust comes about at zero pressure.) It results in the formation of a shock. In this approximation, (9.68) simplifies to *inviscid* (cf. Exercise 4.10) or *viscous* Burgers' equation, given by, respectively,

$$\mathbf{u}_t + (\mathbf{u} \cdot \nabla)\mathbf{u} = \mathbf{0}, \quad \mathbf{u}_t + (\mathbf{u} \cdot \nabla)\mathbf{u} = \nu\Delta\mathbf{u}. \tag{9.78}$$

[2]To see this, recall that (9.76) is invariant under $\mathbf{B} \to \mathbf{B} + \nabla f$ for any sufficiently smooth function f, whereby $\nabla \cdot \mathbf{B} \to \nabla \cdot \mathbf{B} + \Delta f$. The latter expression can be put to zero by a solution f to the Poisson equation $\Delta f = \rho$, $\rho = -\nabla \cdot \mathbf{B}$.

In one dimension, they reduce to

$$u_t + uu_x = 0, \quad u_t + uu_x = \nu u_{xx}. \tag{9.79}$$

Example 9.2. In Burgers' equation (9.79), steeping by the convective term uu_x gives rise to shock formation, in response to initial data $u_0(x) = u(0, x)$ that feature regions with $u_0'(x) < 0$. Consider, for instance, $u_0(x) = \sin x$ on the real line. Consider $U(t) = u(t, X(t))$ along a *characteristic* $X(t)$, satisfying the inviscid equation in (9.79),

$$U' = u_t + X'u_x = 0, \tag{9.80}$$

i.e., $U(t)$ is constant along $X'(t) = U(t)$. For small $t > 0$, tracing back along such characteristic uniquely identifies $u(t, x) = u_0(t, \xi)$. However, two characteristics meet at a finite time, t_S, whenever

$$\xi_2 + u_0(\xi_2)t_S = \xi_1 + u_0(\xi_1)t_S : \ t_S(\xi_1, \xi_2) = -\left[\frac{u(\xi_2) - u(\xi_1)}{\xi_2 - \xi_1}\right]^{-1} \tag{9.81}$$

Thus, $t_S > 0$ whenever the $u_0(\xi)$ is decreasing. Letting ξ_1, ξ_2 be arbitrarily close, the limit $\xi_1, \xi_2 \to \xi$ obtains (Exercise 4.10)

$$t_S(\xi) = -\frac{1}{u_0'(\xi)}, \tag{9.82}$$

which is positive whenever $u_0'(\xi) < 0$.

Some problems are well described by Euler's equations, Stokes' equation or Burgers' equation. In complex fluid fluids, these different regimes may co-exist, i.e., by small and large Reynolds numbers across different scales in boundary layers and macroscopic bulk motion.

Generally, these different flow regimes should be appreciated by examples, their richness being essentially countless [1], further in the presence of, e.g., interfaces between different fluids, (self-)gravity and buoyancy, shocks, heat flux, magnetic fields, ionization and radiation, suspensions in medicine and biology, chemical/nuclear reactions, curved space-time around black holes and in the early Universe—and anisotropy (liquid crystals). Modern developments increasingly derive from high-tech experiments and observations, including numerical simulations.

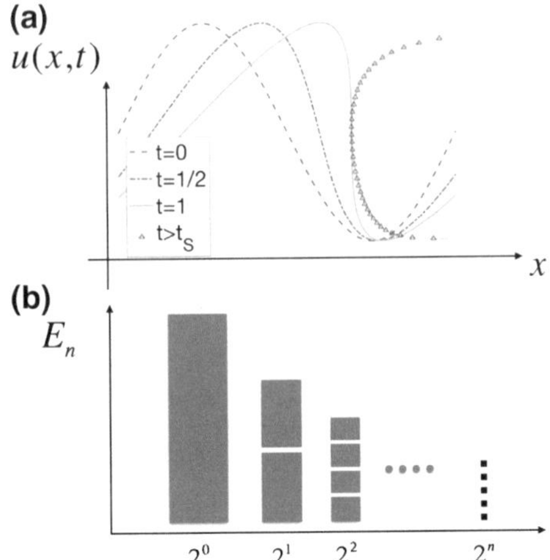

Fig. 9.3 The convective term $\mathbf{u}\cdot\nabla\mathbf{u}$ produces steepening at negative velocity gradients ($t = 0, 1/2, 1$) that may lead to shocks ($t = t_S$) (A). In two or three dimensions, it may produce break-up of large eddy fluid motion (B). The first may appear in dust (pressureless fluids) and traffic flow. Shock formation results from crossing of converging characteristics ($t = t_S$), formally leading to two-valuedness ($t > t_S$). In the second, *a cascade of period doubling in the inertial range* may give rise to a cascade of the latter that, for large n, appears as turbulence, schematically shown as energy density E_n, as a function of eddies of wavelength λ proportional to 2^{-n}. Starting from a driving agent at a macroscopic length scale L, this cascale terminates at some small length scale l at which scale viscosity becomes dominant

9.4.2 Vorticity Equation

In atmospheric problems and in describing transitions between high and low Reynolds number flows near solid boundaries, it is often useful to consider the vorticity field

$$\omega = \nabla\times\mathbf{u}. \tag{9.83}$$

Based on the Navier-Stokes equation (9.72), the ∇ and vector identities

$$\mathbf{u}\cdot\mathbf{u} = \frac{1}{2}\nabla u^2 - \mathbf{u}\times\omega,\quad \mathbf{a}\times(\mathbf{b}\times\mathbf{c}) = (\mathbf{a}\cdot\mathbf{c})\,\mathbf{b} - (\mathbf{a}\cdot\mathbf{b})\,\mathbf{c}, \tag{9.84}$$

where $u^2 = \mathbf{u}\cdot\mathbf{u}$, we can derive the *vorticity equation*

$$\frac{D\omega}{Dt} = (\omega\cdot\nabla)\mathbf{v} - (\nabla\cdot\mathbf{v})\omega + \frac{1}{\rho^2}\nabla\rho\times\nabla p + \nu\Delta\omega, \tag{9.85}$$

where the right hand side brings about variations $(\omega\cdot\nabla)\mathbf{v}$ due to velocity gradients, stretching of vorticity $(\nabla\cdot\mathbf{v})\omega$ due to compressibility (cf. the ballerina effect), a *baroclinic term* $\rho^{-2}\nabla\rho\times\nabla p$ and *diffusion of vorticity* in $\nu\Delta\omega$. For a barotropic fluid, $p = p(\rho)$, $\nabla\rho\times\nabla p = \mathbf{0}$, which also holds true for fluids with constant density. For incompressible fluids, therefore, (9.85) becomes the vorticity transport equation

$$\frac{D\omega}{Dt} = (\omega \cdot \nabla)\mathbf{u} + \nu\Delta\omega. \tag{9.86}$$

In this formulation, explicit appearance of pressure gradients is absent, even though the pressure field is generally non-zero and dynamically relevant in departures from the Stokes flow approximation. For incompressible fluids, however, these pressure gradients do not affect the vorticity distribution.

Example 9.3. For the laminar flow between two plates, as in Fig. 9.2 or Example 9.1., the velocity $u(t, y)\mathbf{i}$ is uniform along the x-direction and the vorticity $\omega(t, y)\mathbf{k}$ is orthogonal along the z-direction. Consequently, both $(\omega \cdot \nabla)\mathbf{u}$ and $(\mathbf{u} \cdot \nabla)\omega = 0$ vanish, whereby (9.86) reduces to the heat equation (4.52) in one dimension,

$$\omega_t = \nu\omega_{yy}. \tag{9.87}$$

Example 9.4. Consider two concentric cylinders or radius $0 < R < R+L$ with a small gap size $L \ll R$. The inner cylinder rotates with angular velocity Ω and the outer cylinder is kept fixed. The gap is filled with a fluid with dynamical viscosity μ and density ρ. The velocity $U = \Omega R$ defines a Reynolds number $\mathrm{Re} = UL/\nu$, $\nu = \mu/\rho$. The torque on the cylinders as a function of ω arising from velocity shear. In a corotating frame of reference, outer cylinder recedes at velocity $V = L\Omega$. When Re is small, the flow is laminar, and the radial velocity gradient $V/L = \Omega$ gives a shear force $f = \mu\Omega$ and torque $fR = \mu\Omega R$ per unit area.

9.4.3 Jeans Instability in Linearized Euler's Equations

Linearizing Euler's equations defines the evolution of small perturbations about a given fluid state. This appears, for instance, in the problem of stars formation in over-dense regions in molecular clouds (Fig. 9.4). This occurs by gravitational collapse in sufficiently cold regions, when gravity wins over thermal pressure. The overall rate of star formation can further be accelerated when, e.g., stellar winds from new-born massive stars compress gas in their surroundings provided that this process is sufficiently cooled by radiation. There may further be a dual role for turbulence, promoting and inhibiting star formation in overdense regions [2].

Fig. 9.4 Shown is a *Hubble Space Telescope* image of N11B in the LMC showing massive stars (*blue* and *white*) on the *left* with additional star formation in the center and on the *right*, indicative of a sequential star formation process. (http://hubblesite.org/newscenter/archive/releases/2004/22/image/a/.)

Without gravity, a local analysis for small amplitude variations will recover sound waves. What, then, is the impact of self-gravity? By dimensional analysis, we anticipate that gravity will be important at low frequency sound waves at angular frequencies below

$$\omega = \sqrt{4\pi G\rho} \tag{9.88}$$

for a mean density ρ in the cloud in view of the Poisson equation for the Newtonian gravitational potential Φ,

$$\Delta\Phi = 4\pi G\rho, \tag{9.89}$$

where G denotes Newton's constant. Below this frequency, gravity wins over thermal pressure, thus setting a critical *Jeans length* defined by the wave length

$$\lambda > c_s\sqrt{\frac{\pi}{G\rho}}, \tag{9.90}$$

where $c_s = \sqrt{\partial P/\partial\rho}$ denotes the sound speed, using $\omega\lambda = 2\pi c_s$. Relatively small mass elements are stable by thermal pressure, whereas relatively large masses are unstable to collapse by self-gravity.

Working on a large scale, these results obtain in the approximation of large Reynolds numbers, using Euler's equations of motion for a compressible fluid,

$$\partial_t u + u\partial_x u = -\frac{1}{\rho}\frac{\partial P}{\partial x} - \frac{1}{\rho}\frac{\partial\Phi}{\partial x}, \quad \partial_t\rho + \partial_x(\rho u) = 0, \quad \frac{\partial^2\Phi}{\partial x^2} = 4\pi G\rho. \tag{9.91}$$

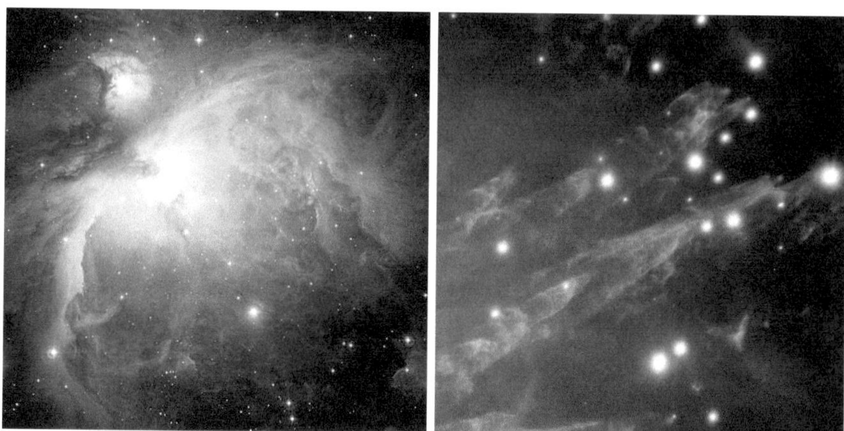

Fig. 9.5 (*Left*) Composite image of the Orion Nebula (M42, NGC 1976): a nearby region of massive star formation ($M \simeq 2000 M_\odot$, $D = 1344 \pm 20$ ly, 85×60 arcmin) captured by the Wide Field Imager on the MPG/ESO 2.2-m telescope at the La Silla Observatory, Chile. (*red*: *red* light, hydrogen gas; *green*: *yellow-green* light; *blue*: *blue* light; *purple*: UV-light.) (ESO/Igor Chekalin (2011)). (*Right*) An IR composite image of the Orion "bullets" (*blue*: Fe II; *orange*: H2 1-0in the wakes), representing 50 arcs across and structure on 0.1 arcs (2 pixel). (Gemini Observatory, 2007)

Linearization to small perturbations about a static (non-moving) background state, we may drop all higher order terms, leaving

$$\partial_t u = -\frac{1}{\rho}\frac{\partial p}{\partial x} - \frac{\partial \delta\Phi}{\partial x}, \quad \partial_t \delta\rho + \rho\partial_x u = 0, \quad \frac{\partial^2 \delta\Phi}{\partial x^2} = 4\pi G\delta\rho, \tag{9.92}$$

where $p \equiv \delta P$, $\delta P = c_s^2 \delta\rho$. The latter two reduce to

$$c_s^{-2}\frac{\partial p}{\partial t} = -\rho\partial_x u, \quad \frac{\partial^2 \delta\Phi}{\partial x^2} = 4\pi G c_s^{-2} p. \tag{9.93}$$

Equivalently, we have the second order wave equation

$$p_{tt} - c_s^2 p_{xx} = 4\pi G\rho p \tag{9.94}$$

with dispersion relation

$$\omega^2 = c_s^2 k^2 - 4\pi G\rho \tag{9.95}$$

for frequencies ω as a function of wave number $k = 2\pi/\lambda$. Evidently, ω is real provided that k is sufficiently large, and ω becomes complex when k satisfies Jeans' instability criterion (9.90).

Star forming regions in molecular clouds must therefore be sufficiently large in linear size and/or overdense to be susceptable to gravitational collapse. A well-studied example is the Orion nebula (e.g., [3]).

Example 9.5. At a distance of $D = 1344$ ly = 411 pc (1 pc = 3.09×10^{18} cm; 1 ly = 0.3062 pc) and angular size of about $\theta = 70$ arcmin = 0.02 rad (1 arcs = $(2\pi/360)/3600 = 4.8 \times 10^{-6}$ rad), the Orion Nebula (Fig. 9.5) has a radius of about

$$R \simeq D\theta = 1.25 \times 10^{19}\ \mathrm{cm}. \tag{9.96}$$

With a mass $M \simeq 2000 M_\odot$, its mean density is about

$$\rho = \frac{3M}{4\pi R^3} = 5.8 \times 10^{-23}\ \mathrm{g\,cm^{-3}} \simeq 35\, m_p\, \mathrm{cm}^{-3}, \tag{9.97}$$

where $m_p = 1.67 \times 10^{-24}$ g is the mass of the proton. Thus,

$$\omega = \sqrt{4\pi G\rho} \simeq 7 \times 10^{-15}\ \mathrm{s^{-1}} \simeq 2 \times 10^{-7}\ \mathrm{yr^{-1}} \tag{9.98}$$

points to a potentially short collapse time on the order of a few Myr. It contains numerous HII regions excited by shock heating, rendering its temperature (and density) strongly non-uniform in the range of 0.01–1eV (1eV = 11,000 K). Since shocks are highly localized, the mean (isothermal) sound speed, $c_s = \sqrt{k_B T/m_p}$ is expected to be determined largely by the lower end of the temperature range,

$$c_s = 3 \times 10^4 \left(\frac{\bar{T}}{10\,\mathrm{K}}\right). \tag{9.99}$$

For the fiducial mean $\bar{T} = 10$ K, our Jeans' criterium (9.90) becomes

$$\lambda > 2.6 \times 10^{19}\ \mathrm{cm} \tag{9.100}$$

which is about the diameter $2R$ above. Given the relatively short time scale of gravitational collapse (9.98), this suggests that the Orion nebula is in a state of persistent, self-regulated star formation, driven by gravitational collapse and inhibited by heating from stars and shocks.

9.5 Kolmogorov Scaling of Homogeneous Turbulence

With sufficient energy input, intermittent and time-variable flows can develop that are quite common, e.g., water flows in rivers, in air flow around buildings, flows in ducts and pipes, and in outflows from astrophysical systems. These high Reynolds number flows can also be found, perhaps unwittingly, around a cyclist on a fast track, around planes and automobiles and in propulsion systems. The first scientific recognition of turbulence is due to Leonardo da Vinci (1452–1519) (e.g., [4]).

Kolmogorov realized that the observed *turbulent flow* represents a response of fluids to a relatively strong energy input at a scale L, that is much larger than a scale l at which viscous dissipation is taking place [5, 6]. In our definition (9.57) for the dynamical viscosity, dissipation is immediate in case of a Stokes' flow, wherein the scales of energy input and dissipative flow structures are equal. Low Reynolds numbers, encountered on small scales, hereby offer a suitable site for dissipation into heat. If the energy input occurs on relatively large scales, however, energy is manifest mostly in vortical motions that are essentially non-dissipative, *except* for those at the smallest scales. In stationary turbulence, then, how is the connection established between energy input at some large scale and energy dissipation a much smaller scale?

The Lagrangian derivative in the Navier-Stokes equation includes the nonlinear convective term $(\mathbf{u}\cdot\nabla)\mathbf{u}$. At high velocities, this term provides a self-interaction whereby large scale structures can transform to small scale structures with no loss in energy, first noticed by Lewis F. Richardson. This is most readily expressed in Fourier space, where it can be seen to produce period doubling, that is canonical for quadratic nonlinearities. In turbulent motion, this process has been recognized as an *energy cascade from large scale eddies to small scale eddies.*[3]

The cascade terminates with small eddies approaching low Reynolds number flows, i.e., in the Stokes regime set by a small scale l, where viscous effects dissipate energy of small eddies into heat.

By the qualitative description above, we are led to consider a continuum of scales λ covering the *inertial range* of turbulent eddies, $l < \lambda < L$, i.e.,

$$\text{Re}_l < \text{Re} < \text{Re}_L \quad (k_{min} < k < k_{max}) \tag{9.101}$$

where $\text{Re}_l = Ul/\nu$, $\text{Re}_L = UL/\nu$, given a fluid velocity U on the macroscopic scale L.

If $\epsilon(k)$, $k = 2\pi/\lambda$, denotes the energy conversion rate at wave number k associated with the cascade process of continuously creating smaller eddies ($k' > k$) out of larger eddies ($k'' < k$), then (Fig. 9.3b)

[3]"... The small eddies are almost numberless, and large things are rotated only by large eddies and not by small ones, and small things are turned by both small eddies and large", *Leonardo Da Vinci,* observing water flow past rocks in river beds.

$$\epsilon(k) = \epsilon_0, \quad k_{min} \leq k \leq k_{max}, \quad k_{min} = \frac{2\pi}{L}, \quad k_{max} = \frac{2\pi}{l}. \tag{9.102}$$

Here, $\epsilon(k_{max})$ refers to viscous energy dissipation into heat at the scale of small eddies.

Thus, Kolmogorov realized that ϵ_0 is the key parameter controlling turbulence, provided that it is large enough for the onset of turbulence. For homogenous, isotropic and stationary turbulence, therefore, a proper starting point is ϵ_0 in (9.102) in terms of the energy dissipation per unit mass, that corresponds to the energy dissipation $\dot{q}$ per unit volume, encountered earlier, divided by the mass density.

Following a Fourier analysis, we may consider the spectral energy density $E(k)$ describing the total energy per unit mass of eddies per wave number k. By the Plancherel formula,

$$\frac{1}{2}\mathbf{u}^2 = \int_0^\infty E(k)dk. \tag{9.103}$$

By dimensional analysis, we have

$$[\epsilon] = \frac{\dot{q}}{\rho} = \mathrm{cm}^2\ \mathrm{s}^{-2} \times \frac{1}{\mathrm{s}}, \quad [E(k)] = \mathrm{cm}^2\mathrm{s}^{-2}\left[\mathrm{k}^{-1}\right] = \mathrm{cm}^3\mathrm{s}^{-2}. \tag{9.104}$$

If $E(k)$ is some function of ϵ and k in the inertial range satisfying $\epsilon = \epsilon_0$, then

$$E(k) = C\epsilon^{\frac{2}{3}}k^b = C\epsilon^{\frac{2}{3}}k^{-\frac{5}{3}} \tag{9.105}$$

by selecting $b = -5/3$ for consistency with (9.104). The Kolmogorov spectrum

$$E(k) \propto k^{-\frac{5}{3}} \ (k_{min} < k < k_{max}) \tag{9.106}$$

now follows by Kolmogorov's insight of a conservative cascade (9.102) in the inertial range above the smallest scale, at which time dissipation sets in.

The power of Kolmogorov scaling (9.106) is in its universality, in high Reynolds number flows in fluids and gases. It appears also, for instance, in broadband temporal fluctuations in gamma-ray rays, believed to originate in ultra-relativistic outflows such as cosmological gamma-ray bursts (Fig. 9.6).

In addition to experiments,[4] studies of (9.106) become increasing accessible to detailed Direct Numerical Simulations (DNS) using the Navier-Stokes equations in various numerical implementations (e.g., [9]).

The Kolmogorov spectrum (9.106) leads to a *two-point correlation function* that, in light of the assumed homogeneity, isotropy and stationarity, simplifies to

$$Q(\xi) = \langle \mathbf{u}(\mathbf{r} + {}_{\text{,}}, t) \cdot \mathbf{u}(\mathbf{r}, t)\rangle = 2\int_0^\infty \frac{\sin k\xi}{k\xi} E(k)dk \tag{9.107}$$

[4]E.g., Fig. 6 in [4], see further [7, 8, 14].

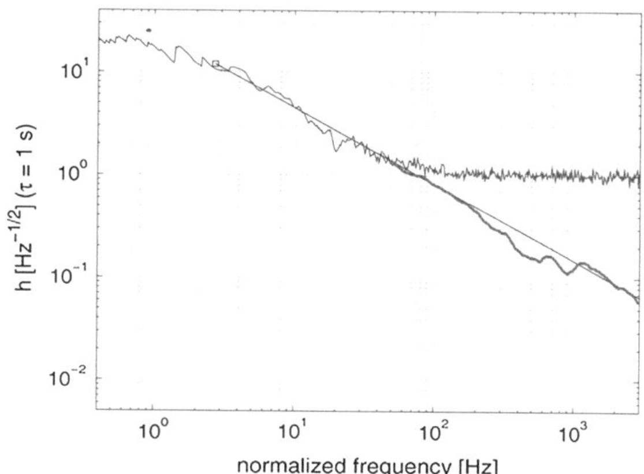

Fig. 9.6 Broadband Kolmogorov spectrum in the prompt emission of bright long gamma-ray bursts in the catalogue of *BeppoSAX*, shown transformed to the comoving frame and identified by application of butterfly filtering of Sect. 6.10. (Reprinted from van Putten, M.H.P.M., Guidorzi, C., & Frontera, F. 2014, ApJ, 786, 146

that is amenable to experimental observations in turbulent flows in the laboratory. Following (9.105), we have

$$Q(\xi) = C\epsilon^{\frac{2}{3}}\xi^{\frac{2}{3}}. \tag{9.108}$$

This relation has been experimentally measured.[5]

The cascade of large to small eddy motion between the limits $k_{min} < k < k_{max}$ entails a finite number of degrees of freedom. To estimate this, we note

$$\epsilon(k_{max}) = \epsilon(k_{min}) = U^2 \times \frac{U}{L} = \frac{U^3}{L}. \tag{9.109}$$

At k_{max}, energy input is converted into heat by molecular viscosity. Complementary to (9.104), $\epsilon = \nu^3 l^{-4}$ since $[\nu] = \mathrm{cm}^2\, s^{-1}$. With $\mathrm{Re}_L = UL/\nu$, it follows that

$$l = \left(\frac{\nu^3}{\epsilon_0}\right)^{\frac{1}{4}} : \ \frac{L}{l} = \mathrm{Re}_L^{\frac{3}{4}}. \tag{9.110}$$

The effective number of degrees of freedom, therefore, scales with $\mathrm{Re}_L^{\frac{9}{4}}$ with an equivalent number of period doublings

$$n = \frac{9}{4}\log_2 \mathrm{Re} \tag{9.111}$$

as illustrated in Fig. 9.3. In fully developed turbulent flows, it can be quite large when $\mathrm{Re}_L \gg 10^3$.

[5]See e.g., Fig. 14 in [10], see further [11].

Table 9.1 Perturbation theory and dimensional scaling

1. For problems with a small parameter ϵ, solutions often can be expanded in a series in ϵ. *Regular* perturbations allow Taylor series expansions in ϵ. *Singular* perturbation do not, and are commonly associated with a change in the order of the problem at hand.
2. Two-timing methods appear in naturally in perturbed harmonic oscillations, e.g., by dissipation or a nonlinearity.
3. The Reynolds number in the Navier-Stokes equation parameterizes the relative importance of convective to diffusive momentum transport. Large Reynolds numbers define the inviscid limit, given by Euler's equations. Small Reynolds numbers define the limit of lubrication theory, given by Stokes equations, giving rise to laminar flow.
4. At large Reynolds numbers, the convective term may produce steeping and/or period doubling that may give rise to, respectively, shock formation and free turbulent motion satisfies Kolmogorov scaling in the inertial range, that terminates in dissipation on small scales with low Reynolds numbers in small eddy motion.

Table 9.1 summarizes this discussion.

9.6 Exercises

9.1. Show that for the damped pendulum equation (9.17), the addition $\omega_1\epsilon$ is suppressed in the second equation of (9.21).

9.2. Consider the *Duffing equation*

$$\ddot{x} + \omega^2 x + \epsilon x^3 = 0. \tag{9.112}$$

Use two-timing (9.19) to derive the nonlinear amplitude-frequency coupling

$$x_0(t_*, \tau) = A_0 \cos\left(\omega t + \frac{3}{8}A^2\epsilon t\right). \tag{9.113}$$

9.3. The gravitational radius (9.42) defines the Schwarzschild radius $R_S = 2R_g$ of the black hole with the mass M after complete gravitational collapse. The Compton wave number $k = Mc/\hbar$, where $\hbar$ denotes the Planck constant, introduces another length scale $\hbar/Mc$. The Planck mass M_p denotes the minimal mass of a black hole. Derive M_p.

9.4. In light of (9.66), consider the second equation in (9.73) in cylindrical geometry with coordinates (r, ϕ), e.g., the flow of coca-cola through a straw, back into a cup. Derive the Poisseuille flow profile $u = u(r)$ in this case as a function of the gravitational acceleration g.

9.5. Derive (9.85) from (9.84). For (9.2), set up the initial value problem to describe the diffusion of vorticity based on (9.86), analogous to the heat equation (4.52).

9.6. Large eddy motions are subject to finite sized effects, e.g., imposed by solid boundaries of a container or the thickness of accretion flows in disks around a central star or compact object. Does a finite size affect an otherwise Kolmogorov spectrum and, if so, how?

9.7. For Euler's experiment dropping a sphere into a bottle of viscous fluid, derive the terminal velocity. Take into account all three forces: the gravitational force downwards, the viscous force upwards and the buoyancy force due to a discrepancy between the density of the sphere and the fluid.

9.8. Reynolds numbers larger than a few thousand, the flow past a sphere becomes turbulent due to instabilities. The resulting *drag force* $F = \frac{1}{2}C_D\rho v^2 A$ is described by a dimensionless *drag coefficient* C_D, where $A = \pi R^2$ denotes the cross-sectional area of a sphere of radius R. $C_D = 0.47$ for a fully turbulent flow. Describe the consequences for timing behavior of the pendulum, starting with a relatively large θ_0 for which the flow past m is turbulent.

9.9. Consider the *forced* pendulum equation

$$\ddot{x} + \alpha\dot{x} + \omega_0^2 x = A\cos(\omega t). \tag{9.114}$$

By Fourier analysis, derive the exact expressions for the amplitude $a = a(\omega/\omega_0)$ and phase lag $\varphi = \varphi(\omega/\omega_0)$ in the response $x(t) = a\cos(\omega t - \varphi)$. Identify the *inertial* and *viscous* limits in the amplitude and phase, corresponding to, respectively $\omega/\omega_0 \gg 1$ and $\omega/\omega_0 \ll 1$. Next, consider the reduced equations

$$\ddot{x} + \alpha\dot{x} = A\cos(\omega t), \quad \alpha\dot{x} + \omega_0^2 x = A\cos(\omega t) \tag{9.115}$$

and discuss their solutions. What are the corresponding limits in (9.114)?

9.10. In our one-body formulation (9.32 and 9.33), points of maximal (*perihelion*) or minimal (*aphelion*) distance satisfy $v = r\dot{\theta} = uj$ with kinetic energy

$$E_k = \frac{1}{2}v^2 = \frac{1}{2}j^2u^2. \tag{9.116}$$

Since H is conserved, H can be evaluated at these *turning points* $\varphi = 0, \pi$

$$H = \mu\left(\frac{1}{2}j^2u_\pm^2 - GMu_\pm\right) = -\frac{\mu G^2M^2}{2j^2}(1-e^2) = -\frac{GM\mu}{2a}, \tag{9.117}$$

using short hands $u_+ = u(0)$ and $u_- = u(\pi)$, giving rise to the *Vis-viva equation*

$$v^2 = 2\mu^{-1}H + \frac{2GM}{r} = GM\left(\frac{2}{r} - \frac{1}{a}\right). \tag{9.118}$$

Derive the orbital average

$$2\bar{E}_k + \bar{U} = \frac{GMe^2}{a(1-e^2)} \geq 0. \tag{9.119}$$

For bound orbits, the orbital average (9.119) is non-negative. When it vanishes, we say the binary is *virialized*. Consider $H = H(e, J)$ for constant J. Show that keeping $j^2 = GMa(1-e^2)$ fixed, H decreases as e becomes small: circular (virialized) orbit is a minimum energy state for a given angular momentum, i.e., orbits may circularize by angular momentum preserving dissipative interactions.[6]

9.11. Consider a modified Newtonian potential

$$U = -\frac{M}{r} + A\ln\left(\frac{r}{r_0}\right), \tag{9.120}$$

where the new term expresses the presence of dark matter and/or a modification of Newton's theory in the limit of weak gravitational attraction (small accelerations on the scale of cH_0 or less, where c is the velocity of light and H_0 is the Hubble parameter.) Following Sect. 9.3, apply perturbation theory to derive precessional motion in (9.120).

9.12. Consider the Jeans' criterion for gravitational collapse in spherical geometry. Show that (9.88) is the frequency of radial oscillations in a stratified atmosphere, assuming constant background temperature. Derive (9.90) by equating the thermal

[6] A number of binaries of compact objects show a residual ellipticity away from zero, even though the time scale of circularization would predict essentially circular orbits. The residual ellipticity may be understood to result from the dissipation-fluctuation theorem applied to tidal interactions [12, 13].

and gravitational energies within a radius λ. Derive (9.90) by equating the sound speed to the free fall time from a radius λ. Calculate the Jeans length for the ISM, e.g., $c_s = 10$ km s^{-1}, $\rho = 10^{-24}$ g cm^{-3}. Express the Jeans length as a function of c_s and the mass density relative to these fiducial values.

References

1. Van Dyke, M., 1982, *An Album of Fluid Motion* (Parabolic Press).
2. Shetty, R., & Ostriker, E.C., 2012, ApJ, 754, 2; McKee, C.F., & Ostriker, E.C., 2007, ARAA, 45, 565.
3. Colgan, S.W.J., Schultz, A.S.B., Kaufman, M.J., Erickson, E.F., & Hollenbach, D.J., 2007, ApJ, 671, 536.
4. Ecke, R., 2005, "The Turbulence Problem," Los Alamos Science, 29, 124.
5. Kolmogorov, A.N., 1941, Dokl. Akad. Nauk SSSR, 30(4); *trans.* 1991, Proc. Roy. Soc. London A., 434, 9.
6. Kolmogorov, A. N., 1941, Compt. Rend. Acad. Sci. USSR, 30, 301; ibid. Compt. Rend. Acad. Sci. USSR, 32, 16.
7. Champagne, F. H., 1978, J. Fluid Mech. 86, 67; Grant, L., Stewardt, R., & Moillieta, W., 1962, J. Fluid Mech. 12, 241.
8. Saddouchi, S., & Veeravalli, S.V., 1994, J. Fluid Mech., 268, 333.
9. Foias, C., Holm, D.D., & Titi, E.S.., 2001, Physica D, 152–153, 505.
10. Neumann, M., et al., 2009, Flow Meas. Instrum., 20, 252.
11. Benedict, L.H., Gould, R.D., 1999, Exp. Fluids, 26, 3818.
12. Phinney, E.S., 1992, Phil. Trans. R. Soc. Lond. A, 341, 39.
13. Lanza, A., & Rodonò, M., 2001, A&A, 376, 165.
14. Grant, L., Stewardt, R., & Moillieta, W., 1962, J. Fluid Mech., 12, 241.

Part III
Selected Topics

Chapter 10
Thermodynamics of N-body Systems

We next discuss some elements self-gravitating many body systems.[1] The equations of motion satisfy an action principle, that facilitates the set up of initial value problems for the evolution of systems with arbitrarily many particles, as a system of equations that is second order (Euler-Lagrange) or first order (Hamiltonian) in time, assuming no dissipation. When the number of particles becomes large, the resulting systems do not permit direct numerical integration and their behavior is to be interpreted in terms of statistical properties. This has motivated the design of various numerical methods of integration to deal with realistic numbers of stars and galaxies in astronomy and cosmology. For dense stellar clusters, weak gravitational scattering tends to be sufficient to establish near thermal equilibrium, however. In this event, a thermodynamic approach can provide an effective description.

10.1 The Action Principle

Classical mechanics appears to derive from magic: the classical trajectories of particles satisfy the action principle

$$\delta S = 0, \quad S = \int_0^T L(t)dt, \quad L(t) = L(x(t), \dot{x}(t)), \tag{10.1}$$

[1] We shall assume gravitation according to Newton's law. For astronomical systems, however, this may entail weak gravitational accelerations on the scale of or less than the cosmic scale cH_0 of the velocity of light c times the Hubble parameter H_0. In this regime, Newton's law has not yet been established by first principle laboratory experiments.

M.H.P.M. van Putten, *Introduction to Methods of Approximation in Physics and Astronomy*, Undergraduate Lecture Notes in Physics,
DOI 10.1007/978-981-10-2932-5_10

where L refers to the Lagrangian, that leads to the Euler-Lagrange equations (cf. Exercise 5.10)

$$\frac{d}{dt}\frac{\partial L}{\partial \dot{x}} - \frac{\partial L}{\partial x} = 0, \tag{10.2}$$

which defines a $2nd$ order ordinary differential equation. We obtain an initial value problem with

$$x(0) = x_0, \quad \dot{x}(0) = x_1, \tag{10.3}$$

describing the initial coordinate positions and coordinate velocities of the particle.

Example 10.1. $L = E_k - U$ is the difference of kinetic energy and potential energy. For a pendulum, we have

$$L = \frac{1}{2}ml^2\dot{\theta}^2 - mgl(1 - \cos\theta) \tag{10.4}$$

and (10.2) becomes

$$\ddot{\theta} + \omega^2 \sin\theta = 0, \quad \omega^2 = \frac{g}{l}. \tag{10.5}$$

It describes the *nonlinear pendulum equation* in view of the sine function. By a Taylor series expansion about $\theta = 0$, $\sin\theta = \theta - \frac{1}{6}\theta^3 + O(\theta^5)$, we have

$$\ddot{\theta} + \omega^2\theta = \frac{\omega^2}{6}\theta^3 + O(\theta^5). \tag{10.6}$$

Noether's theorem shows conserved quantities as the result of symmetries in L. If

$$\frac{\partial L}{\partial x_i} = 0 \tag{10.7}$$

for some i, then

$$\frac{d}{dt}\frac{\partial L}{\partial \dot{x}_i} = 0: \quad \frac{\partial L}{\partial \dot{x}_i} = \text{const.} \tag{10.8}$$

If x_i is ignorable—a symmetry of the problem—it comes with a conserved quantity. In (10.8), this quantity is momentum.

10.2 Momentum in Euler-Lagrange Equations

The second order equations of motion for a point particle m with kinetic energy E_k moving in an external potential U derives from $L = E_k - U$ by (10.2). We define the momentum

$$p = \frac{\partial L}{\partial x} \tag{10.9}$$

and obtain and equation for its time rate-of-change

$$\frac{d}{dt}p = -\frac{\partial U}{\partial x} \tag{10.10}$$

resulting from a force $F = -\partial U/\partial x$.

Example 10.2. Consider again Newton's problem of an apple falling from a tree in the Earth's gravitational acceleration g. The kinetic and potential energies satisfy

$$E_k = \frac{1}{2}mv^2, \;\; U = mgh. \tag{10.11}$$

With motion and force is along the $z-$axis with unit normal $\mathbf{i}_z$,

$$\mathbf{v} = \frac{dh}{dt}\mathbf{i}_z, \;\; E_k = \frac{1}{2}m\dot{h}^2, \;\; \frac{\partial U}{\partial z} = mg\mathbf{i}_z. \tag{10.12}$$

In this event, our coordinate x in (10.9–10.10) is the height h, giving a momentum and the familiar result

$$p = \frac{\partial L}{\partial \dot{h}} = m\dot{h}, \;\; m\ddot{h} = -mg. \tag{10.13}$$

Taking the apple far out into space, g approaches zero and L reduces to $L = E_k$. It follows immediately from (10.10) that

$$\frac{d}{dt}p = 0: \;\; p(t) = p_0, \tag{10.14}$$

showing that momentum is constant. This is a general property. Whenever L does not depend on a coordinate, the associated momentum is conserved.

Example 10.3. Consider (10.10) for Kepler's problem of planetary motion in circular orbits expressed in polar coordinates (r, θ) about the Sun of mass M. In the approximation of negligible radial motion, the kinetic energy reduces to

$$E_k = \frac{1}{2}mr^2\dot{\theta}^2, \tag{10.15}$$

giving

$$L = \frac{1}{2}mr^2\dot{\theta}^2 + \frac{GM}{r}. \tag{10.16}$$

Taking our coordinate x in (10.9–10.10) to be θ, we obtain

$$p_\theta = \frac{\partial L}{\partial \dot{\theta}} = mr^2\dot{\theta}, \quad \frac{d}{dt}p_\theta = 0, \tag{10.17}$$

since $U = -GM/r$ is independent of θ. The momentum p_θ is recognized to be *angular momentum* $J = mj$, where j denotes the specific angular momentum

$$j = r^2\dot{\theta} = r^2\frac{2\pi}{P} = 2\frac{dA}{dt} \tag{10.18}$$

for an orbit with period P. By (10.17), conservation of p_θ, j is constant and orbital motion traces out a constant rate of change of surface area.

10.3 Legendre Transformation

The Euler-Lagrange equations of motion represent a second order (system of) differential equation(s). A common procedure for casting it in first order form is by a transformation of variables

$$(x, \dot{x}) \rightarrow (q, p), \quad p = \frac{\partial L}{\partial \dot{x}}. \tag{10.19}$$

Here, $\partial L/\partial \dot{x}$ serves as a new independent variable in passing from $\dot{x}$ to p. To illustrate this step, let us turn to functions of one variable.

If $f(x)$ is convex, i.e., $f''(x) > 0$, then $f'(x)$ is monotonically increasing whereby it is invertible, i.e.:

$$s(x) = f'(x) \tag{10.20}$$

defines a 1-1 map between x and s. We are therefore at liberty to consider $f(x) = f(x(s)) \equiv f^*(s)$. Consider further

$$g(s) = sx - f(x), \tag{10.21}$$

where $x = x(s)$. It is instructive to write (10.21) in symmetric form (e.g., [1])

$$g(s) + f(x) = sx, \tag{10.22}$$

where we are at liberty to consider $s = s(x)$ or $x = x(s)$. With (10.20) in hand, differentiation of (10.22) with respect to s gives its counter part

$$x(s) = g'(s) \tag{10.23}$$

and differentiation twice, also with respect to x, gives the slopes

$$\frac{ds}{dx} = f''(x), \quad \frac{dx}{ds} = g''(s), \tag{10.24}$$

the product of which equals 1, following our assumption that $f''(x)$ is nonzero.

10.4 Hamiltonian Formulation

Given $L = L(q, \dot{q})$ (following $q = x$), the Hamiltonian formulation is a first order system obtained following a Legendre transform, given by

$$H = \dot{q}p - L(q, \dot{q}), \quad p \equiv \frac{\partial L}{\partial \dot{q}}. \tag{10.25}$$

We note a total variation

$$dH = pd\dot{q} + \dot{q}dp - \frac{\partial L}{\partial q}dq - \frac{\partial L}{\partial \dot{q}}d\dot{q} = \left(p - \frac{\partial L}{\partial \dot{q}}\right)d\dot{q} + \dot{q}dp - \frac{\partial L}{\partial q}dq. \tag{10.26}$$

The first term on the right hand side vanishes by our definition of p in (10.25). For solutions satisfying the Euler-Lagrange equations, we further have

$$\frac{\partial L}{\partial q} = \frac{d}{dt}\frac{\partial L}{\partial \dot{q}} = \dot{p}. \tag{10.27}$$

Consequently, (10.26) reduces to

$$dH = \dot{q}dp - \dot{p}dq. \tag{10.28}$$

showing that

$$H = H(q, p). \tag{10.29}$$

The Legendre transformation (10.25) hereby obtains a reformulation in the canonical variables (q, p) of position and momentum, instead of coordinate position and velocity $(q, \dot{q})$. Moreover, (10.28) gives the Hamiltonian system of equations

$$\dot{p} = -\frac{\partial H}{\partial q}, \quad \dot{q} = \frac{\partial H}{\partial p}, \tag{10.30}$$

that define an initial value problem specified by initial data $p(0) = p_0, q(0) = q_0$.

An extension to N particles is straightforward:

$$\dot{p}_{ni} = -\frac{\partial H}{\partial q_{ni}}, \quad \dot{q}_{ni} = \frac{\partial H}{\partial p_{ni}} \quad (n = 1, 2, \ldots N, i = x, y, z) \tag{10.31}$$

with associated initial positions and momenta for each of the particles at $t = 0$. This is a system of $6N$ equations.

10.5 Globular Clusters

To examplify Hamiltonian systems, we consider the N-body problem of globular clusters comprising dense systems of $N \sim 10^4 - 10^6$ stars. There are some 140 of them around our Milky Way, some listed in Table 10.1. If m_i denotes the mass of each star, the Hamiltonian is given by the *sum* of the kinetic and gravitational binding energies,

$$H = \Sigma_{i=1}^{N} \frac{p_i^2}{2m_i} - \Sigma_{i>j} \frac{Gm_i m_j}{r_{ij}}, \quad r_{ij} = \left| \mathbf{r}_j - \mathbf{r}_i \right|. \tag{10.32}$$

The ages Milky Way globular clusters approach the age of the Universe. They are *open self-gravitating* systems with the peculiar property of having *negative specific heat*. Some energy is emitted into the galactic halo by escaping high-velocity stars, in response to which the globular cluster contracts conform the Virial Theorem (Fig. 10.1).

A remarkable feature of escaping stars is the formation of tidal streams, apparent in a number of globular clusters. Because these systems are open, they gradually evaporate by escaping high velocity stars, perhaps about one per 10^5 years. The resulting lifetime as governed by the *Ambartsumian-Spitzer* and *Kelvin-Helmholtz* time scales, to be explored below.

Table 10.1 Data on selected globular clusters derived from the catalogue of Harris, W.E., 1996. AJ 112, 1487 (revised 2010)

Name[a]	$D^{\rm b}$ [kpc]	M_c [$10^6 M_\odot$]	$r_t^{\rm c}$ [pc]	t_{rh} [Gyr]	r_h [pc]	σ [km/s]
NGC 104[1]	4.5(7.4)	1.45	64	3.5	4.1	16
Pal 4[1]	109(111)	0.0541	100	2.6	16	1.6
Pal 5[1]	23.2(18.6)	0.0284	101	6.6	18	1.1
Pal 14[1]	73.0(71.6)	0.0200	118	10	26	0.8
NGC 5139[1]	5.2(6.4)	2.64	104	12	7.6	16
NGC 5904[1]	7.5(6.2)	0.834	52	2.5	3.9	13
NGC 4590[1]	10.3(10.2)	0.306	89	1.9	4.5	7.3
NGC 5466[1]	16.0(16.3)	0.133	162	5.7	11	3.1
NGC 6254[1]	4.4(4.6)	0.225	30	0.8	2.5	8.4
NGC 6121[1]	1.7[d](5.9)	0.225	19	0.9	2.1	9.0
NGC 6752[2]	4.0(5.2)	0.364	52	0.7	2.2	11
NGC 7078[2]	10.4(10.4)	0.984	62	2.1	3.0	16

[1]King morphology, [2]Post Core-Collapse morphology [2]
[a]NGC 104 = 47 Tuc, NGC 5139 = ω Centauri, NGC 5904 = M5, NGC 4590 = M68, NGC 6121 = M4, NGC 6254 = M10, NGC 7078 = M15
[b]Distances to the Sun (galactic center) with uncertainties of 6% [3]
[c]For the old clusters of the Milky Way, r_t is much larger than the radius of the cluster
[d][4]

The dynamics of globular clusters is described by (10.32). Out of this mathematical expression for the total energy comes to life with the pristine beauty[2] M80 and NGC 6752 in Fig. 10.2.

As gravitationally bound open systems, globular clusters have $H < 0$ satisfying the Virial Theorem,

$$2\bar{E}_k + \bar{U} = 0, \tag{10.33}$$

where $\bar{E}_k$ and $\bar{U}$ denote the mean kinetic and gravitational binding energy of the stars. *Virialization* tends to be extremely fast (Fig. 10.1).

Small angle gravitational scattering between the stars tends to preserve *thermal equilibrium*, subject to occasional ejection of stars into the galactic halo and a finite heat flux outwards from excess heat in the core. The escapers present a singular perturbation away from perfect thermodynamic equilibrium. In the large N limit, they have the following properties.

1. *Negative specific heat.* Escapers have a total energy $h > 0$ and deposit h in kinetic energy in the halo of the host galaxy. By conservation of total energy, $\delta H = -h < 0$. By the Virial Theorem, $H = \frac{1}{2}N\bar{U}$, so that $\delta U = -2N^{-1}h < 0$.

[2]"If one cannot see gravity acting here, he has no soul." R.P. Feynman, 1995, *Six Easy Pieces*, (Perseus Books), Chap. 5.

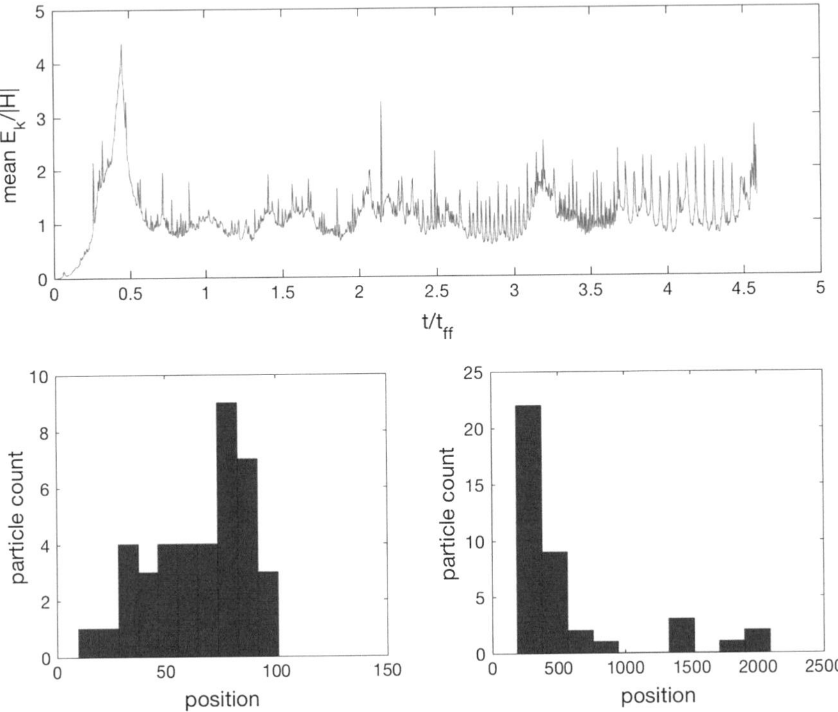

Fig. 10.1 Virialization in self-gravitating N-body problem ($N = 40$) takes place on about one-half the free fall time scale t_{ff}. Initially, the particles were spread out approximately uniformly within a sphere of dimensionless radius 100 with zero momenta. Subsequently, the cluster gradually spreads by developing a high velocity tail in its velocity distribution

Due to escapers, therefore, the cluster gradually contracts, *heating* up as $\delta \bar{E}_k = -\frac{1}{2}\delta\bar{U} = N^{-1}h > 0$.

2. *Gravitothermal instability.* Core collapses occurs in a finite time by thermal runaway in heat extraction with negative specific heat. The time-scale of core-collapse (CC) is short compared to the evaporative lifetime [5]. Indeed, globular clusters appear with both smooth King morphology (KM) and Post Core Collapse morphology (PCC) (e.g., [6]);
3. *Tidal tails.*[3] Escapers from the cluster tend to leave beyond the tidal radius r_t,[4] leaving an S-shaped tidal streams, oriented relative to the L_1 and L_2 Lagrange points of the cluster in orbit about the host galaxy, due to angular momentum conservation (Fig. 10.3). Ultimately, these streams disperse in the galactic halo.

[3]E.g., Table 10.1 in [7].

[4]The tidal sphere has the effect of lowering the potential barrier for stars to escape in the direction L_1 and L_2 [8].

Fig. 10.2 (*Left*) Globular cluster M80 (NGC 6093, in Scorpius, $D \sim 10$ kpc, $M \sim 10^{5.53} M_\odot$, age ~ 12.5 Gyr) with King morphology. (*Right*) Globular cluster NGC 6752 (in Paco, $D \sim 4$ kpc, $M \sim 10^{5.33} M_\odot$, age ~ 11.8 Gyr) with Post Core Collapse morphology (PCC)

Small angle scattering of stars within the cluster are mainly two-body encounters, producing deflections in hyperbolic trajectories of one star passing another, say, of mass m. This process is described by a (positive) total energy H (different from (10.32)) and net angular momentum l. In polar coordinates (r, φ), we have

$$H = \frac{1}{2}\dot{r}^2 + \frac{1}{2}r^2\dot{\phi}^2 - \frac{Gm}{r} = \frac{1}{2}v^2, \;\; l = r^2\dot{\phi}, \tag{10.34}$$

where v denotes the velocity at infinity. A Möbius transformation $u = b/r$ with $u = u(\varphi)$ for an impact parameter b (distance at periastron), transforms (10.34) into

$$u'' + u = \frac{Gmb}{l^2} : \;\; u = \frac{Gmb}{l^2}\left(1 + \frac{\sin\varphi}{\sin(\Delta\varphi/2)}\right). \tag{10.35}$$

The initial (i) and final states (f) satisfy $u_i = u_f = 0$ ($r_{i,f} = \infty$) with

$$\dot{r}_{i,f} = \mp v, \;\; \varphi_i = \pi + \frac{\Delta\varphi}{2}, \;\; \varphi_f = -\frac{\Delta\varphi}{2}. \tag{10.36}$$

By angular momentum conservation, $\dot{r} = -bu^{-2}u'\dot{\varphi} = -(l/b)u'$, i.e.,

$$\dot{r}_f = \frac{Gmb}{l^2}\frac{\cos(\Delta\varphi/2)}{\sin(\Delta\varphi/2)} \times \frac{l}{b} \simeq \frac{2Gmb}{b^2 v\Delta\varphi} = v \tag{10.37}$$

for weak interactions with small $\Delta\varphi$ and relatively minor velocity variations along the trajectory, so that $b\dot{\varphi} \simeq -v$, so that $l = r^2\dot{\varphi} = -bv$. It follows that

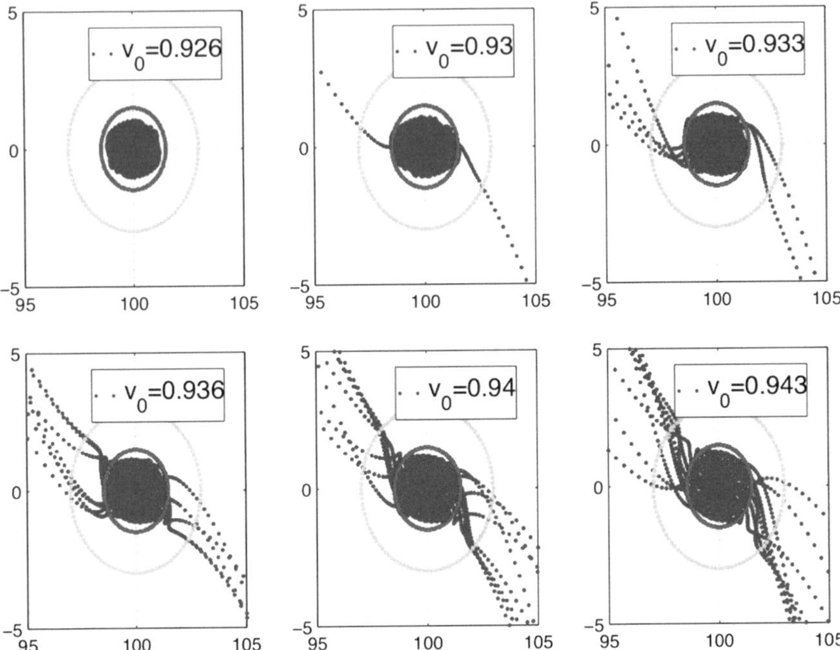

Fig. 10.3 Shown is the formation of S-shaped tails produced in the evaporation of a cluster of 30 equal mass stars. The initial velocities v_0 shown are normalized to $\sqrt{2M/r_0}$, isotropically in all directions in the tidal field of a host galaxy, where v_0 increments of 0.5% in each panel. The tidal sphere (*solid circle*) forms an effective threshold for stars to escape. The results illustrate the critical onset of anisotropic evaporation by escapers into tails of high velocity stars. The results obtain by direct numerical integration of the Hamiltonian equations of motion. (Reprinted from van Putten, M.H.P.M., 2012, NewA, 17, 411.)

$$|\Delta\varphi| \simeq \frac{2Gm}{bv^2}. \tag{10.38}$$

In crossing a globular cluster, a star experiences numerous small deflections. On average, the sum will be zero but the dispersion σ of the net deflections will be nonzero conform the theory of random walks. Simplifying the interaction with stars in the cluster by interactions with the projection of the cluster into a disk with density $\Sigma = N/(\pi R^2)$, we have

$$\delta\sigma^2 = \Sigma(2\pi b\delta b)(\Delta\phi)^2 = \frac{N}{\pi R^2}(2\pi b\delta b)\left(\frac{2Gm}{bv^2}\right)^2, \tag{10.39}$$

and hence

$$\sigma^2 = \frac{8N}{R^2}\left(\frac{Gm}{v^2}\right)^2 \log\left(\frac{R}{b_{min}}\right), \tag{10.40}$$

where $b_{max} = R$. A natural scale for b_{min} obtains from the maximal (in absolute magnitude) binding energy on par with kinetic energy in circular motion i.e.,

$$\frac{Gm}{b_{min}} = v^2. \tag{10.41}$$

By (10.33), $Nmv^2 = GM^2/R$, i.e., $v^2 = GmN/R$, so that

$$\frac{R}{b_{min}} = N, \quad \frac{Gm}{Rv^2} = N^{-1}. \tag{10.42}$$

Our expression (10.40) becomes $\sigma^2 = 8N^{-1}\log N$ after one *crossing time*

$$t_c = \frac{2R}{v} = \frac{2R^{\frac{3}{2}}}{\sqrt{NGm}}. \tag{10.43}$$

How long does it take for a significant net deflection to develop? The variance σ^2 scales linearly with time. We define one *relaxation time* t_{relax} to be the time for a large variance of deflection angles to develop from multiple scatterings,

$$\frac{t_{relax}}{t_c}\sigma^2 = 1, \tag{10.44}$$

that is

$$t_{relax} = \frac{R^{\frac{3}{2}}N^{\frac{1}{2}}}{4\sqrt{Gm}\log N}. \tag{10.45}$$

In the literature, (e.g., [9]), various estimates for (10.45) are given that generally agree within a factor of two. For the intended leading order estimates, (10.45) is more than accurate.

An individual star i in the globular cluster is typically gravitationally bound to all other stars (unless it is an escaper), i.e., its total energy is negative:

$$H_i = \frac{p_i^2}{2m_i} - \Sigma_{j\neq i}\frac{Gm_im_j}{|\mathbf{r}_i - \mathbf{r}_j|} < 0. \tag{10.46}$$

By frequent small angle scattering, globular clusters preserve near-thermal equilibrium by relaxation. Exchange of energy and momentum nevertheless can result in some stars acquiring $H_i > 0$. As an open system, globular cluster hereby slowly evaporates by ejection of high velocity stars. To calculate the rate of this evaporation process, consider the total binding and kinetic energy

$$U = -\Sigma_{j>i}\frac{Gm^2}{\left|\mathbf{r}_j - \mathbf{r}_i\right|}, \quad E_k = \Sigma_i \frac{1}{2} m v_i^2 \tag{10.47}$$

subject to (10.33). The average binding and kinetic energy per star are

$$\bar{U} = -\frac{1}{N}\Sigma_{j>i}\frac{Gm^2}{\left|\mathbf{r}_j - \mathbf{r}_i\right|} = -\frac{1}{2N}\Sigma_{j\neq i}\frac{Gm^2}{\left|\mathbf{r}_j - \mathbf{r}_i\right|} \simeq -\frac{Gm^2N}{2R}, \ \bar{E}_k = \frac{E_k}{N}, \tag{10.48}$$

The kinetic energy ϵ_k of an escaping star i is such that it exceeds to binding energy

$$u_i = -\Sigma_{j\neq i}\frac{Gm^2}{\left|\mathbf{r}_j - \mathbf{r}_i\right|} \tag{10.49}$$

to all other stars. Let $\bar{u}$ denote the average of the u_i over all stars $i = 1, 2, \ldots N$, i.e.,

$$\bar{u} = 2\bar{U} \tag{10.50}$$

by (10.48–10.49). Then by (10.48), the condition for escape is

$$\epsilon_k > -\bar{u} = -2\bar{U} = 4\bar{E}_k. \tag{10.51}$$

We thus obtain the *Ambartsumian condition* for the escape velocity to be twice the mean velocity of the stars in the cluster,

$$v_e = 2\bar{v}. \tag{10.52}$$

A general frame work for the evaporative evolution of a globular cluster is the system of two ordinary differential equations, describing the time rate-of-change of N and H,

$$\frac{dN}{dt} = -N\tau_{AS}^{-1}, \quad \frac{dH}{dt} = H\tau_{KH}^{-1} \tag{10.53}$$

with the *Ambartsumian-Spitzer* time scale τ_{AS} and the *Kelvin-Helmholtz* time scale τ_{KH} defined as multiples of the relaxation time,

$$\tau_{AS} = f_N^{-1} t_{relax}, \quad \tau_{KH} = f_H^{-1} t_{relax}. \tag{10.54}$$

Here, f_N and f_H denote the fractional changes in N and H, respectively, per unit of relaxation time, due to the escaping stars.

10.6 Coefficients of Relaxation

To estimate f_N and f_H, we build on Ambartsumian [10]. The kinetic energies of escapers is considered to be in the tail of a Boltzmann distribution (10.55), defined by velocities larger than the escape velocity. For an idealized globular cluster with equal mass stars, that may be appropriate for old but not young clusters, the kinetic energy of the stars below the escape velocity effectively satisfies the Boltzmann distribution. Consider the number density of stars n, $dN = nd^3v$, satisfying

$$n(\epsilon) \propto e^{-\epsilon/k_BT}, \quad \epsilon = \frac{1}{2}m(v_x^2 + v_y^2 + v_z^2), \quad \frac{3}{2}k_BT = \bar{\epsilon}. \tag{10.55}$$

Extending Ambartsumian's estimate for f_N, we now include f_E as follows:

$$f_N = \frac{\int_2^\infty e^{-\frac{3}{2}s^2} 4\pi s^2 ds}{\int_0^\infty e^{-\frac{3}{2}s^2} 4\pi s^2 ds} \simeq \frac{1}{135}, \quad f_E = \frac{\int_2^\infty s^2 e^{-\frac{3}{2}s^2} 4\pi s^2 ds}{\int_0^\infty s^2 e^{-\frac{3}{2}s^2} 4\pi s^2 ds} \simeq \frac{1}{29}, \tag{10.56}$$

where $s^2 = \epsilon/\bar{\epsilon}$, f_E denotes the fraction of total kinetic energy exceeding the threshold for escape. Let us now look at the following two processes.

1. *Core-Collapse.* Escapers from the core represent heat flux outwards to the cluster envelope. Because U attains a minimum for stars at the center, they effectively deposit a net kinetic energy into the envelope with minor change in potential energy. f_E signifies excess heat in the core that cannot be retained, which is hereby transported outwards essentially uninhibited,

$$f_H \simeq f_E \simeq 4.7 f_N. \tag{10.57}$$

   This inevitable result will be core-collapse by negative specific heat.
2. *Cluster Evaporation.* Total escape represents a net particle and energy flux to infinity. The latter is somewhat reduced in overcoming the gravitational binding energy to the cluster as a whole. The energy deposited at infinity equals initial kinetic energy minus the minimum kinetic energy required for escape,

$$f_H = \left(\frac{f_E}{f_N} - 4\right) f_N \simeq 0.7 f_N. \tag{10.58}$$

By (10.57–10.58), core-collapse and evaporation are driven by dramatically different net luminosities,. f_H defines the Kelvin-Helmholtz time scale (lifetime based on energy extraction, keeping N fixed) and f_N defines the Ambartsumian-Spitzer time scale (lifetime based on escapers with zero net energy deposition at infinity). Comparing f_H and f_N, we have

$$\text{core}: \ \tau_{KH} < \tau_{AS}, \quad \text{cluster}: \ \tau_{KH} > \tau_{AS}. \tag{10.59}$$

Overall, evolution based on f_H and f_N combined, we have

$$\tau_{evap} >> \tau_{CC}. \quad (10.60)$$

Detailed results obtain from numerical integration of the system of two ordinary differential equations (10.54) with (10.57–10.58). (Uncertainties in (10.59) largely stem from the degree to which our collisionless interactions are indeed modeled by weak two-body scattering, ignoring the effect of compact binaries.) Assuming further the Virial Theorem and (10.48),

$$H = -\frac{GN^2m}{4R}, \quad (10.61)$$

so that the coefficients (10.56) define a closed system (10.53–10.54) of ordinary differential equations for (R, N), that is readily integrated. The result gives the following estimates for the time to core-collapse [11]

$$t_{CC} = 13.85\, t_{relax}, \quad t_{CC} = 20.53\, t_{rh}, \quad (10.62)$$

where t_{relax} is the relaxation time derived from R and t_{rh} is the relaxation time derived from r_h ($R = 1.3\, r_h$). This result (10.62), derived from thermodynamic arguments, agrees with detailed large N-body simulations.

Example 10.4. Based on Table 10.1, $N = M_c/\bar{m}$ for a globular cluster with mean stellar mass of, e.g., $\bar{m} = \frac{1}{3}M_\odot$. Tidal tails are created over a typical period $T = T_7 10^6$ yr of tens million years since the last crossing of the galactic disk. Scaled to one-quarter of the period $P = P_6$ Myr of the orbit around the galactic center, the associated outflow of evaporating stars satisfies

$$n \simeq \frac{PNf_N}{4t_{relax}} = 0.275 \times N_5 P_6 \left(\frac{t_{rh}}{10^9\ \mathrm{yr}}\right)^{-1} \quad (10.63)$$

with $N = 10^5 N_5$.

By angular momentum conservation, bipolar outflows of stars emanate from L_1 and L_2, deflecting into leading and trailing streams inside and outside the orbit of the cluster [12] Produced by high velocity escapers, it appears with a correlation of radial separation to cluster core temperature, apparent in an angular separation

$$\alpha = \frac{\Delta r}{D_{gc}} = 3.4\frac{\sigma}{D_{gc}\Omega} = 0.55\, \sigma_1 D_4^{1/2}\ \mathrm{deg}, \quad (10.64)$$

where $\Omega = \sqrt{M_g/D^3}$ with $M_g = 3 \times 10^{11} M_\odot$ [13] denotes the orbital period of the cluster about the galactic center at a distance $D_{gc} = D_4$ 10 kpc and $\sigma = \sigma_1$ 1 km s^{-1}.

Table 10.2 Thermodynamics of N-body systems

1. Globular clusters are self-gravitating N-body systems that are essentially virialized. They develop a near-thermal velocity distribution on a relaxiation time scale, defined by small angle gravitational scattering. By a negative specific heat, their cores are susceptable to a gravitothermal catastrophe.
2. Escapers are in the tail of a Boltzmann velocity distribution with positive total energy. Their escape represents a singular perturbation away from a thermal distribution.
3. Evaporation by escapers carries off a net kinetic luminosity out to infinity (into the halo of their host galaxy), described by a system of two ordinary differential equations comprising conservation of total number of stars and total energy.
4. Numerical integration shows that gravitothermal core-collapse takes place after about twenty relaxation times.

Table 10.2 summarizes this discussion.

10.7 Exercises

10.1. Give the Lagrangian for a two-dimensional motion of the pendulum, including motion about the vertical axis. Show that Noether's theorem implies conservation of angular momentum about this vertical axis. For small deflections, obtain solutions to the equations of motion.

10.2. Use Noether's theorem in the Lagrangian for the trajectory of a canon ball, ignoring friction, to show that horizontal momentum is conserved. Obtain the distance traversed when it reaches ground level as a function of maximal height in its trajectory.

10.3. Consider a globular cluster orbiting a host galaxy at a distance R. Expand the Newtonian binding energy to host in terms of Legendre polynomials. In a spherical coordinate system (r, θ, φ) centered at the host galaxy, identify the tidal interaction with the second Legendre polynomial $P_2(\cos\theta)$, where θ is the poloidal angle relative to $\mathbf{R}$. ($\theta = 0$ at L_2 and $\theta = \pi$ at L_1.) Obtain an estimate of the tidal radius r_t and apply this to NGC104 (Table 10.1).[5]

10.4. M80 and NGC 6752 are two globular clusters of the Milky Way with $N = 3\times10^5, r_h = 0.61$ pc and, respectively, $N = 2\times10^5$ and $r_h = 1.91$ pc. [15]. Here, the *half-light radius* r_h is related to the virial radius by $R_{vir} = 1.3\,r_h$. For the relaxation time of the cluster, we may choose R_{vir} or r_h, giving $t_{rh} = kt_{relax}$. Following t_{relax} in (10.45), what is the conversion factor k? Calculate t_{relax} following (10.45) for M80 and NGC 6752. The time to core collapse is given (10.62). Calculate t_{CC} for M80 and NGC 6752. How do the results compare with the classification of M80 and NGC 6752 having KM and, respectively, PCC morphology? You may assume each to have the age of the Universe.

10.5. For the two-body problem in Newtonian mechanics, derive an equivalent one-body Hamiltonian of a test particle about the center of mass. Derive Kepler's third law from this reformulation.

10.6. Saturn's rings consist of cm to m sized rock-like particles. Although the total mass of Saturn's rings is small, they are self-gravitating.[6] Write down the Hamiltonian of one particle in orbit around Saturn, obtain the Hamiltonian equations of motion and integrate these using the Maxima *rk* Runge-Kutta integration method for a few orbits. Show the results for circular and non-circular orbits. Consider two particles of equal mass, now including their gravitational interaction. Determine initial conditions chosen such that these two particle form a small binary. Use *rk* to calculate the orbit of this binary around Saturn.

10.7. Using Runge-Kutta integration, perform a direct numerical simulation of a globular cluster of $N \leq 10$ stars with initially zero velocities. Plot the total kinetic energy in the cluster as a function of time and interpret the result. [*Hint.* The initial evolution shows equilibration following a first bounce.]

References

1. Zia, R.K., Redish, E.F., & McKay, S.R., 2009, arXiv:0806.1147v2.
2. Chernoff, D.F., & Djorgovski, S., 1989. ApJ 339, 904.
3. Chaboyer, B., 2007, Proc. IAU Symp. 248, Eds. Jin, W.J., Platais, I., Perryman, M.A.C.
4. Richer, H.B. et al., 2004. AJ 127, 2771.
5. Lynden-Bell, D., Wood, R., 1968, MNRAS, 138, 495; Lynden-Bell, D., Eggleton, P.P., 1980, MNRAS, 191, 483; Lynden-Bell, D., 1988, arXiv/cond-mat/9812172; Antonov, V.A., 1962, Vest. Leningrad Univ. 7, 135 (IAU Symp. 113, 525 (1995).
6. Chernoff, D., & Djorgovski, S., 1989, ApJ, 339, 904.
7. van Putten, M.H.P.M., 2012, New A, 17, 411.

[5] For a recent discussion, see, e.g., [14].

[6] Possibly, there is further a soft magnetic moment-magnetic moment repulling force between them, induced by Saturn's 1 Gauss magnetic field, nearly perfectly aligned with its spin axis and orthogonal to Saturn's disk, that tends to contribute to stability.

8. Hayli, A., 1967. Bull. Astron. 2, 67.
9. Aarseth, S.J., 2003, *Gravitational N-Body Simulations* (Cambridge: Cambridge University Press).
10. Ambartsumian, V., 1938, Uch. Zap. LGU, 22, 19; *in* Dynamics of Star Clusters, Princeton, 1984, eds. J. Goodman & P. Hut (IAU Symposium, No. 113), 1985, p. 521.
11. van Putten, M.H.P.M., 2012, NewA, 17, 411.
12. Heggie, D.C., 2001. In: Deiters, S., Fuchs, B., Spurzem, R. Just, A., Wielen R. (Eds.), *Dynamics of Star Clusters and the Milky Way*, ASP Conference Series, p. 228.
13. Odenkirchen, M. et al., 1997. New A 2, 477.
14. Gajda, G., & Lokas, E.L., 2016, ApJ, 819, 20.
15. Harris, W.E., 2010, Catalogue of parameters for Milky Way Globular Clusters, http://physwww.mcmaster.ca/harris/mwgc.dat.

Chapter 11
Accretion Flows onto Black Holes

Accretion is a general phenomenon in the interaction of astrophysical objects with their environment. Accretion onto white dwarfs, neutron stars and black holes in particular may give rise to multi-wavelength continuous and transient emission. The most extreme accretion scenarios are envisioned following birth of compact objects in core-collapse of massive stars. Accretion is a complex process of mass flow. Quite generally, in fall of angular momentum rich matter gives rise to disks. Accretion may continue provided angular momentum diffuses outwards, perhaps by a combination of turbulent viscosity and outflows.

In the limit of low angular momentum, mass may fall directly onto the central object without the formation of an accretion disk. This problem has been worked out for adiabatic flows in spherical symmetric flows by Bondi [1] and in cylindrical symmetry by Hoyle and Lyttleton [2]. These solutions provide leading order approximations to mass capture by gravitational focusing applicable to compact objects in a variety of settings, e.g., subject to fallback matter in core-collapse of massive stars, stellar winds from a binary companion, or when moving through a molecular cloud or the insterstellar medium.

In the mass transfer between two stars in a binary stellar system, accretion flows from a donor are naturally angular momentum rich and tend to form an accretion disk around the acceptor. In this process, high energy emissions may be produced particularly in the deep potential well around neutron stars or black holes.

Here, we include a discussion on the theory of thin accretion disks around black holes, to highlight the importance of anomalous diffusion to account for the observed high accretion disk luminosities. This is commonly modeled by turbulence [3] and possibly so by magnetic instabilities [4].

The extremely deep potential well around black holes allows no light escape from beyond their event horizon. This was originally envisioned by John Michell and Pierre-Simon Laplace based on Newton's corpuscular theory of light [5]. Remark-

M.H.P.M. van Putten, *Introduction to Methods of Approximation in Physics and Astronomy*, Undergraduate Lecture Notes in Physics,
DOI 10.1007/978-981-10-2932-5_11

ably, this predicts the correct radius of a non-rotating Schwarzschild black hole following the Hamiltonian

$$H = \frac{1}{2}v^2 - \frac{GM}{r} \tag{11.1}$$

of a particle of unit mass at a distance r from a central mass M. At a radius R, $H = 0$ gives the escape velocity

$$v_e = \sqrt{\frac{2GM}{R}}. \tag{11.2}$$

Particles with $v < v_e$ are trapped within. With an initial velocity c, light remains trapped within the *Schwarzschild radius*

$$R_S = 2R_g, \quad R_g = \frac{GM}{c^2} = 1.5 \times 10^5\,\mathrm{cm}\left(\frac{M}{M_\odot}\right), \tag{11.3}$$

where R_g is a commonly used definition for the *gravitational radius* of an object of mass M. Even though (11.1) is a classical Hamiltonian in Newton's flat and non-relativistic spacetime, (11.3) is the exact result predicted by general relativity, upon identifying $4\pi R_S^2$ with the area of the black hole event horizon.

In reality, the surface of a black hole is an event horizon at infinite redshift.[1] Black holes can grow without bound in accumulating mass by accretion. With no or negligible radiative back reaction, this process is different from accretion onto the hard surface of a neutron star. Accreting back holes can hereby appear sub-luminous.

Various methods exist for determining the mass of black holes, based on the orbital motion of accompanying gas or stars or by causality constraints applied to time variable emission. For instance, the extragalactic source PKS 2155-304 observed by H.E.S.S. showed short time scale variability on the order of 10 min in a >200 GeV flare [8]. A 10 min time scale can be realized in a nucleus, provided its linear size is no larger than the corresponding light crossing distance, i.e.,

$$R_S \le t_{var}c, \quad M \le \frac{c^3 t_{var}}{2G} \sim 10^7 M_\odot. \tag{11.4}$$

The theory of general relativity introduced embedding of Newton's theory of gravitation in a four-covariant theory, allowing an extension to strong gravity around black holes (Fig. 9.1).

It should be mentioned that the Einstein equations are mixed hyperbolic-elliptic, which also predicts the entirely new phenomena of gravitational wave motion. The

[1] In the classical limit, ingoring evaporation. Black holes evaporate by photon emission one-by-one that, in preserving unitarity, involves an astronomical amount of computation [6]. Subject to the Margolus-Levitin bound on quantum computation (1998, Physica D, 120, 88), this process is slow, one photon every few thousand light crossing time scales [7].

detection of black hole binary coalescence GWB150914 [9] has now made direct detection of gravitational wave observations a practical reality, taking us a major step beyond the indirect detection in the decay of the Hulse-Taylor binary pulsar system PSR 1913+16 [10]. Recent measurements by the LAGEOS II and Gravity Probe B [11] satellites provided a test for frame dragging around the Earth, confirming an essential feature (of the elliptic part of) general relativity to within measurement error. Together, these experimental results confirm remarkable non-Newtonian behavior of spacetime in general relativity.

The most extreme accretion flows may be found in energetic supernovae associated with the birth of neutron stars and black holes in core-collapse of massive stars. In this event, accretion flows reach high densities, as inferred from the $>10\,\mathrm{MeV}$ neutrino burst in SN1987A. Around black holes, the ratio of disk mass to the mass of the central object is expected to reach up to 1% in models of extreme events associated with gamma-ray bursts. If so, any non-axisymmetry in these accretion flows is inevitably luminous in gravitational radiation.

11.1 Bondi Accretion

A remarkable result of Bondi's study is the existence smooth solutions in spherically symmetric accretion, starting from a finite temperature, zero velocity medium at large distances. These solutions are *adiabatic* flows (with no shocks) at special accretion rates, given a density ρ_0 and adiabatic sound speed a_0 at infinity for an adiabatic index $1 \leq \gamma \leq 5/3$.

Starting point is the specific enthalpy of adiabatic flow along streamlines,

$$h = \frac{1}{2}v^2 + \frac{a^2}{\gamma - 1} - \frac{GM}{r} = \frac{a_0^2}{\gamma - 1}. \tag{11.5}$$

The time-independent radial flow velocity v is governed by Euler's equation of motion

$$v\partial_r v = -\frac{1}{\rho}\frac{\partial P}{\partial r} - \frac{GM}{r^2}, \tag{11.6}$$

where r denotes the radial distance to a central mass M, P and ρ are the fluid pressure and, respectively, density, satisfying a polytropic equation of state

$$P = K\rho^\gamma, \tag{11.7}$$

where $\gamma = 1 + 1/n$ in terms of the polytropic index n. The associated adiabatic sound speed is $a = \sqrt{\gamma P/\rho}$. Conservation of baryon number implies that the accretion rate

$$\dot{M} = 4\pi r^2 \rho v \tag{11.8}$$

is constant throughout.

In considering smooth solutions, we allow radial derivatives of all relevant flow quantities. To this end, we consider (11.6) and the logarithmic derivate of (11.8), giving

$$v' + \frac{a^2}{v\rho}\rho' + \frac{GM}{vr^2} = 0, \quad \frac{\rho}{v}v' + \rho' + \frac{2}{r}\rho = 0. \tag{11.9}$$

As a system of equations for two unknowns, its solutions are

$$\begin{pmatrix} v' \\ \rho' \end{pmatrix} = \frac{v^2}{v^2 - a^2}\begin{pmatrix} \frac{a^2}{v\rho} & -1 \\ -1 & \frac{\rho}{v} \end{pmatrix}\begin{pmatrix} \frac{2\rho}{r} \\ \frac{GM}{vr^2} \end{pmatrix}, \tag{11.10}$$

that is

$$v' = \frac{2a^2 - \frac{GM}{r}}{v^2 - a^2} r^{-1} v, \quad \rho' = \frac{\frac{GM}{r} - 2v^2}{v^2 - a^2} r^{-1} \rho. \tag{11.11}$$

The vanishing of the determinant in (11.10) defines a critical point, which is the sonic point where the inflow velocity equals the sound velocity. For smooth solutions to exist, the numerators of (11.11) will have to vanish simultaneously, i.e.,

$$v_s^2 = a_s^2 = \left(\frac{2}{5 - 3\gamma}\right) a_0^2, \quad r_s = \frac{5 - 3\gamma}{4}\frac{GM}{a_0^2}, \tag{11.12}$$

where a_0 denotes the sound velocity at infinity, where we assume $v = 0$. In meeting these conditions (11.12), we fix the accretion rate, i.e., the accretion rate assumes critical values for smooth solutions satisfying

$$\dot{M} = 4\pi r_s^2 \rho_s v_s = 4\pi\lambda \left(\frac{GM}{a_0^2}\right)^2 \rho_0 a_0 = 4\pi\lambda R_g^2 c \rho_0 \left(\frac{c}{a_0}\right)^3 \propto a_0^{-3}, \tag{11.13}$$

where $R_g = GM/c^2$, ρ_0 denotes the density at infinity and

$$\lambda = \frac{1}{4}\left(\frac{2}{5 - 3\gamma}\right)^{\frac{5-3\gamma}{2(\gamma-1)}}, \tag{11.14}$$

taking into account $1 \leq \gamma \leq \frac{5}{3}$. It assumes values of order unity as shown in Fig. 11.1. Note that the accretion rate decays with the third power of the sound speed at infinity: accretion prefers cold over hot gas.

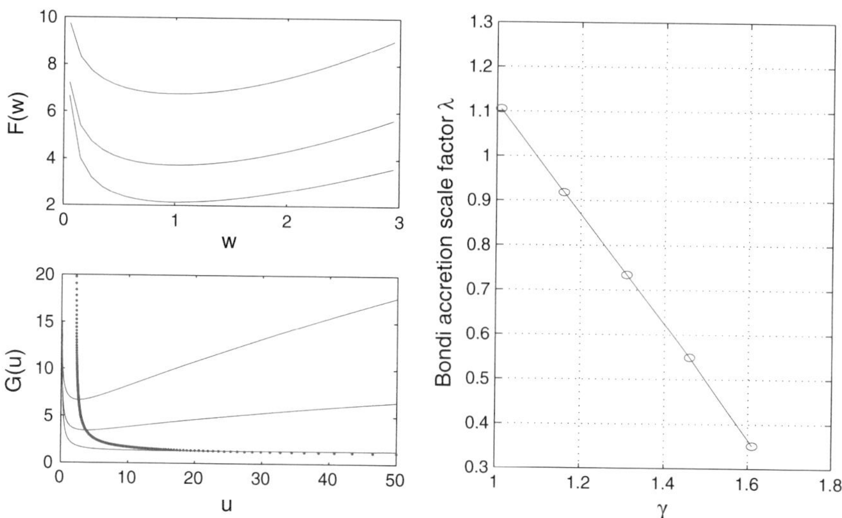

Fig. 11.1 (*Left*) Shown are $F(w)$ and $G(u)$ for the values $\gamma = 1.16, 1.31, 1.61$ (*top* to *bottom*). $F(w)$ is minimal at the sonic point $w = 1$. The corresponding locus of the minima (*bold line*) of $G(w)$ are at $u_s = 4/(5-3\gamma)$. (*Right*) The ratio of these minima defines the Bondi accretion scale factor λ, which serves as a bound on the accretion rate in a smooth flow

Example 11.1. The Bondi accretion rate (11.13) can be scaled as

$$\dot{M} = \dot{m}_{-3}\left(\frac{2}{5-3\gamma}\right)^{\frac{5-3\gamma}{2(\gamma-1)}}\left(\frac{M}{M_\odot}\right)^2\left(\frac{\rho_0}{1\,\mathrm{g\,cm^{-3}}}\right)\left(\frac{0.001\,c}{a_0}\right)^3, \quad (11.15)$$

where $\dot{m}_{-3} = 0.001\,M_\odot\,\mathrm{s}^{-1}$. If a fraction $\epsilon = \epsilon_{-2}1\%$ of $\dot{M}c^2$ converts to X-ray radiation, $L_X \sim \epsilon_{-2}10^{49}\,\mathrm{erg\,s}^{-1} \sim \epsilon_{-2}10^{11}L_{Edd}$ for the scaling parameters at hand, where $L_{Edd} = 1.26\times10^{38}\,\mathrm{erg\,s}^{-1}$ is the Eddington luminosity. Radiative backreaction is hereby prohibitive for accretion onto a neutron star, but not for accretion onto a black hole.

It turns out that (11.13) is the maximal attainable accretion rate in adiabatic flows, based on the dimensionless implicit formulation for the Mach number $w = v/a$ as a function of the dimensionless inverse distance $u = GM/(ra_0^2)$ given by

$$F(w) = kG(u). \quad (11.16)$$

With $\alpha = (\gamma-1)/(\gamma+1)$, we have

$$F(w) = w^{-2\alpha}\left(\frac{1}{2}w^2 + \frac{1}{\gamma-1}\right), \quad G(u) = u^{-4\alpha}\left(u + \frac{1}{\gamma-1}\right). \tag{11.17}$$

Here, $k = \lambda^{-2\alpha}$ is a constant set by the accretion rate. In view of $1 < \gamma < 5/3$, $0 < \alpha < \frac{1}{4}$, the functions

$$F(w) = \frac{1}{2}w^{\frac{4}{\gamma+1}} + \frac{1}{\gamma-1}w^{-2\alpha}, \quad G(u) = u^{\frac{5-3\gamma}{\gamma+1}} + \frac{1}{\gamma-1}u^{-4\alpha} \tag{11.18}$$

divergence for small and large arguments (Fig. 11.1). Qualitatively similar to $f(x) = x^{-1} + x$ $(x > 0)$, the functions $F(w)$ and $G(u)$ define a sonic point of Mach number 1 at their minimum, $F'(w) = G'(u) = 0$, given by $w = 1$ at $u = 4/(5-3\gamma)$, whereby

$$F(w) \geq F_s = \frac{1}{2\alpha}, \quad G(u) \geq G_s = \frac{1}{4\alpha}\left[\frac{5-3\gamma}{4}\right]^{-\frac{5-3\gamma}{\gamma+1}}. \tag{11.19}$$

For (11.16) to have a global solution (connecting infinity $u = 0$ to the center $u = \infty$), $kG(u)$ must exceed the lower bound (11.19) *everywhere*. That is, $kG(u_s) \geq F_s$, which puts an upper bound on the accretion rate scale factor λ by

$$k \geq k_s = \frac{F_s}{G_s} = 2\left[\frac{5-3\gamma}{4}\right]^{\frac{5-3\gamma}{\gamma+1}}. \tag{11.20}$$

Example 11.2. To illustrate numerical continuation (Sect. 7.6), Fig. 11.2 is shows a computation of two branches of global solutions ($k > k_s$) to (11.16) by root finding in two-dimensions. Consider the roots of

$$Z(w, u) = 0, \tag{11.21}$$

where

$$Z(w, u) = F(w) - kG(u). \tag{11.22}$$

For a given k satisfying (11.20), roots of (11.21) define flow solutions of interest that approach the transonic point in the limit as k approaches k_s. In using Newton's method, we choose an initial guess for w at an upstream position $u < u_s$, e.g., a small value like $u = 0.1$, and iterate

$$w_{n+1} = w_n - \frac{Z(w_n, u)}{Z_w(w_n, u)}, \quad Z_w(w, u) = \frac{\partial Z(w, u)}{\partial w}. \tag{11.23}$$

By quadratic convergence of Newton's method, w_n converges rapidly to a limit point $w^* = w_*(u)$ in a few iterations. For instance, for $u = 1.25$ with $k = 1.04\,k_s$, we find

$$\begin{array}{ll} w_n: & 0.5000\ 0.1217\ 0.2093\ 0.2592\ 0.2677\ \ \ \ \ \ 0.2679, \\ Z_n: & -0.5851\ 0.9451\ 0.2752\ 0.0358\ 7.66\times10^{-4}\ 3.6582\times10^{-7}, \end{array} \quad (11.24)$$

For a choice of step size h, we next choose a neighboring point $u + h$ with initial guess $w^*(u)$ from the previous iteration. A few iterations of (11.23) now produces $w^*(u+h)$. Repeating this procedure step-by-step obtains $w^*(u+nh)$, $n = 1, 2, \ldots$, thus producing a continuous branch. For the initial choice of initially subsonic value $w = 0.5$, a subsonic branch obtains. Alternatively, for an initial choice of supersonic value $w = 1.5$, a supersonic branch obtains.

The sonic point is numerically unstable as a *bifurcation point* across which continuation can choose between a subsonic or supersonic branch on $r < r_s$ $(u > u_s)$. The result is physically determined by the boundary condition at $r = 0$, i.e., a hard surface or a surface of maximal inflow in case of a black hole event horizon.

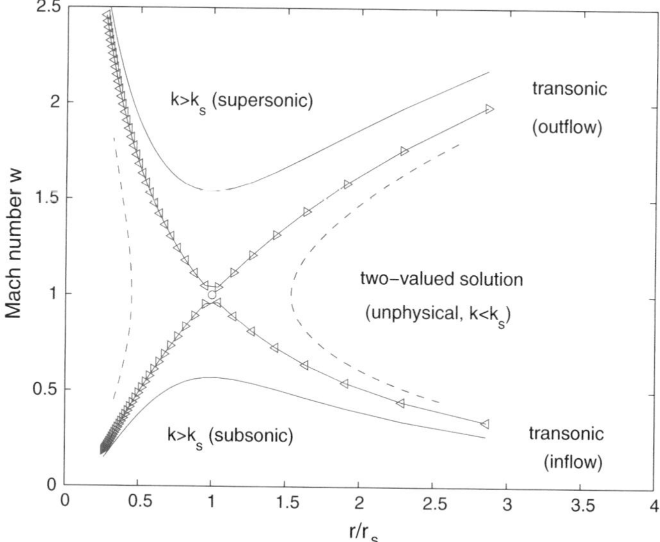

Fig. 11.2 Solutions to Bondi's equations of spherical accretion include transonic in- and outflows, above and below which are super- and subsonic solutions ($k > k_s$). Additional fallback and bouncing solutions to the left and right of the sonic point are unphysical two-valued velocity solutions. The results shown obtain by numerical continuation with $\gamma = 1.2$

In case of non-global solutions at sub-critical accretion rates ($k < k_s$), numerical continuation can similarly be applied to the inverse of (11.23), i.e.,

$$u_{n+1} = u_n - \frac{Z(w, u_n)}{Z_u(w, u_n)}, \quad Z_u(w, u) = \frac{\partial Z(w, u)}{\partial u}. \tag{11.25}$$

It produces two additional bouncing solutions, shown by unphysical two-valued solution branches to the left and right of the sonic point.

Example 11.3. In the formation of a neutron star or black hole in core-collapse of a massive star, Bondi accretion may be compared with free fall upon considering the sonic radius r_s relative to the radius R of the progenitor. Bondi accretion will be relatively slow when $r_s/R \ll 1$. Prior to collapse, the progenitor was in equilibrium with a thermal energy

$$E_{th} = Mc_s^2 \tag{11.26}$$

with $c_s^2 = GM/(4R)$ based on Sect. 4.2.1. In the approximation $a_0 \simeq c_s$, and using (11.12), it follows that

$$\frac{r_s}{R} \simeq 5 - 3\gamma. \tag{11.27}$$

Thus, r_s falls within the star whenever $4/3 < \gamma < 5/3$. As γ approaches 4/3, most of mass infall is expected to be subsonic. In contrast, r_s falls outside the star when $1 < \gamma < 4/3$, in which case infall is essentially supersonic. These two cases point to a slow respectively fast fall in of the stellar envelope, relative to the free fall time scale.

11.2 Hoyle-Lyttleton Accretion

An interesting variation to the spherical Bondi accretion is the accretion of gas flow captured by gravitational focusing from the environment.[2]

In the comoving frame of the star, flow appears upstream with asymptotic velocity v_0. In a spherical coordinate system (r, θ, φ) aligned with the direction of motion to the right, streamlines upstream are deflected downstream towards the semi-infinite axis $\theta = 0$. Away from the $\theta = 0$, we treat the flow in the limit of zero pressure (cf. Burgers' equations of motion of Sect. 9.4.1.) The enthalpy

[2]Historically, this problem was treated before Bondi's spherical accretion.

$$h = \frac{1}{2}\dot{r}^2 + \frac{1}{2}r^2\dot{\theta}^2 - \frac{GM}{r} = \frac{1}{2}v_0^2 \tag{11.28}$$

is constant along streamlines satisfying

$$u = \frac{GM}{j^2}(1 + \cos\theta) - \frac{v_0}{j}\sin\theta \ \ (0 < \theta < \pi), \tag{11.29}$$

taking into account the boundary condition $u = 1/r \to 0$ as θ approaches π with $u = u(\theta)$,

$$\dot{r} = -ju' \to -v_0 \tag{11.30}$$

as gas is coming from the left as seen in the frame of the star. Here,

$$j = r^2\dot{\theta} = -bv_0 \tag{11.31}$$

denotes the specific angular angular momentum that, associated with the impact parameter b, labels a specific streamline.

By gravitational focusing, streamlines cross the semi-infinite line $\theta = 0$, where our approximation (11.28) breaks down and a fluid dynamical treatment with internal energy due to pressure induced by compression of the gas is to be included. We model this as being local to the $\theta = 0$, where the angular momentum of the fluid elements is reduced to zero about $\theta \simeq 0$, giving rise to out- or inflow of fluid elements with positive, respectively, negative enthalpy. Given the reduced enthalpy h_* (11.32), this corresponds to fluid elements that are unbound ($h_* > 0$) and, respectively, gravitationally bound ($h_* < 0$) to the star.

Reduction of angular momentum to zero at $\theta = 0$ by collision and compression appears in a vanishing of the poloidal velocity $r\dot{\theta}$ at $\theta = 0$. Thus, $\theta = 0$ contains a *stagnation point* at some radius $r_* > 0$ from the star, that may be derived from the reduced specific energy

$$h_*(r) = \frac{1}{2}v_0^2 - \frac{GM}{r}, \ \ \dot{r} = v_0 \ \ (\theta = 0), \tag{11.32}$$

where the second equation follows by (11.29) and (11.31). The stagnation point is at the root

$$h_*(r_*) = 0: \ \ r_* = \frac{2GM}{v_0^2}. \tag{11.33}$$

Thus, gas along streamlines crossing θ beyond (before) the stagnation point escapes (falls back). From (11.29), the associated streamline (labeled by angular momentum) is

$$r_* = \frac{j_*^2}{2GM}, \tag{11.34}$$

and hence

$$j_*^2 = 2GMr_* = \frac{4G^2M^2}{v_0^2}. \tag{11.35}$$

Streamlines entering the stagnation point carry a specific angular momentum j_* given by r_* times $v_\theta = r\dot{\theta} = ju$, just before the hydrodynamical interactions set in. From (11.31), we have $j_* = b_* v_0$, where b_* is the critical impact parameter, measured by separation to $\theta = \pi$ at large distances upstream. That is, we have

$$b_* = \frac{j_*}{v_0} = \frac{2GM}{v_0^2}. \tag{11.36}$$

The mass rate of accretion now follows from the capture area πb_*^2:

$$\dot{M}_{HL} = \pi b_*^2 v_0 \rho_0 = 4\pi \frac{G^2M^2}{v_0^3}\rho_0 = 4\pi R_g^2 c\rho_0 \left(\frac{c}{v_0}\right)^3 \propto v_0^{-3}. \tag{11.37}$$

Its asymptotic behavior is similar to (11.13). It let Bondi to suggest the general result

$$\dot{M}_{BHL} = \pi b_*^2 v_0 \rho_0 = 4\pi \frac{G^2M^2}{(v_0^2 + a_0^2)^{3/2}}\rho_0 \tag{11.38}$$

to apply to bodies moving with arbitrary velocity v_0 (including zero) in media with asymptotic density and sound speed ρ_0 and a_0, respectively.

Example 11.4. Similar to Example 11.2, the Hoyle-Lyttleton accretion rate (11.37) can be scaled as

$$\dot{M} = \dot{m}_{-3} \left(\frac{M}{M_\odot}\right)^2 \left(\frac{\rho_0}{1\text{ g cm}^{-3}}\right) \left(\frac{0.001\,c}{v_0}\right)^3, \tag{11.39}$$

and the same conclusions apply. Upon moving into the envelope of a companion star, a neutron star is expected to grow slowly by Eddington limited accretion, whereas a black hole may grow rapidly.

The case of bodies moving supersonically is different, in that a shock front will form in the accretion flow. It effectively combines the free stream Hoyle-Lyttleton accretion flow upstream and Bondi flow downstream [12].

11.3 Accretion Disks

The basic principle of accreting accretion disks is mass flow *inwards* due to angular momentum transport *outwards* (Fig. 11.3). The latter is generally considered to be dissipative by internal anomalous diffusion, possibly augmented by outflows in disk winds or astrophysical jets and, in the most extreme cases, gravitational radiation [13].

To illustrate the limitations of molecular diffusion, we recall the scaling of kinematic viscosity in the theory of random walks,

$$\nu = \lambda c_s, \quad c_s = \sqrt{\frac{k_B T}{m}}, \tag{11.40}$$

by mean free path length λ (between atoms in air or protons in fully ionized plasmas alike) and the isothermal sound speed c_s in a disk of temperature T. What is the implied luminosity of accretion flows driven by molecular diffusion?

Consider a disk with scale height $H = H(r) \ll r$, where r denotes the radial distance to the central object. The vertical component of gravitational acceleration imparted by a central mass M satisfies

$$g_z = -\frac{GM}{r^2}\sin\theta = -\frac{GMz}{r^3} \tag{11.41}$$

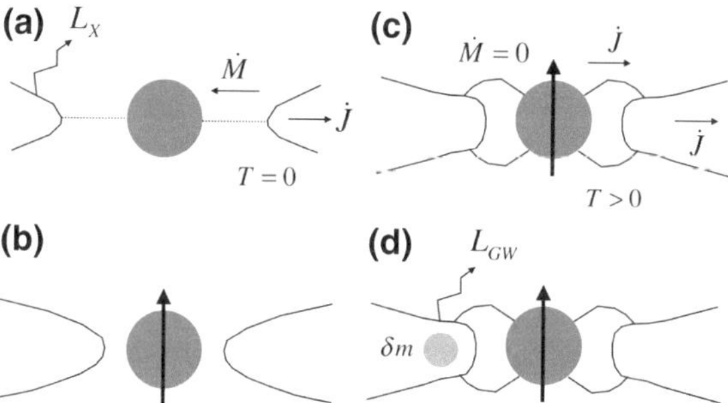

Fig. 11.3 **a** The theory of thin accretion disk describes accretion in response diffusive angular momentum transport outwards subject to zero torque boundary conditions at the inner radius of the disk. Viscous heating produces X-rays, as matter falls into the deep potential well around a black hole. **b** It may extend down to the Inner Most Stable Circular Orbit (ISCO) of a rotating black hole, whose spin energy represents a potentially large energy reservoir for additional interactions. **c** A backreaction may appear through an inner torus magnetosphere, acting back onto the disk to suspend accretion. **d** Gravitational wave emission from non-axisymmetric mass motion induced by thermal instabilities may last for the lifetime of black hole spin, at balance between heating by dissipation and cooling in gravitational radiation. (Adapted from van Putten, M.H.P.M., 1999, Science, 284, 115; 2001, Phys. Rev. Lett., 87, 091101.)

where $\sin\theta = z/r$ denotes the poloidal angle relative to the equatorial plane. In the isothermal approximation, balance of pressure $p = \rho c_s^2$ with g_z,

$$\frac{1}{\rho}\frac{\partial p}{\partial z} = g_z, \tag{11.42}$$

gives rise to a density ρ and a vertically integrated column density

$$\Sigma(r) = \int_{-H(r)}^{H(r)} \rho(r,z)dz \tag{11.43}$$

satisfying

$$\rho(r,z) = \rho_0 e^{-z^2/H^2}, \quad H(r) = \sqrt{2}rc_s/v_K, \tag{11.44}$$

where H denotes the scale height. In near-Keplerian orbital motion, $v_K = r\Omega = \sqrt{GM/r}$ at radius r, the disk will be thin whenever $v_K \gg c_s$, i.e., in a relatively cold disk with supersonic orbital velocities.

A Keplerian disk shows differential rotation: matter at inner radii rotate faster than further out. Following Example 9.4., transport of angular momentum outwards can ensue by viscous shear between concentric surface elements, here by

$$r\frac{d\Omega}{dr} = -\frac{3}{2}\Omega. \tag{11.45}$$

The associated viscous force on cylindrical cross sections gives rise to an azimuthal torque upon including an additional factor r,

$$\tau_\phi = r\int_{-H}^{H} \rho\nu\left(r\frac{d\Omega}{dr}\right)2\pi r dz = 2\pi\nu\Sigma r^3\frac{d\Omega}{dr} \tag{11.46}$$

In a stationary state, transport of angular momentum outwards by τ_ϕ and convective angular momentum inwards with accretion rate $\dot{M}$ is constant throughout the disk:

$$\tau_\phi + \dot{M}\Omega r^2 = C, \tag{11.47}$$

where C is some constant. A complete boundary value problem for this first order differential equation obtains with one boundary condition.

Consider a *zero stress* boundary condition at the inner radius $r = r_{in}$, putting $C = 0$. Here, we ignore potentially powerful interactions by losses in gravitational radiation and gain by coupling to black hole spin (Fig. 11.3).
Integration gives a solution to $H(r)$ via $\Sigma(r)$ in (11.43),

$$3\pi\nu\Sigma = \dot{M}\left[1 - \left(\frac{r_{in}}{r}\right)^{\frac{1}{2}}\right]. \tag{11.48}$$

At large radii, it shows the asymptotic radial drift velocity

$$v_r = -\frac{3\nu}{2r}. \tag{11.49}$$

Energy dissipation per unit volume satisfies

$$\dot{q} = \frac{1}{\rho\nu}\left(\rho\nu r\frac{d\Omega}{dr}\right)^2 = \rho\nu\left(r\frac{d\Omega}{dr}\right)^2, \tag{11.50}$$

and hence the dissipation per unit surface area at radius r is

$$j_q = \int_{-H}^{H}\dot{q}dz = \Sigma\nu\left(r\frac{d\Omega}{dr}\right)^2 = \frac{3GM\dot{M}}{4\pi r^3}\left[1-\left(\frac{r_{in}}{r}\right)^{\frac{1}{2}}\right], \tag{11.51}$$

where we used $rd\Omega/dr = -(3/2)\Omega$ in Keplerian motion. The total luminosity of the disk follows by integration,

$$L = 2\pi\int_{r_{in}}^{\infty} j_q r dr = \frac{GM\dot{M}}{2r_{in}} = \frac{1}{4}\left(\frac{R_s}{r_{in}}\right)\dot{M}c^2. \tag{11.52}$$

In view of (11.48), the disk luminosity (11.52) is proportional to ν.

Let us express ν relative to some fiducial values of our model parameters. In the theory of random walks, mean free path length is defined by $\lambda n\sigma_c = 1$, where σ_c is the cross section of pp collisions. The typically proximity between two protons is determined by a balance between their repulsive Coulomb interaction and the kinetic energy set by temperature. That is, $\sigma_c = \pi r_c^2$, where

$$\frac{e^2}{r_c} = k_B T : \ r_c = \frac{e^2}{k_B T}, \quad \sigma_c = \frac{\pi e^4}{(k_B T)^2}. \tag{11.53}$$

That is,

$$\lambda = \frac{1}{n\sigma_c} = \frac{(k_B T)^2}{n\pi e^4}, \tag{11.54}$$

and hence

$$\nu \sim \lambda\sqrt{\frac{k_B T}{m_p}} \sim \frac{(k_B T)^{\frac{5}{2}}}{n\pi e^4\sqrt{m_p}}. \tag{11.55}$$

For characteristic values $n = 10^{16}\,\mathrm{cm}^{-3}$, $T = 10^6\,\mathrm{K}$ and $e = 4.8\times10^{-10}\,\mathrm{esu}$, $k_B = 1.38\times10^{-16}\,\mathrm{erg\,K}^{-1}$, we have

$$\sigma_c = 8.75 \times 10^{-18}\text{cm}^2,\ \lambda = 11\ \text{cm},\ \ c_s = 9 \times 10^6\ \text{cm s}^{-1}, \tag{11.56}$$

so that

$$\nu = \lambda c_s \sim 10^8 \left(\frac{T}{10^6\ \text{K}}\right)^{\frac{5}{2}} \left(\frac{n}{10^{16}\ \text{cm}^{-3}}\right)^{-1}\ \text{cm}^2\ \text{s}^{-1}. \tag{11.57}$$

Then (11.48) with $\Sigma \simeq 2H\rho$, $\rho = nm_p$ gives the estimate

$$\dot{M} \simeq 3\pi\Sigma\nu \simeq 6\pi m_p c_s H/\sigma_c = \frac{24}{e^4}(k_B T)^3 \frac{GM}{c^3}\left(\frac{r}{R_S}\right)^{\frac{3}{2}} \tag{11.58}$$

at $r \gg r_{in}$. Scaled to characteristic values, we have

$$\begin{aligned} \dot{M} &= 6 \times 10^3 \left(\frac{T}{10^6\ \text{K}}\right)^3 \left(\frac{r}{R_S}\right)^{\frac{3}{2}} \left(\frac{M}{M_\odot}\right)\ \text{g s}^{-1} \\ &\simeq \text{few} \times 10^{-14}\dot{M}_{Edd}\left(\frac{T}{10^6\text{K}}\right)^3 \left(\frac{r}{R_S}\right)^{\frac{3}{2}} \end{aligned} \tag{11.59}$$

in terms of the Eddington accretion rate $\dot{M}_{Edd} = L_{Edd}/c^2 = 10^{17}(M/M_\odot)\,\text{g s}^{-1}$ associated with the Eddington luminosity $L_{Edd} = 10^{38}(M/M_\odot)\,\text{erg s}^{-1}$.

For the active nuclei that we observe—in galactic X-ray binaries and extragalactic AGN—*the estimate* (11.59) *is essentially negligible.*

The above led Shakura & Sunyaev [3] to propose an anomalous viscosity mediated by turbulent eddies,

$$\nu = \alpha H c_s, \tag{11.60}$$

where α is a dimensionless constant of order unit. In this *α-disk model*, ν is vastly greater than that predicted by molecular diffusion, allowing for Eddington limited accretion rates. As mentioned in the introduction, the turbulent origin of (11.60) may be due to instabilities in magnetic fields.

11.4 Gravitational Wave Emission

Non-axisymmetric accretion flows may produce and may be partially driven by gravitational wave emission, potentially relevant to the most extreme accretion processes in catastrophic such as core-collapse of massive stars and accretion disks formed by tidal break-up around a companion black hole.

A starting point is the quadrupole formula of gravitational wave luminosity from binaries such as the Hulse-Taylor binary is a system of two neutron stars. It consists of masses $M_{1,2}$ of the two neutron stars: $M_{NS} \simeq 1.44 M_\odot$, semimajor axis $a = 2 \times 10^{11}$ cm, period: $P = 7.75$ h, and ellipticity: $e = 0.617$. The angular velocity $\omega =$

$2\pi/P$ satisfies

$$\omega^2 = \frac{G(M_1+M_2)}{(2a)^3} \simeq \frac{2GM_{NS}}{8a^3}, \quad \omega^2 = \frac{R_S}{4a^3}, \quad R_S = \frac{2GM_{NS}}{c^2}. \tag{11.61}$$

In the second expression, we converted to geometrical units, wherein mass ($M_{NS} \to R_S$) and time ($t \to ct$) are expressed in cm. Similarly, the associated binding energy is

$$U = -\frac{GM_1M_2}{2a} \to U = -\frac{R_S^2}{2a}. \tag{11.62}$$

By the Virial Theorem, binding energy is about twice the kinetic energy (equal in their time-averaged values),

$$-U \simeq 2E_k = R_S v^2 \simeq R_S(\omega a)^2. \tag{11.63}$$

According to general relativity, energy produces a strain amplitude in space time, in gravitational waves of transverse elliptical deformation of circles (*linear polarization*, in the plane of the binary) or rotation of ellipsoidal deformations (*circular polarizations*, along the axis of rotation of the binary). These strains reflect a rotating gravitational tidal field in the binary. The dimensionless strain amplitude h measured at a distance r from the source will be proportional to the dimensionless ratio defined by the energy in the tidal field (in geometrical units) relative to r, as in the calculation of small angles. By (11.63), we have the wave amplitude

$$\hat{h}(r) \sim \frac{E_k}{r} \simeq \frac{R_S}{2r}(\omega a)^2. \tag{11.64}$$

The luminosity in $h(r,t) = \hat{h}re^{ikr-i\omega t}$ follows the standard expressions for a transverse wave,

$$L_{GW} \sim 4\pi r^2 \left(\frac{\partial h}{\partial t}\right)^2 \sim R_S^2 a^4 \omega^6 \tag{11.65}$$

and should be compared with the exact quadrupole formula for gravitational wave emission from two masses M_1 and M_2 in a circular binary ($e=0$),

$$L_{GW} = \frac{32}{5}\tilde{\mu}^2 a^4 \omega^6 = \frac{32}{5}(\omega\mu)^{\frac{10}{3}}, \ \mu = \frac{M_1^{\frac{3}{5}} M_2^{\frac{3}{5}}}{(M_1+M_2)^{\frac{1}{5}}}, \ \tilde{\mu} = \frac{M_1M_2}{M_1+M_2} \tag{11.66}$$

in units of

$$L_0 = \frac{c^5}{G} = 3.6 \times 10^{59} \text{ erg s}^{-1} = 1.8 \times 10^5 \, M_\odot c^2 \text{ s}^{-1} \tag{11.67}$$

For an equal mass binary, $\mu = 2^{-\frac{6}{5}} R_S$, so that (11.65–11.66) agree within a factor of $\frac{32}{5} \times 2^{-\frac{12}{5}} = 1.21$, which is better than anticipated.

For the Hulse-Taylor binary, we note its large ellipticity. It hereby radiates appreciably at higher harmonics, described by an additional factor in the Peters & Mathews formula [14],

$$L_{GW} = \frac{32}{5}\mu^2 a^4 \omega^6 F(e), \quad F(e) = \frac{1 + \frac{73}{24}e^2 + \frac{37}{96}e^4}{(1-e^2)^{\frac{7}{2}}}. \tag{11.68}$$

For the Hulse-Taylor binary, the ellipticity $e = 0.617$ increases the luminosity to

$$L_{GW} = 7.35 \times 10^{31}\,\mathrm{erg\,s^{-1}}. \tag{11.69}$$

The quadrupole formula (11.66) is a general result that can be applied also to a mass inhomogeneities δm in accretion disks. In this event, $M_1 = M$ and $M_2 = \delta m$ gives

$$L_{GW} = \frac{32}{5}\left(\frac{M}{R}\right)^5\left(\frac{\delta m}{M}\right)^2 \sim \text{few} \times 10^{51}\left(\frac{4M}{R}\right)^5\left(\frac{\delta m}{10^{-3}M}\right)^2\,\mathrm{erg\ s^{-1}} \tag{11.70}$$

for a inhomogeneities at a radius R. Scaling on the right hand side refers to an estimate for $\delta m \simeq 10\% M_D$ for a disk $M_D \simeq 1\% M$ around a rotating black hole of mass M.

Rotating black holes are believed to form in core-collapse of relatively high mass stars.

Example 11.5. Rotating black holes are subject to the Kerr constraint

$$J \leq \frac{GM^2}{c}. \tag{11.71}$$

Consider black hole formation produced in core-collapse of a rotating star, here modeled by a uniform density ρ_0 and constant angular momentum Ω throughout. In this event, a central region of the star out to a radius r contains a mass and angular momentum satisfying

$$M(r) = \frac{4\pi}{3}\rho_0 r^3, \quad J(r) = \frac{2}{5}M(r)r^2\Omega. \tag{11.72}$$

For prompt collapse, (11.71) imposes a *minimum* mass according to

$$r \geq \frac{3}{10\pi}\left(\frac{c\Omega}{G\rho_0}\right). \tag{11.73}$$

At this radius, fall back matter has a specific angular momentum of at most $j \simeq \Omega r^2$. Continuing in the approximation of Newtonian gravity, this matter stalls against an angular momentum barrier at radius r_* satisfying $j = c\sqrt{R_g(r)r_*}$, where $R_g = GM/c^2$, i.e.,

$$\frac{r_*}{R_g} \simeq \left(\frac{15}{8\pi}\right)^2 < 1. \tag{11.74}$$

Following prompt collapse, accretion continues unabated before an accretion disk first forms. When it does, the black hole is non-extremal. Continuing accretion will be driven by viscosity and outflows in disk winds, that may further be subject to feedback from the black hole (11.3) [15].

Rapidly spinning black holes are candidate inner engines to long gamma-ray bursts and superluminous supernovae, leaving slow spin remnants following forceful spin-down against massive accretion disks. A powerful black hole spin-disk connection may prolong emission (11.70), (11.3), the time scale of which is set by the inverse ratio of disk mass to M [15]. This opens the prospect that some nearby supernovae are luminous in gravitational waves [16]. This question may be investigated by future observations with LIGO-Virgo and KAGRA.

11.5 Mass Transfer in Binaries

It is perhaps not surprising to find stars born in multiples with orbital motion absorbing an excess in angular momentum, permitting gravitational collapse to a star. These binary stellar systems may be observable as *visual binaries*, (single or double-line) *spectroscopic binaries*, *photometric binaries* by periodic variability of fluxes or colors, or (partial or full) *eclipsing binaries*. Stars in wide binaries (or isolated stars) evolve in their hydrogen burning phase as main sequence stars with a lifetime

$$T_{MS} = 13\left(\frac{M}{M_\odot}\right)^{-5/2} \text{Gyr}. \tag{11.75}$$

Example 11.6. T_{MS} of stars born with $M \leq 1M_\odot$ is on the order of the age of the Universe of about 13.7 Gyr or longer. Those with higher mass can have remarkably short lifetimes, e.g., 400 Myr for $M = 4M_\odot$ or 40 Myr for $M = 10M_\odot$.

Following (11.75), evolution proceeds on relatively short time scales with the burning of heavier elements, starting with helium. Ultimately, core burning ceases with the formation of an iron core. The final fate of the star depends on the mass of the iron core thus formed, which will be a function of the initial mass of the star at birth. In the absence of nuclear burning, the core is supported against gravitational collapse by electron-degenerate pressure *up to a certain point*. While stars with $M \leq 4M_\odot$ are believed to end as white dwarfs, relatively more massive cores produced by $M > 4M_\odot$ will experience continued collapse to a (proto-)neutron star. Core-collapse of massive stars is believed to trigger supernovae, that may produce neutron stars ($4M_\odot < M < 10M_\odot$) or stellar mass black holes ($M > 10M_\odot$). Based on Salpeter's initial mass function for new born stars,

$$\psi(m)dm = m^{-2.35} dm \text{ pc}^{-3} \text{ yr}^{-1} \tag{11.76}$$

estimates for the population density of white dwarfs, neutron stars and stellar black holes obtain as remnants of stellar formation within a Hubble time ($M \geq 1M_\odot$),

$$n_{WD} = 0.1 \text{ pc}^{-3}, \ n_{NS} = 0.02 \text{ pc}^{-3}, \ n_{BH} = 0.0008 \text{ pc}^{-3} \tag{11.77}$$

by integration of (11.76) over the intervals $M_\odot \leq M \leq 4M_\odot$, $4M_\odot \leq M \leq 10M_\odot$ and, respectively, $M \geq 10M_\odot$.

When stars are born in close binaries, the evolution is expected to be different from those on the main sequence due to interactions.

Overall, we distinguish three classes of compact binaries: *detached*, *semi-detached* and *contact* binaries. Their photospheres are then, respectively, inside their Roche lobes, inside and coincident with their Roche lobes, and both coincident or exceeding their Roche lobes forming a common stellar envelope. For circular orbits, the two Roche lobes in a binary system meet a fixed first Lagrange point L_1 as seen in a co-rotating frame of the binary.

Close binaries may experience mass transfer that generally prodices accretions disks as discussed in Sect. 11.3. If semi-detached, this transfer proceeds through the inner Lagrange point L_1 by Roche lobe overflow. It may proceed naturally in evolution towards equal mass stars, lowering their gravitational binding energy. Evolution in the opposite direction requires energy input from one of the two stars. As the binary tightens or when both stars expand, a common stellar envelope phase may follow.

The Roche lobes in a binary is an equipotential surface passing through the first (*inner*) Lagrange point L_1 (Fig. 11.4), defined as a saddle point in the Roche potential

$$\Phi(\mathbf{r}) = -\frac{GM_1}{|\mathbf{r} - \mathbf{r_1}|} - \frac{GM_2}{|\mathbf{r} - \mathbf{r_2}|} - \frac{1}{2}(\omega \times \mathbf{r})^2, \tag{11.78}$$

where the first two terms give the potential binding energies to the stars of mass M_1 and M_2 at positions $\mathbf{r_1}$ and $\mathbf{r_2}$, respectively. The third term contributes the centrifugal force at the distance **œ** to the axis of rotation in the force $\mathbf{F} = -\nabla\Phi(\mathbf{r})$.

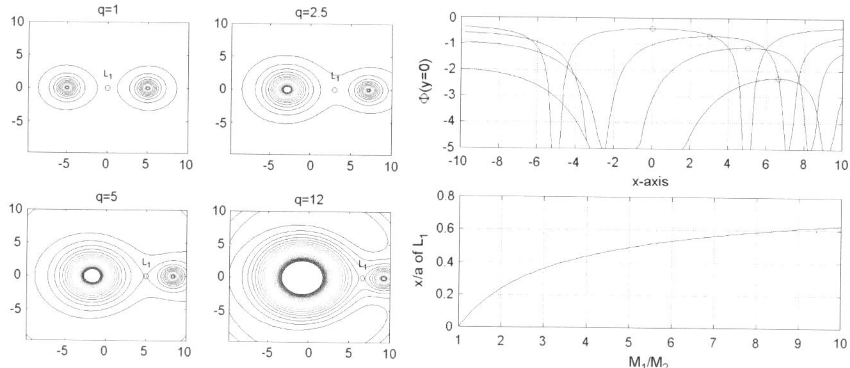

Fig. 11.4 (*Left*) Equipotential surfaces of $\Phi(\mathbf{r})$ and the location of the Lagrange point L_1 at the saddle point between M_1 and M_2 for various mass ratios M_1/M_2. (*Right*) The locus of L_1 moves towards the relatively less massive star with increasing M_1/M_2

In the full dynamical equations of motion of a mass-less test particle (in the *restricted three-body problem*), $\mathbf{F}$ is balanced by inertia of the two stars and Coriolis forces,

$$\frac{d^2\mathbf{r}}{dt^2} = \mathbf{F} - \omega \times \mathbf{v} \tag{11.79}$$

where $\mathbf{v}$ denotes the velocity as measured in the co-rotating frame. We commonly choose the angular velocity of the latter to be constant, defined by the orbital period P in Kepler's law:

$$|\omega| = \Omega = \frac{2\pi}{P}, \quad \Omega^2 = \frac{G(M_1 + M_2)}{a^3}. \tag{11.80}$$

The orbit has latus semi-rectum $p = a(1 - e^2)$ with ellipticity e and semi-major axis a. The motion in the orbital plane is conveniently described in polar coordinates (r, θ) by

$$r = \frac{a(1 - e^2)}{1 + e\cos\theta}. \tag{11.81}$$

Thus, $1/p$ equals the orbital mean of $1/r$ when $e < 1$. For circular orbits, $e = 0$ and $a = a_1 + a_2$ equals the binary separation, given by the sum of distances of each to the center of mass. Because L_1 is a saddle point, Roche lobes take the shape of droplets. Relevant to mass transfer is the effective radius R_L of an equivalently spherical star of the same volume, relative to a. The effective radius of M_1 satisfies [17] (Fig. 11.5)

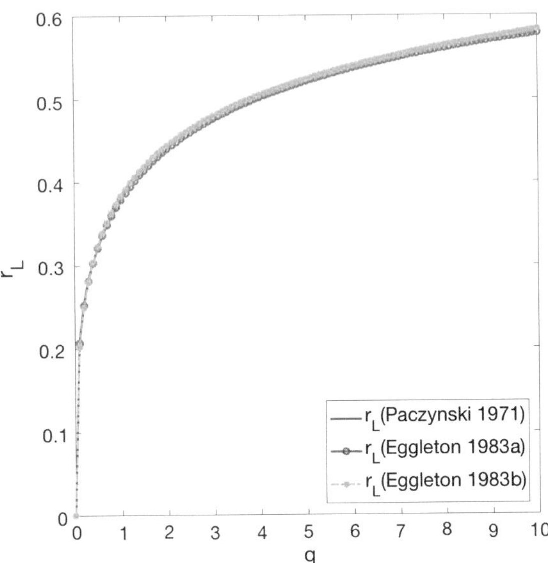

Fig. 11.5 Approximate formulas for the effective radius $r_L = R_L/a$ as a function of $q = M_1/M$

$$\frac{R_L}{a} \simeq \max\left\{\frac{2}{3^{4/3}}\left(\frac{M_1}{M}\right)^{\frac{1}{3}}, 0.38 + 0.2\log_{10} q\right\}, \tag{11.82}$$

and, to a better approximation [18]

$$\frac{R_L}{a} = r_L, \quad r_L = \frac{0.49q^{2/3}}{0.6q^{2/3} + \ln\left(1 + q^{1/3}\right)}, \quad q = \frac{M_1}{M_2}. \tag{11.83}$$

The ratio R_L/a increases with the mass ratio q. For instance, $R_L = 0.3789a$ for an equal mass binary ($M_1 = M_2$), increasing to $R_L = 0.5782a$ when $M_1/M_2 = 10$ and decreasing to $R_L = 0.2068a$ when $M_1/M_2 = 0.1$.

In a semi-detached binary, conservative mass transfer in Roche lobe overflow preserves the total mass $M = M_1 + M_2$ and total orbital angular momentum J, ignoring exchanges of orbital and stellar angular momentum by tidal interactions. In this event,

$$J = \mu j = \frac{M_1 M_2}{M}\sqrt{GMa(1-e^2)} = x(1-x)M^{\frac{3}{2}}\sqrt{Ga(1-e^2)}, \tag{11.84}$$

where we put $x = M_1/M$. Note that the parabola $x(1-x)$ assumes a maximum at $x = 1/2$. If the orbit does not change shape (e remains constant), mass transfer from the more to less massive member causes a tightening, i.e., the major semi-axis a decreases and the binary becomes more compact. In this process, the total energy (9.32)

$$H = -x(1-x)\frac{GM^2}{2a} \tag{11.85}$$

decreases and assumes a minimum at $x = 1/2$. Increasing e may ameliorate this energy loss. Opposite conclusions hold mass transfer from the less to more massive member. This process requires forcing by one of the stellar members and/or circularization.

For circular orbits, (11.84) can be written as

$$\frac{a}{a_*} = \left(\frac{M^2}{M_1M_2}\right)^2 = \frac{(1+q)^4}{q^2}, \quad a_* = \frac{J^2}{GM^3}, \tag{11.86}$$

where $a_0 = 16a_*$ denotes the minimum binary separation when $M_1 = M_2$. Differentiating (11.84) with respect to time keeping J constant ($\dot{J}/J = 0$) gives

$$\frac{\dot{a}}{a} = \frac{2(M_1 - M_2)}{M_1M_2}\dot{M}_1 = 2\frac{\dot{M}_1}{M_1}(q-1), \quad q = \frac{M_1}{M_2}. \tag{11.87}$$

Given the rate of change of the effective radius $R_L = ar_L$, $r_L = r_L(q)$, we have

$$\frac{\dot{R}_L}{R_L} = \frac{r_L'}{r_L}\frac{\dot{q}}{q} + \frac{\dot{a}}{a} = 2\left(\frac{1+q}{2}\frac{r_L'}{r_L} + q - 1\right)\frac{\dot{M}_1}{M_1}. \tag{11.88}$$

Here $r_L' > 0$, and hence R_L of the more massive star will be decreasing with M_1 whenever $q > 1$.

The *stability* of conservative mass transfer depends on the rate at which the radius of the donor star changes relative to $\dot{R}_L$. Evidently, mass transfer is stable about the equilibrium $q = 1$. However, stability is not ensured when $q > 1$. If R_L decreases faster than the radius of the donor star, mass transfer proceeds increasingly faster and a run-away may occur. It does not preclude recovering stability as q approaches 1. The governing physical parameter of the donor star is the logarithmic derivative [19]

$$\xi = \frac{d\log R}{d\log M}. \tag{11.89}$$

Generally, we distinguish different parameter regimes of mass transfer, $\dot{M}_1 = M_1/\tau$ according to different time scales τ. The system may be thermally stable with slow mass transfer, e.g., evolving on its nuclear burning time scale. Or it may be thermally unstable while dynamically stable (on time scale of the binary period). Or it may be both thermally and dynamically unstable, when nothing stops the system from rapidly shrinking towards common envelope state.

Since (9.32) is expressed in terms of the velocity difference v and the separation distance r between the two members of the binary, it is *Galilean invariant*. If one member goes supernova, mass is lost effectively instantaneously. If the supernova is spherically symmetric, v is the same just before and after the explosion. If the progenitor binary is also circular, then $r = a$ and $v^2 = GM/a$. For H to remain negative after the explosion, mass lost must satisfy $\Delta M < M/2$, i.e., no more than 50% of the total mass may be ejected.

Table 11.1 highlights some of the main issues.

11.6 Exercises

11.1. State the dimensions of the following quantities in cgs units.
(a) Angular velocity ω
(b) Hubble radius R_H
(c) Mass density ρ_c at the center of the Sun
(d) Boltzmann constant k_B
(e) Kinematic viscosity ν

11.2. Conversion to *geometrical units* of mass, energy and time with length is according to the formulae

$$M \to \frac{GM}{c^2}, \quad E \to \frac{GE}{4}, \quad t \to ct, \tag{11.93}$$

where G is Newton's constant and c is the velocity of light. Show that G and c become dimensionless in geometrical units and determine the result for the following quantities:
(a) Accretion rate $\dot{M}$
(b) Surface mass density Σ
(c) Planck constant $\hbar$
(d) Kinematic viscosity ν

11.3. Derive the following estimate for the orbital period of a star of mass M_1, radius R_1 in a semi-detached binary with mass ratio $q = M_1/M_2$:

$$P \simeq 0.35 \left(\frac{R_1^3}{M_1}\right)^{\frac{1}{2}} \left(\frac{2}{1+q}\right)^{0.2} \text{day}. \tag{11.94}$$

11.4. Consider an isotropic stellar wind from M_1 with a companion binary of mass M_2, i.e., $\dot{M}_1 < 0$ and $\dot{M}_2 = 0$. Compute

$$\alpha = \left(\frac{\dot{J}}{J}\right)_{wind}. \tag{11.95}$$

Table 11.1 Overview of axisymmetric and non-axisymmetric accretion

1. Let $R_S = 2GM/c^2$ denote the Schwarzschild radius of a star of mass M. The Bondi-Hoyle-Lyttleton mass accretion rate onto a compact object with velocity v_0 in a medium with density ρ_0 and sound velocity a_0 satisfies

$$\dot{M}_{BHL} = \pi b_*^2 v_0 \rho_0 = 4\pi R_S^2 c \rho_0 \frac{c^3}{(v_0^2 + a_0^2)^{3/2}}. \quad (11.90)$$

Since $\dot{M} \propto v_0^{-3}, a_0^{-3}$, the accretion rate is suppressed in the limit of large velocities or high temperature environments. Conversely, compact objects will be relatively bright when moving slowly in low temperature media.

2. Accretion flows satisfy nonlinear equations of conservation of energy-momentum and mass. Formulated implicitly as

$$Z(x, y) = 0, \quad (11.91)$$

solution branches can be found by numerical continuation using Newton's method. For a choice of x and an initial guess y_0, consider the sequence

$$y_{n+1} = y_n - \frac{Z(x, y_n)}{Z_y(x, y_n)}. \quad (11.92)$$

If convergent, the limit $y_* = y_*(x)$ is a solution to (11.91). It can be used as an initial guess to (11.92) to calculate $y_*(x + h)$. Repeating this procedure obtains a branch $y_*(x_m)$ $(m = 1, 2, \cdots)$ away from bifurcation points, where $Z_y = 0$. Similarly, branches $x = x(y)$ can be calculated.

3. Accretion flows occur following mass transfer in stellar binaries following Roche lobe overflow. The result is typically binary evolution on a secular time scale. Accretion onto the acceptor is mediated by angular momentum transfer outwards by diffusion and outflows.

4. Extreme accretion flows may occur during growth of black holes newly formed in core-collapse of massive stars. In this event, (a) accretion onto may be driven by gravitational wave emission from non-axisymmetric accretion flow, generally producing ascending chirps; (b) descending chirps may derive from non-axisymmetric waves in high density matter about the ISCO spinning down a rapidly rotating black hole sustained for the lifetime of rapid spin.

Derive an expression for $\dot{a}/a$. How does the binary period evolve?

11.5. Show that $P^2 = a^3/(M_1 + M_2)$ expresses the period in years of a binary with major semi-axis a in A.U. and masses $M_{1,2}$ expressed in units of $M_\odot$.

11.6. Calculate the luminosity in gravitational waves for the orbital motion of Jupiter around the Sun. Here, $M_J = 2 \times 10^{30}$ g and $R_J = 7.8 \times 10^{13}$ cm.

11.7. Accretion disks are a natural outcome of mass transfer in binaries, when one of the two stars fills its Roche lobe. In deriving a leading order description of this processes, consider the following and state the relevant quantities without any equations.

- Give three or more time scales in the problem of mass transfer and associated stellar evolution.
- In Roche lobe overflow of the donor feeding mass to the acceptor, state two principally conserved quantities.
- Why does mass transfer give rise to an accretion disk around the acceptor, in particularly around compact object such as a white dwarf, neutron star or black hole?
- What governs the mass accretion rate *onto* the acceptor?
- In accretion onto neutron stars or black holes, what are potentially observable differences?

11.8. State two radiation channels by which we can directly probe the inner-most workings of core-collapse supernovae. For supernovae at large distances, which of these two channels is preferred?

11.9. Express (11.69) as a fraction of the Solar luminosity $L_\odot$ in electromagnetic radiation.

References

1. Bondi, H., 1952, MNRAS, 112, 195; Shapiro, S.L., & Teukolsky, S.A., 1983, *Black Holes, White Dwarfs and Neutron Stars* (John Wiley & Sons); Ryden, B.S., Lecture notes AST825, http://www.astronomy.ohio-state.edu/~ryden/ast825/ch8.pdf
2. Hoyle, F. & Lyttleton, R. A.: 1939, Proc. Cam. Phil. Soc. 35, 405; Bondi, H. & Hoyle, F., 1944, MNRAS, 104, 273
3. Shakura, N.I., & Sunyaev, R.A., 1973, A&A, 24, 337.
4. Balbus, S.A., & Hawley, J.F., 1991, ApJ, 376, 214; Hawley, J.F., & Balbus, S.A., 1991, ApJ, 376, 223.
5. Montgomery, C., Orchistron, W., & Whittingham, I., 2009, J. Astron. History and Heritage, 12, 90.
6. van Putten, M.H.P.M., 2015, arXiv:1506.08075
7. Bekenstein, J.D., & Mukhanov, V.F., 1995, Phys. Lett. B, 360, 7
8. Aharonian, F., Akhperjanian, A.G., Bazer-Bachi, A.R., et al., 2007, ApJ, 664, L71.

9. Abbott, B.P., et al., 2016, Phys. Rev., Lett., 116, 061102.
10. Russel A. Hulse and Joseph H. Taylor received the Nobel Prize in Physics, 1993, for their discovery; Weisberg, J. M., & Taylor, H., in Radio Pulsars, eds. M. Bailes, D. J. Nice & S. E. Thorsett, San Francisco: APS 2003 (Conf. Series CS 302), p. 93.
11. Ciufolini, I., & Pavlis, E.C., 2004, Nature, 431, 958; Everitt, C.W.F., et al., 2011, Phys. Rev. Lett., 106, 221101.
12. Hunt, R., 1971, MNRAS, 154, 141
13. Levinson, A., van Putten, M.H.P.M., Guy, P., 2015, 812, 124.
14. Peters, P. C., & Mathews, J., 1963, Phys. Rev., 131, 435.
15. van Putten, M.H.P.M., & Della Valle, M., 2016, MNRAS, 464, 3219.
16. van Putten, M.H.P.M., & Levinson, A., 2012, *Relativistic Astrophysics of the Transient Universe* (Cambridge: Cambridge University Press).
17. Paczynński, B., 1967, Acta Astron., 17, 287
18. Eggleton, P.P., 1983, ApJ, 268, 368; Benacquista, M., 2013, *An Introduction to the Evolution of Single and Binary Stars* (Springer-Verlag)
19. Hjellming, M. S. & Webbink, R.F. 1987, ApJ, 318, 794; Soberman, G.E., Phinney, E.S., & van den Heuvel, E.P.J. 1997, A&A, 327, 620

Chapter 12
Rindler Observers in Astrophysics and Cosmology

In setting up a Hamiltonian system to describe the motion of binaries, globular clusters and galaxies alike, we commonly start with Newton's second law with inertial mass m equal to rest mass m_0. We here revisit this premise in a cosmological setting, wherein Rindler and cosmological horizons may collude.

Einstein's theory of relativity introduces (apparent or event) horizons to Rindler observers, of black holes and, on the largest scale, in cosmology. Largely due to the equivalence principle, these horizon surfaces share some common properties in causality and thermodynamics, described by entropy and temperature according to area and surface gravity,[1] where entropy varies with the distribution of mass within.[2]

Here, we discuss the finite temperature of Rindler horizons based on the moving mirror problem of Birrell and Davies (1982) [2], as it highlights a common origin in pair creation due to a change in basis of the Hilbert spaces defining radiative states in asymptotically flat spacetimes. This temperature is, in fact, a thermodynamic quantity as it recovers Newtonian inertia, of Rindler observers whose horizon falls within the cosmological horizon.

12.1 The Moving Mirror Problem

Quantization of fields representing free particle motion is commonly given in momentum space. For massless fields describing, e.g., photons, we introduce countably infinite states for each momentum by means of creation and destruction operators $\hat{a}^\dagger$ and, respectively, $\hat{a}$ associated with a choice of vacuum $\mid 0 \,\rangle$ satisfying

---

[1]Internal surface gravity of the cosmological horizon.

[2]According to position information $I = 2\pi\Delta\varphi$ in a holographic approach, where $\Delta\varphi$ represents distances expressed in Compton phase on the basis of unitarity in particle propagators; [1].

M.H.P.M. van Putten, *Introduction to Methods of Approximation in Physics and Astronomy*, Undergraduate Lecture Notes in Physics,
DOI 10.1007/978-981-10-2932-5_12

$$\hat{a} \mid 0 \rangle = 0, \tag{12.1}$$

i.e., a vacuum defined as the empty state from which no particles can be removed. A general formalism for quantizing a realistic field requires completeness. For free massless fields in Minkowski spacetime coordinatized by x^b, we can associate $\hat{a}^\dagger$ and $\hat{a}$ with momentum states of the form $e^{ik_a x^a}$. From the theory of Fourier transforms, we know that these harmonic states provide a complete basis. For plane waves in 1+1 Minkowski spacetime, for instance, we have $e^{i(kx-\omega t)}$ with wave number k and angular frequency ω subject to the dispersion relation $\omega = ck$. This carries over to standing waves on finite domains by the theory of discrete Fourier transforms.

The linear span of such basis defines a *Hilbert space*, which preserves all the nice properties of normed linear vector spaces in finite dimensions. Evidently, such Hilbert space lacks manifest covariance in depending on a choice of spacetime foliation, here associated with (12.1). Different foliations in Cauchy surfaces of constant time readily gives rise to a different set of basis functions, bringing along their own creation and destruction operators. A similar dependency in the orthogonality of functions on a choice of domain was encountered in Chap. 6: $\{z^n\}_{n=0}^{\infty}$ form a complete basis for analytic functions on unit disk, that are orthogonal on S^1 in $\mathbb{C}$ but not on $[-1, 1]$. Even so, a basis transformation exists that transforms them into Legendre functions $P_n(x)$ that are orthogonal on $[-1, 1]$.

As one-step (up and down) creation and annihilation operators, $\hat{a}$ and $\hat{a}^\dagger$ satisfy

$$[\hat{a}, \hat{a}^\dagger] = 1. \tag{12.2}$$

An alternative choice of Hilbert space to quantize a given field brings along a second set of creation and annihilation operators $\hat{b}$ and $\hat{b}^\dagger$ subject to the same. By some linear transformation of basis functions associated with (12.2),

$$\hat{b} = \alpha\hat{a} + \beta\hat{a}^\dagger, \;\; \hat{b}^\dagger = \alpha^*\hat{a}^\dagger + \beta^*\hat{b} \tag{12.3}$$

and imposing (12.2) also on the $(\hat{b}, \hat{b}^\dagger)$, the *Bogoliubov coefficients* α and β satisfy

$$|\alpha|^2 - |\beta|^2 = 1. \tag{12.4}$$

We are hereby at liberty to employ the hyperbolic factorization $\alpha = e^{i\theta_1}\cosh\lambda$ and $\beta = e^{i\theta_2}\sinh\lambda$. In (12.3), α and β are linear operators; we suppress summation associated with each $(\hat{b}, \hat{b}^\dagger)$ over all basis elements of the Hilbert space of $(\hat{a}, \hat{a}^\dagger)$.

Importantly, the vacuum state (12.1) associated with $(\hat{a}, \hat{a}^\dagger)$ is no longer empty according to $(\hat{b}, \hat{b}^\dagger)$,

$$\hat{b} \mid 0 \rangle = \left(\alpha\hat{a} + \beta\hat{a}^\dagger\right) \mid 0 \rangle = \beta\hat{a}^\dagger \mid 0 \rangle, \tag{12.5}$$

showing

$$\langle\, 0 \mid \hat{b}^\dagger\hat{b} \mid 0 \,\rangle = |\beta|^2 \langle\, 0 \mid \hat{a}\hat{a}^\dagger \mid 0 \,\rangle = |\beta|^2 \langle\, 0 \mid [\hat{a}, \hat{a}^\dagger] \mid 0 \,\rangle = 1. \tag{12.6}$$

The vacuum state $| \, 0' \, \rangle$ of $(\hat{b}, \hat{b}^\dagger)$,

$$\hat{b} \, | \, 0' \, \rangle = 0, \tag{12.7}$$

is hereby inequivalent to (12.1).

The moving mirror problem described by a Rindler trajectory (Fig. 6.5) describes just such transformation between in- and outstates X and Y. The X and Y domains define Hilbert spaces at radiation states at null-infinity that are related by the non-trivial map imposed by the mirror. In choosing a uniform distribution of phase on Y, we readily can define positive and negative energy states. On this basis, we define $| \, 0 \, \rangle$ as the state that contains no positive energy particles, while prohibiting negative energy states (Sommerfeld radiation boundary conditions). As shown in Example 6.5, this implies negative energy states in X, by projection on its Hilbert space of Fourier modes. In particular, the true vacuum (empty) states of X and Y are incommensurable. According to (6.31) and (12.4),

$$|\beta|^2 \left(e^{2\pi\alpha} - 1\right) = 1 : \quad |\beta|^2 = \frac{1}{e^{2\pi\alpha} - 1}. \tag{12.8}$$

Upon restoring summation in (12.3), $|\beta|^2$ refers to the photon number flux per unit time per unit frequency passing through Y. For a complete discussion the reader is referred to the original work of Birrell and Davies (1982).

Following Exercise 6.20,

$$2\pi\alpha = ka, \quad k = \frac{2\pi}{\lambda} \tag{12.9}$$

in (6.31), where a denotes the acceleration of the mirror in Fig. 6.5 and λ is the wave length in the uniform distribution of phase over Y. With the dispersion relation $\omega = ck$ of photons in vacuum, $2\pi\alpha = \hbar\omega/k_B T$ in terms of the *Unruh temperature*

$$k_B T = \frac{a\hbar}{2\pi c} \tag{12.10}$$

of essentially thermal radiation according to (12.8). Here, k_B is the Boltzmann constant, $\hbar$ is Planck's constant and c is the velocity of light.

12.2 Implications for Dark Matter

The result highlights a universal property of essentially thermal radiation from event horizons according to their surface gravity, including those of black holes[3] and de Sitter space [4].

[3] [3], that shares the same map between X and Y by ray-tracing as shown in Fig. 6.5.

Event horizons formed in gravitational collapse give rise to a finite lifetime of black holes. Over the course of photon emission one-by-one at a rate [5]

$$\nu \simeq 40 \left(\frac{M}{M_\odot}\right)^{-1} \text{Hz}, \tag{12.11}$$

a black hole of mass M produces a finite random walk due to minute kicks in response to this emission. (*A black hole is never at rest.*) Over the course of complete evaporation, the corresponding image by photon detections on the celestial sphere possibly restores unitarity at 4 bits per photon after complete evaporation [6]. A given total mass-energy M imposes a bound on the number of bit flips per second by the Margolus-Levitin bound for a given total mass-energy M [7]. The slow emission (12.11) is consistent with the resulting limit on computation [5, 8].

In de Sitter spacetime, a cosmological event horizon appears at a distance $R_H = c/H$ (Fig. 12.1), in (8.71) with exponentially growing scale $a(t) = a_0 e^{H_0 t}$ for a Hubble parameter H_0. Analogous to Fig. 6.5, Fig. 12.1 shows a logarithmic phase map from a null surface element Y in the future to a null surface element X in the past associated with an event horizon H at a Hubble radius

$$R_H = \frac{c}{H} \tag{12.12}$$

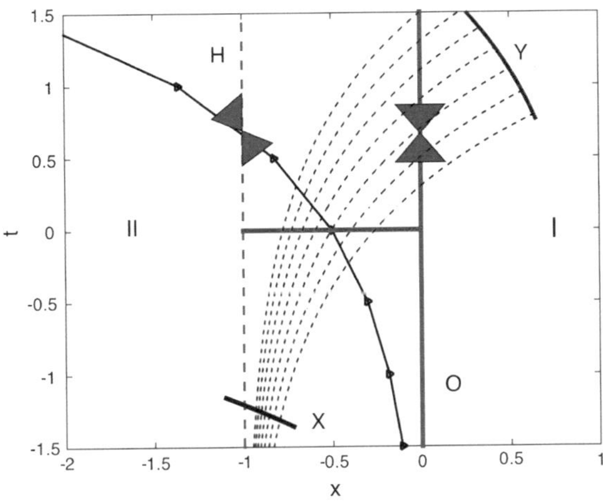

Fig. 12.1 In de Sitter space, there is an outgoing Hubble flow of galaxies (curved trajectory) passing through a cosmological event horizon H across a spacelike distance (*thick horizontal line*) R_H to an inertial observer $\mathcal{O}$. A map of light rays with uniform phase distribution across a null-surface element Y maps into a logarithmic distribution onto a null-surface element X close to H. The result is thermal radiation of the same origin as produced by a moving mirror at constant acceleration shown in Fig. 6.5

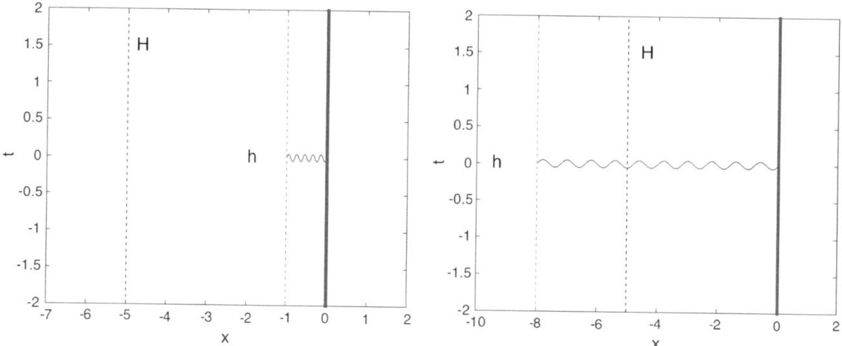

Fig. 12.2 A Rindler observer $\mathcal{O}$ is trailed by an event horizon h at constant distance ξ, whose inertia satisfies a thermodynamic potential associated with h. *Strong* and *weak* gravity limits (*left* and, respectively, *right panel*) are defined by ξ relative to the distance R_H to the cosmological horizon H seen by inertial observers

(in a three-flat cosmology). Thus, (12.10) attains the Gibbons-Hawking temperature

$$k_B T = \frac{\hbar H_0}{2\pi} \tag{12.13}$$

of de Sitter space.

For Rindler observers in de Sitter space, R_H introduces a characteristic scale for accelerations a, according to whether the Rindler horizon at $\xi = c^2/a$ (cf. Sect. 1.5.1) falls inside or beyond the cosmological horizon, i.e., the cases $a > a_{dS}$ and $a < a_{dS}$, $a_{dS} = cH$, where (Fig. 12.2).

$$\frac{cH_0}{2\pi} \simeq 1\text{Å}\,\mathrm{cm\,s}^{-2} \tag{12.14}$$

for the present value H_0 of the Hubble parameter.

The result of Example 1.9 is no coincidence, but reflects on the nature of inertia. On a screen at distance $\Delta\varphi = k\xi$ measured in total Compton phase, unitary encoding of particle position requires an information $I = 2\pi\Delta\varphi$, where $k = mc/\hbar$ is the Compton wave number of a particle of mass m. At the Unruh temperature (12.10) observed by Rindler observers, the associated change in entropy $dS = -dI$ identifies inertia in Newton's law as an entropic force [9]:

$$F = mT\frac{dS}{d\xi} = -mT\frac{dI}{d\xi} = -m\frac{\hbar a}{2\pi c}(2\pi k) = ma. \tag{12.15}$$

Inertia hereby has a thermodynamic potential $U = mc^2$ associated with the horizons of Rindler charts of Minkowski spacetime (Appendix C.3). For a discussion on Newtonian gravity from entropic forces, see [10].

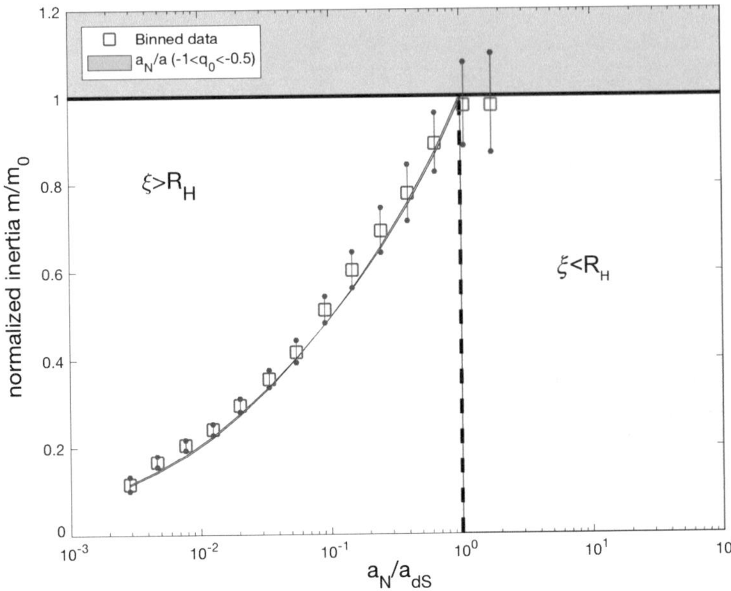

Fig. 12.3 Accelerations $a = a_{dS}$ define a transition to weak gravity (4.99) as Rindler and cosmological horizon collude. For $a < a_{dS}$, inertia m drops below the Newtonian value m_0. This transition is sharp as manifest in galaxy rotation curves, here plotted as m/m_0 as a function of Newtonian gravitational acceleration a_N/a_{dS} based on baryonic matter content. Shown are binned data accompanied by 3σ uncertainties. The *curved green band* is the theoretical curve covering $-1 < q_0 < -0.5$ with $H_0 = 73\,\text{km}\,\text{s}^{-1}\,\text{Mpc}^{-1}$. (Reprinted from van Putten, M.H.P.M., 2017, ApJ, 837, 22. Binned data from Lelli, F., McGaugh, S.S., Schombert, J.M., Pawlowski, M.S., 2017, ApJ, 836, 152 http://astroweb.case.edu/SPARC/RARbins.mrt.)

The cosmological scale of acceleration (12.14) is readily attained on galactic scales beyond distances (cf. (8.82))

$$r_t = \sqrt{R_g R_H}\,:\ \ r_t = 4.7\,\text{kp}\,M_{11}^{1/2} \tag{12.16}$$

where R_g denotes the gravitational radius of the galaxy of mass with $M = 10^{11} M_\odot$.

At distances $r > r_t$ $(a < a_{dS})$, the Rindler horizon of orbiting stars and gas formally falls beyond the cosmological horizon. Thus, R_H introduces a cut-off upon integration of (12.15) by causality, that suppresses inertia whenever $a < a_{dS}$. Figure 12.3 clearly shows this transition in rotation curve data of low redshift spiral galaxies.

If the observed non-Newtonian behavior in galaxy rotation curves originate in suppressed inertia and dark matter is present on cosmological scales, then clustering of the latter is limited to scales of galaxy clusters. If so, the putative dark matter particle must be extremely light [11]. Existing dark matter experiments may well

Table 12.1 Overview Rindler inertia in cosmology

1. In Minkowski spacetime, inertia of a particle at acceleration a has a thermodynamic potential mc^2 associated with its Rindler horizon at a distance $\xi = c^2/a$. When ξ falls within the cosmological horizon, the inertia m is constant and equal to the rest mass m_0 of the particle.

2. In three-flat cosmologies with Hubble parameter H_0, the Rindler horizon colludes with the cosmological horizon at $R_H = c/H_0$ when $a_N = a_{dS}$, $a_{dS} = cH_0$. By causality, R_H acts as a cut-off at $a_N < a_{dS}$, effectively reducing m by a factor R_H/ξ below m_0. The transition to $m < m_0$ is *sharp* in the first derivative across $a_N = a_{dS}$, manifest in rotation curve data of a large number of galaxies shown in Fig. 12.3.

3. At extremely low accelerations $a_N << a_{dS}$, the asymptotic behavior in $a = a_N m_0/m$ captures Milgrom's empirical law of acceleration, $a = \sqrt{a_0 a_N}$ [13]. In the limit of a de Sitter cosmology, Milgrom's parameter satisfies $a_0 = cH_0/(\sqrt{2}\pi)$[9].

4. There is no need for dark matter on galactic scales. Any dark matter clustering is hereby limited to galaxy clusters. If so, the putative dark matter particle is extremely light, expected to produce null-results in laboratory dark matter detector experiments.

consistently produce a null-result.[4] The associated dynamical dark energy may be probed by redshift dependence in a_{dS} [12].

Table 12.1 highlights some of the main issues.

12.3 Exercises

12.1. Derive (12.11) from Hawking radiation. Why are black holes never at rest?

12.2. In geometrical units, derive the surface gravity of a black hole horizon $a_H = 1/(8\pi M)$. [*Hint*. Identify the total mass energy of a test particle suspended above the event horizon with a suitably defined thermodynamic potential in the Schwarzschild line-element.]

[4]If so, the significance of which will be similar to the experiment of A.A. Michelson (1852–1932) and E.W. Morley (1938–1923).

12.3. Derive (12.16).

12.4. For three-flat cosmological in FRW cosmologies, derive

1. The extrinsic surface gravity analogous to that of black hole event horizons;
2. The intrinsic surface gravity based on the Gauss-Bonnet theorem;
3. When are intrinsic and extrinsic surface gravity equal?

References

1. van Putten, M.H.P.M., 2015, IJMP, 24, 1550024.
2. Birrell, N.D., & Davies, P.C.W., 1982, *Quantum Fields in Curved Space*, Cambridge University Press, Cambridge.
3. Hawking, S.W., 1975, Commun. Math. Phys., 43, 199; Wald, R.M., Commun. Math. Phys., 1975, 45, 9.
4. Gibbons G. W., Hawking S. W., 1977, Phys. Rev. D, 15, 2738).
5. Spallicci, D.A.M., & van Putten, M.H.P.M., 2016, IJMMP, 13, 1630014.
6. van Putten, M.H.P.M., 2015, IJMP-D, 24, 155024.
7. Margolus, N., & Levitin, L.B., 1998, Physica D 120, 188.
8. van Putten, M.H.P.M., 2015, arXiv:1506.08075.
9. van Putten, M.H.P.M., 2017, ApJ, 837, 22.
10. Verlinde, E., 2011, JHEP, 4, 29.
11. van Putten, M.H.P.M., 2015, MNRAS, 450, L48.
12. van Putten, M.H.P.M., 2016, ApJ, 824, 43.
13. Milgrom, M., 1983, ApJ, 270, 365.

Appendix A
Some Units and Constants

Physical Constants

Black body constant	$\alpha = \pi^2 k^4/15c^3\hbar^3 = 7.56 \times 10^{-15}$ erg cm^{-3} °K^{-4}
Stefan-Boltzmann constant	$\sigma = \pi^2 k^4/60\hbar^3 c^2 = 5.67 \times 10^{-5}$ g sec^{-3} °K^{-4}
Bekenstein-Hawking entropy	$S_H/A = kc^3/4G\hbar = 1.397 \times 10^{49}$ cm^{-2}
Bohr radius	$a_0 = \hbar^2/m_e e^2 = 0.529 \times 10^{-8}$ cm
Boltzman constant	$k_B = 1.38 \times 10^{-16}$ erg °K^{-1}
	$1/k_B = 1160$ °K/eV
Critical magnetic field	$B_c = m_e^2 c^3/e\hbar = 4.43 \times 10^{13}$ G
Compton wavelength	$\lambda_c/2\pi = \hbar/m_e c = 3.86 \times 10^{-11}$ cm
Velocity of light	$c = 2.99792458 \times 10^{10}$ cm/s
Newton's constant	$G = 6.67 \times 10^{-8}$ cm^3 g^{-1}s^{-2}
	$\kappa = (16\pi G/c^4) = 2.04 \times 10^{-24}$ s cm$^{-1/2}$ g$^{-1/2}$
Planck's constant	$\hbar = 1.05 \times 10^{-27}$ erg s
Planck energy	$E_p = l_p c^4/G = 2.0 \times 10^{16}$ erg $= 1.3 \times 10^{19}$ GeV
Planck density	$\rho_p = l_p^{-2} c^2/G = 5.2 \times 10^{93}$ g cm^{-3}
Planck length	$l_p = (G\hbar/c^3)^{1/2} = 1.6 \times 10^{-33}$ cm
Planck mass	$m_p = l_p c^2/G = 2.2 \times 10^{-5}$ g
Planck temperature	$T_p = E_p/k_B = 1.4 \times 10^{32}$ K
Planck time	$t_p = l_p/c = 5.4 \times 10^{-44}$ s
Electron charge	$e = 4.80 \times 10^{-10}$ esu
Electron volt	1 eV $= 1.60 \times 10^{-12}$ erg
Electron mass	$m_e = 9.11 \times 10^{-28}$ g
	$m_e c^2 = 0.511$ MeV
Fine structure constant	$\alpha = e^2/\hbar c \simeq 1/137$
Proton mass	$m_p = 1.67 \times 10^{-24}$ g
	$m_p c^2 = 938.2592(52)$ MeV
Neutron mass	$m_n c^2 = 939.5527(52)$ MeV
	$= m_p c^2 + 2.31 \times 10^{-27}$ g
	$= m_p c^2 + 1.29$ MeV/c^2
Rydberg constant	$m_e e^4/2\hbar^2 = 13.6$ eV
Thomson cross section	$8\pi e^4/3m_e^2 c^4 = 0.665 \times 10^{-24}$ cm^2

M.H.P.M. van Putten, *Introduction to Methods of Approximation in Physics and Astronomy*, Undergraduate Lecture Notes in Physics,
DOI 10.1007/978-981-10-2932-5

Some Astronomical and Cosmological Constants

Second of arc (")	4.85×10^{-6} rad
Astronomical unit (A.U.)	1.50×10^{13} cm
Light year (ly)	0.946×10^{18} cm
Parsec (pc)	3.26 ly $= 3.09 \times 10^{18}$ cm
Solar mass ($M_\odot$)	1.99×10^{33} g
Distance to Virgo	16.5 ± 0.1 Mpc[1]
Hubble constant (H_0)	67.4 ± 1.4 (km/s) Mpc^{-1} (Planck[2])
Closure density (ρ_c)	$\frac{3\pi H^2}{8\pi G} = 9.4 \times 10^{-30}$ g
de Sitter temperature	$\frac{\hbar H_0}{2\pi k_B} = 2.7 \times 10^{-30}$ K.

[1] Mei, S., Blakeslee, J.P., Côté, P., et al., 2007, ApJ, 655, 144

[2] Planck Collaboration: Ade, P.A.R., Aghanim, N., Armitage-Caplan, C., et al., 2014, A&A, 571, A16

Appendix B
$\Gamma(z)$ and $\zeta(z)$ Functions

The $\Gamma(z)$ and Riemann zeta function $\zeta(z)$ are defined as

$$\Gamma(z) = \int_0^\infty t^{z-1}e^{-t}dt, \quad \zeta(z) = 1 + \frac{1}{2^z} + \frac{1}{3^z} + \cdots \quad (\text{Re}(z) > 1). \tag{B.1}$$

The latter satisfies Euler's identity,

$$\zeta(z) = \Pi\left(1 - p^{-z}\right)^{-1} \tag{B.2}$$

over all prime numbers p. The Riemann-zeta function is defined as its analytic extension to the complex plane. $\zeta(z)$ features a pole at $z = 0$, trivial zeros at $z_k = -2, -4, \ldots$ and an infinite number of non-trivial zeros in the strip $0 < \text{Re}(z_k) < 1$ ($\text{Re}(z_k) \equiv \frac{1}{2}$ if the Riemann hypothesis holds true). They satisfy a symmetry about $\text{Re}z = \frac{1}{2}$, given by

$$\zeta(z)\chi(z) = \zeta(1-z)\chi(1-z), \quad \chi(z) = \pi^{\frac{z}{2}}\Gamma\left(\frac{z}{2}\right). \tag{B.3}$$

The relation $\Gamma(z+1) = z\Gamma(z)$ shows that $\Gamma(n+1) = n!$ for integers $n \geq 0$. It also satisfies Euler's symmetry,

$$\Gamma(z)\Gamma(1-z) = \frac{\pi}{\sin(\pi z)}. \tag{B.4}$$

Its asymptotic properties give the Stirling formula

$$n! \simeq \sqrt{2\pi n}n^n e^{-n} \ (n \gg 1), \tag{B.5}$$

that provides an excellent approximation to within a few percent for n greater than a few.

M.H.P.M. van Putten, *Introduction to Methods of Approximation in Physics and Astronomy*, Undergraduate Lecture Notes in Physics,
DOI 10.1007/978-981-10-2932-5

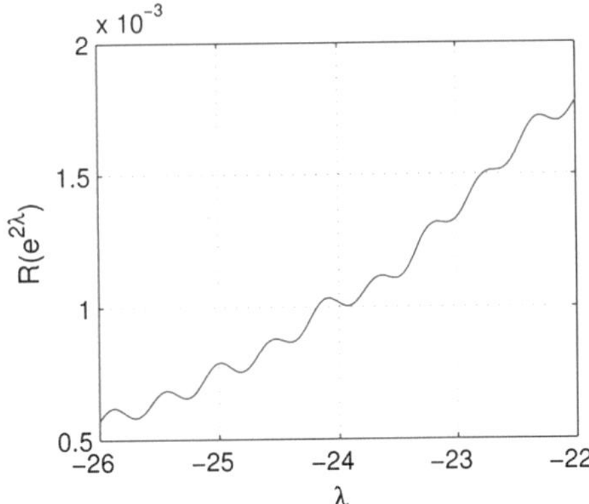

Fig. B.1 Asymptotic behavior in the residual $R(x) = \Phi(x) - (1/2)\gamma(1/2)$ of the regularized prime number sum $\Phi(x)$, $x = e^{2\lambda}$, by explicit numerical summation over the first 10^{12} prime numbers in quad precision. The oscillatory behavior represents the first 21 non-trivial zeros of $\zeta(z)$

As defined by Euler's identity, the Riemann-zeta function—and hence its non-trivial zeros—is deeply related to the prime number distribution. This can be made explicit by considering the regularized sum

$$\varphi(x) = \sum e^{-p^2\pi x} \ln p \tag{B.6}$$

over all primes p and the asymptotic behavior of (Fig. B.1)

$$\Phi(x) = x^{-\frac{1}{4}} \left[1 - 2\sqrt{x}\varphi(x)\right] \tag{B.7}$$

for small $x > 0$. Let

$$Z(\lambda) = \sum \gamma(z_k) e^{-\lambda\left(z_k - \frac{1}{2}\right)}, \quad \gamma(z) = \pi^{-\frac{z}{2}} \Gamma\left(\frac{z}{2}\right) \tag{B.8}$$

where the sum is over all non-trivial zeros z_k of $\zeta(z)$.

It can be shown that $\Phi(x)$ satisfies the small x asymptotic expansion [1]

$$\Phi(x) = \frac{1}{2}\gamma\left(\frac{1}{2}\right) + Z\left(\ln\sqrt{x}\right) + \frac{1}{3}\gamma\left(\frac{1}{3}\right) x^{\frac{1}{12}} + o\left(x^{\frac{1}{12}}\right). \tag{B.9}$$

If $\Phi(x)$ is bounded in the limit as x approaches zero, then the number of non-trivial zeros z_k off the line $\mathrm{Re} z_k = \frac{1}{2}$ cannot be finite or the Riemann hypothesis is true.

In its own right, (B.6–B.7) provide novel functions for benchmarking prime number generators and high precision numerical summation procedures.

Appendix C
Free Fall in Schwarzschild Spacetime

For reference, we revisit some explicit expressions for the radial geodesics of test particles falling onto Schwarzschild black holes, starting at rest from some finite distance away from the event horizon. Illustrative results in Kruskal coordinates are highlighted. The expressions are given as mathematical solutions in the classical Schwarzschild solution, not taking into account the fact the black hole event horizon is evaporating. The true nature of free fall onto black hole event horizons is believed to be fundamentally different from these classical solutions associated with a diminishing of the Kretchmann scalar (C.14) (below) due to evaporation by Hawking radiation [2]. This, however, falls outside the scope of this brief summary of classical results on radial trajectories, in the approximation of a time-invariant mass-energy M at infinity.

C.1 Radial Geodesics

We set out to consider radial geodesic motion of test particles in the Schwarzschild line-element (e.g., [3])

$$ds^2 = -\alpha^2 dt^2 + \frac{dr^2}{\alpha^2} + r^2 d\Omega^2, \quad \alpha^2 = 1 - \frac{2M}{r}. \tag{C.1}$$

in spherical coordinates (t, r, θ, φ). Since $(\partial_t)^b = (1, 0, 0, 0)$ is a Killing vector, the velocity four-vector

$$u^b = \left(\frac{dt}{d\tau}, \frac{dr}{d\tau}, 0, 0\right) = \left(\dot{t}, \dot{r}, 0, 0\right) \tag{C.2}$$

M.H.P.M. van Putten, *Introduction to Methods of Approximation in Physics and Astronomy*, Undergraduate Lecture Notes in Physics,
DOI 10.1007/978-981-10-2932-5

satisfies conservation of energy

$$e = -u_t = \alpha^2 u^t = \alpha^2 \dot{t}. \tag{C.3}$$

The normalization condition $u^c u_c = -1$ hereby satisfies

$$-1 = -\alpha^2 \dot{t}^2 + \alpha^{-2} \dot{r}^2 = -\alpha^{-2} e^2 + \alpha^{-2} \dot{r}^2. \tag{C.4}$$

Consider a particle at rest, $\dot{r} = 0$, at some coordinate distance $2M < R \leq \infty$. Then $e^2 \leq 1$. With $2M/R = 1 - e^2$, we write

$$\dot{r}^2 = e^2 - \alpha^2 = e^2 - 1 + \frac{2M}{r}. \tag{C.5}$$

C.2 Drop from Infinity ($e = 1$)

If $e = 1$, the particle is at rest at infinity. Consider the dimensionless variable and parameter

$$u = \frac{r}{R}, \quad \epsilon = \frac{2M}{R}, \tag{C.6}$$

where R refers to the distance at which $\tau = 0$. Thus, $\tau < 0$ when $u > 1$ and $\tau > 0$ when $u < 1$. For a particle in radial free fall, we have

$$R\dot{u} = -\sqrt{\frac{2M}{r}} = -\sqrt{\epsilon}\frac{1}{\sqrt{u}}. \tag{C.7}$$

Integration gives the equivalent expressions

$$\tau = \frac{2R}{3\sqrt{\epsilon}}\left(1 - u^{\frac{3}{2}}\right), \quad u = \left(1 - \sqrt{\epsilon}\frac{2\tau}{3R}\right)^{\frac{2}{3}}. \tag{C.8}$$

The coordinate time t obtains by integration of energy conservation (C.3), i.e.,[1]

$$dt = \frac{1}{1 - \epsilon u^{-1}} \frac{1}{\dot{r}} dr = -\frac{R}{\sqrt{\epsilon}}\left(1 + \frac{\epsilon}{u - \epsilon}\right)\sqrt{u}\, du. \tag{C.9}$$

[1] Explicit independence of R obtains with $v = r/2M$: $dt/dr = -v\sqrt{v}/(v-1)$.

Substituting $u = x^2$ gives $du = 2xdx$ and

$$\begin{aligned} \frac{dt}{dx} &= -\frac{2R}{\sqrt{\epsilon}}\left(1+\frac{\epsilon}{x^2-\epsilon}\right)x^2 = -\frac{2R}{\sqrt{\epsilon}}\left(x^2+\epsilon+\frac{\epsilon^2}{x^2-\epsilon}\right) \\ &= -\frac{R}{\sqrt{\epsilon}}\left[2x^2+2\epsilon+\epsilon^{\frac{3}{2}}\left(\frac{1}{x-\sqrt{\epsilon}}-\frac{1}{x+\sqrt{\epsilon}}\right)\right]. \end{aligned} \tag{C.10}$$

The result is

$$t = \frac{R}{\sqrt{\epsilon}}\left(\frac{2}{3}\left(1-x^3\right)+2\epsilon(1-x)+\epsilon^{\frac{3}{2}}\log\left[\frac{(x+\sqrt{\epsilon})(1-\sqrt{\epsilon})}{(x-\sqrt{\epsilon})(1+\sqrt{\epsilon})}\right]\right), \tag{C.11}$$

i.e., by $\operatorname{arccoth} u = (1/2)\log\frac{u+1}{u-1}$,

$$t = \frac{2R}{\sqrt{\epsilon}}\left(\frac{1}{3}\left[1-\left(\frac{r}{R}\right)^{\frac{3}{2}}\right]+\epsilon\left(1-\sqrt{\frac{r}{R}}\right)+\epsilon^{\frac{3}{2}}\left[\operatorname{arccoth}\left(\frac{r}{2M}\right)^{\frac{1}{2}}-\operatorname{arccoth}\frac{1}{\sqrt{\epsilon}}\right]\right). \tag{C.12}$$

The expression (C.12) brings out explicitly an expansion in $\sqrt{\epsilon}$, where the leading term is the Newtonian result and the remaining two higher order terms represent the relativistic contributions. To see that dt/dr is independent of R, evident in (C.7), write (C.12) as

$$t(r) = 2M\left[T\left(\frac{R}{2M}\right)-T\left(\frac{r}{2M}\right)\right],\ T(v) = 2\left[\frac{1}{3}v^{\frac{3}{2}}+\sqrt{v}-\operatorname{arccoth}\sqrt{v}\right]. \tag{C.13}$$

C.3 Rindler Spacetime and the Kruskal Extension

The geodesics of test particles above describe the trajectories of free falling observers. In the Schwarzschild line-element, these geodesics can be integrated down to but not across the event horizon, since coordinate time t becomes singular as r reaches $2M$. Even so, the tidal forces experienced about the event horizon remain finite, as indicated by the square root Q of the Kretschmann scalar [4]

$$Q = \sqrt{R_{abcd}R^{abcd}} = 4\sqrt{3}\frac{M^3}{r^3}, \tag{C.14}$$

where R_{abcd} denotes the Riemann tensor of the Schwarzschild metric.

In fact, observed causal structure of spacetime generally depends on the observer's state of the acceleration, made explicit in Rindler spacetimes (e.g., [5])

$$ds^2 = -x^2dt^2 + dx^2 \tag{C.15}$$

A coordinate transformation ($t = i\theta, x = r$) would bring (C.15) into the trivial line-element $ds^2 = dr^2 + r^2 d\theta^2$ of the two-dimensional plane in polar coordinates. The latter is manifestly flat. Being two-dimensional, its Riemann tensor $R_{abcd} = Rg_{a[c}g_{d]b}$ with $R = R(\theta, r) \equiv 0$. By analytic continuation, $R(t, x) \equiv 0$, whereby the Rindler line-element (C.15) is flat. An explicitly coordinate transformation brining it to the Minkowski line-element in (T, X):

$$X = x \cosh t, \ \ T = x \sinh t: \ \ ds^2 = -dT^2 + dX^2. \tag{C.16}$$

Trajectories of constant x have a velocity four-vector

$$u^b = \left(\dot{T}, \dot{X}\right) = (x\kappa \cosh(\kappa\tau), x\kappa \sinh(\kappa\tau)). \tag{C.17}$$

With the normalization $u^c u_c = -1$, particles at constant Rindler coordinate have a constant *acceleration* $\kappa = x^{-1}$, that appear as hyperbolic orbits in Minkowski spacetime. Light cone's therein emanate from the origin as event horizons for accelerating observers. Their distance to this event horizon is measured as a space like interval in surfaces Σ_t of constant Rindler time t. According to (C.15), at constant x, this distance is exactly x, i.e., the inverse of the acceleration κ of the particle. Thus, (C.15) describes a flat parameter space of accelerations with event horizons at distances κ^{-1}. For these reasons, made explicit by (C.15) and (C.16), the Rindler space and its event horizons may be viewed as a property of the tangent space of a physical space-time manifold. To further illustrate Rindler coordinates, we note that geodesic observers, corresponding to a constant Minkowski coordinate X, satisfy

$$t = \tanh^{-1}(\kappa\tau), \ \ x = \kappa^{-1}\sqrt{1 - \kappa^2\tau^2}. \tag{C.18}$$

This transformation shows that $x = 0$ corresponds to the light cone $T^2 = X^2$ $(X > 0)$ of Minkowski spacetime.

For Schwarzschild spacetimes, the event horizon at $r = 2M$ hereby applies to observers at constant acceleration outside, i.e., those at constant coordinate Schwarzschild radius r. By the equivalence principle, the same is *absent* for observers with zero acceleration, i.e., observer freely falling towards the black hole. To these observers, it becomes of interest to have at hand a maximal extension of the Schwarzschild space time, that extends regularly across $r = 2M$. This is achieved by the *Kruskal* metric,

$$ds^2 = \frac{16M^2}{A}\left(dX^2 - dT^2\right) + r^2 d\Omega^2, \ \ A = \frac{r}{2M}e^{\frac{r}{2M}}. \tag{C.19}$$

The geodesic of an observer falling onto the black hole now reaches the location of the event horizon at Schwarzschild radius $r = 2M$ (seen by a non-geodesic observer at

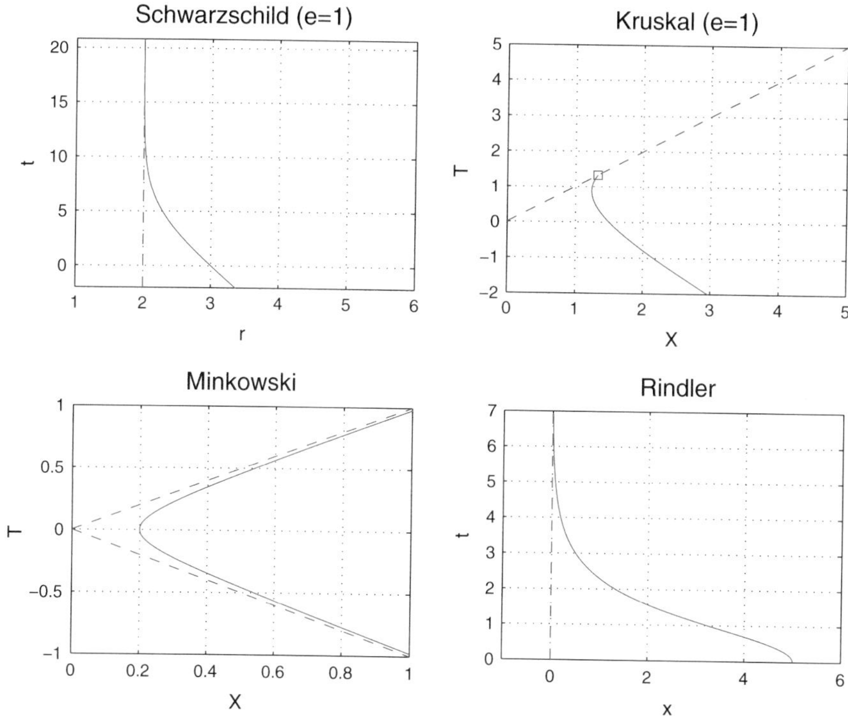

Fig. C.1 (*Top panels*) Shown is a geodesic of a test particle dropped from infinity ($e = 1$) in Schwarzschild coordinate (r, t) (*left panel*) and Kruskal coordinates (X, T). The particle reaches the event horizon in infinite coordinate time t, yet at a finite eigentime τ and a finite Kruskal time T. (*Bottom left*) Shown is a trajectory of a non-geodesic observer at constant acceleration. Its invariant distance to the light cone in Minkowski spacetime is equal to the inverse of its acceleration. (*Bottom right*) The trajectory of a geodesic observer (at constant Minkowski coordinate X) approaches the event horizon $X = T$ in infinite Rindler time t

fixed $r > 2M$) at a finite Kruskal time T.[2] This may be seen by detailed inspection of the explicit coordinate transformation between (X, T) and the Schwarzschild coordinates (r, t),

$$X = \alpha A^{\frac{1}{2}} \cosh\left(\frac{t}{4M}\right), \quad T = \alpha A^{\frac{1}{2}} \sinh\left(\frac{t}{4M}\right). \tag{C.20}$$

Figure C.1 shows a geodesic of a particle coming from infinity in its various coordinate representations.

[2] $M = M(t)$ evolves by Hawking radiation, commonly expressed as a function of time-at-infinity t in the Schwarzschild line-element with associated evolution of the Kretschmann scalar according to (C.14). While $M(t)$ appears slowly evolving to observers at large distances, transformed to the Kruskal time T, an observer in free fall observes a rapidly evolving Kretschmann scalar as a result of $M = M(T)$ as $r(T)$ approaches $2M(T)$. In discussing classical radial trajectories, however, we assume M to be constant.

C.4 Drop from Finite Distance ($e < 1$)

We next consider the problem for finite R, described by $e^2 < 1$. Then $\dot{r} = 0$ at $r = R$. With $\epsilon = 1 - e^2$, write

$$\dot{r}^2 = e^2 - \alpha^2 = \frac{2M}{r} - \frac{2M}{R} = \frac{2M}{R}\left(1 - \frac{r}{R}\right)\frac{R}{r} = \epsilon\frac{1-u}{u}. \tag{C.21}$$

For particles falling in, (C.21) gives

$$\dot{u} = -\frac{\sqrt{\epsilon}}{R}\sqrt{\frac{1-u}{u}}: \quad \frac{\sqrt{u}\,\dot{u}}{\sqrt{1-u}} = -\frac{\sqrt{\epsilon}}{R}. \tag{C.22}$$

Integration of (C.22) gives

$$\frac{\tau}{R} = \frac{1}{\sqrt{\epsilon}}\left[\frac{\pi}{2} - \arcsin\sqrt{u} + \sqrt{u(1-u)}\right]. \tag{C.23}$$

With $u = \sin^2\phi$, the result can be stated neatly as

$$\frac{\tau}{R} = \frac{1}{\sqrt{\epsilon}}\left[\frac{\pi}{2} - \phi + \sin\phi\cos\phi\right]. \tag{C.24}$$

For $u = 0$, it obtains the well-known free fall timescale in Newtonian gravity,

$$\tau_{ff} = \frac{\pi R^{\frac{3}{2}}}{2\sqrt{2M}}. \tag{C.25}$$

The infall is relativistic. The associated coordinate time t as seen by observers at infinity obtains by integration of (C.3). With $d\tau = dr/\dot{r} = du/\dot{u}$, we have

$$dt = \frac{ed\tau}{1 - \frac{2M}{r}} = \frac{eu}{u-\epsilon}\frac{du}{\dot{u}} = -\frac{Reu\sqrt{u}\,du}{\sqrt{\epsilon}\sqrt{1-u}(u-\epsilon)}. \tag{C.26}$$

With aforementioned $u = \sin^2\phi$, we note that

$$\frac{\epsilon}{\sin^2\phi - \epsilon} = -U'(\phi), \quad U(\phi) = \gamma^{-1}\mathrm{arccoth}\left(\gamma\tan\phi\right), \quad \gamma = \sqrt{\frac{1-\epsilon}{\epsilon}}. \tag{C.27}$$

We note that $U(\phi) \geq 0$ is decreasing on $\phi_0 < \phi \leq \frac{\pi}{2}$ with $U\left(\frac{\pi}{2}\right) = 0$, where $\tan\phi_0 = \sqrt{\epsilon/(1-\epsilon)}$ refers to reaching the event horizon at $u = \epsilon$ ($r = 2M$). Writing (C.26) as

$$
\begin{aligned}
dt &= -\tfrac{Re}{\sqrt{\epsilon}}\left(1+\tfrac{\epsilon}{u-\epsilon}\right)\tfrac{\sqrt{u}}{\sqrt{1-u}}du = -\tfrac{2Re}{\sqrt{\epsilon}}\left(1+\tfrac{\epsilon}{\sin^2\phi-\epsilon}\right)\sin^2\phi\, d\phi \\
&= -R\tfrac{2e}{\sqrt{\epsilon}}\left(\sin^2\phi+\epsilon+\tfrac{\epsilon^2}{\sin^2\phi-\epsilon}\right)d\phi.
\end{aligned}
\tag{C.28}
$$

is readily integrated from $\pi/2$ down to ϕ ($\sin\phi > \sqrt{\epsilon}$),

$$
\begin{aligned}
\tfrac{t}{R} &= \tfrac{e}{\sqrt{\epsilon}}\left[(1+2\epsilon)\left(\tfrac{\pi}{2}-\phi\right)+\sin\phi\cos\phi+2\epsilon U(\phi)\right] \\
&= \left(\tfrac{t}{R}\right)_N + 2e\sqrt{\epsilon}\left(\tfrac{\pi}{2}-\phi+U(\phi)\right),
\end{aligned}
\tag{C.29}
$$

where, using (C.24), the first term on the right hand side is

$$
\begin{aligned}
\left(\tfrac{t}{R}\right)_N &= e\left(\tfrac{\tau}{R}\right) = \tfrac{e}{\sqrt{\epsilon}}\left[\left(\tfrac{\pi}{2}-\phi\right)+\sin\phi\cos\phi\right] \\
&= \tfrac{e}{\sqrt{\epsilon}}\left[\arccos\sqrt{u}+\sqrt{u(1-u)}\right].
\end{aligned}
\tag{C.30}
$$

Figure C.2 shows some illustrative examples. Setting $e = 1$, (C.30) recovers the non-relativistic, Newtonian limit $t = \tau$. The relativistic expression (C.29) shows that *the divergent coordinate time behavior as r approaches* $2M$ ($\phi \to \phi_0$) *is due to* $U(\phi)$.

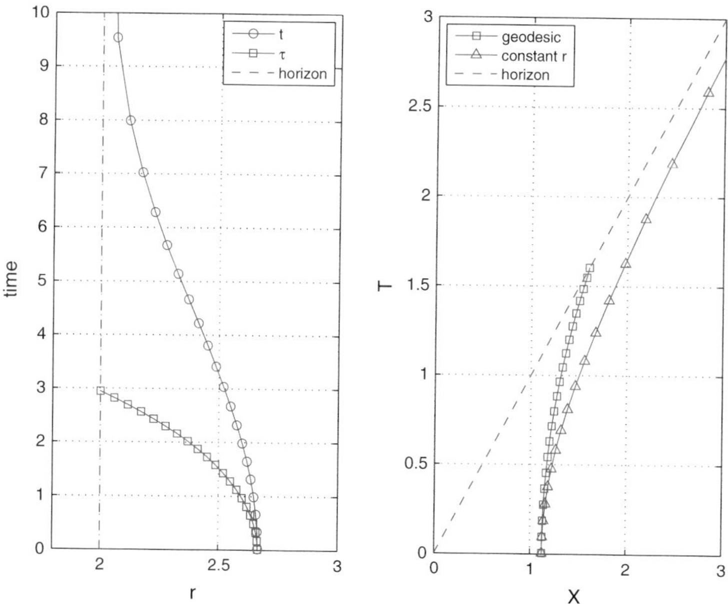

Fig. C.2 (*Left panel*) Shown is a geodesic ($e = 0.5$, $r_0 = 2.667$, $t_0 = 0$) in the Schwarzschild metric coordinates (r, t) along with its eigentime τ. (*Right panel*) The same geodesic is shown in Kruskal coordinates (X, T). Note the counter-intuitive behavior of a positive curvature (towards increasing X) as the particle leaves its initial position $X = 1.1245$ at $T = 0$

C.5 Isotropic Coordinates

We next seek to describe the Schwarzschild line-element by surfaces Σ_t of constant coordinate time t with vanishing vanishing extrinsic curvature tensor K_{ij}. The three-metric h_{ij} on Σ_t are so-called *time-symmetric data*, like turning points in a pendulum reaching maximal deflection. Space-times like Schwarzschild black holes represent eternal turning points, like a pendulum at maximal deflection.[3]

In case of a vanishing extrinsic curvature tensor, the Hamiltonian energy constraint on h_{ij} reduces to

$$^{(3)}R = 0, \tag{C.31}$$

where $^{(3)}R$ denotes the Ricci scalar of the three-metric h_{ij} of Σ_t. Using a conformal scaling of the three-metric $h_{ij} = \phi^4 g_{ij}$, we have

$$^{(3)}R_h = {}^{(3)}R_g - 8\phi^{-1}\Delta_g\phi \tag{C.32}$$

where Δ_g denotes the Laplacian associated with the scaled metric g_{ij}.

Brill and Lindquist (1963) exploit (C.32) by choosing $g_{ij} = \delta_{ij}$ to be the flat metric described by the Kronecker δ-symbol satisfying [7]

$$\delta_{ij} = 1 \ \ (i = j), \ \ \delta_{ij} = 0 \ \ (i \neq j). \tag{C.33}$$

In this event, (C.31) reduces to

$$\Delta\phi = 0, \tag{C.34}$$

where Δ denotes the Laplacian associated with the Euclidean plane described by (C.33). With the asymptotic condition $\phi \sim 1$ at large distances, solutions to (C.34) are a linear superposition of 1 plus Green's functions of the 3-flat Euclidean plane. In particular, we have

$$\phi = 1 + \frac{M}{2\rho} \tag{C.35}$$

in the spherical coordinates

$$ds^2_{(3)} = \delta_{ij}dx^i dx^j = d\rho^2 + \rho^2 d\theta^2 + \rho^2 \sin^2\varphi. \tag{C.36}$$

[3]For a discussion on the thermodynamic potential at the turning points in the motion of black hole binaries, see [6].

Upon further inspection, the solution

$$ds^2 = -N^2dt^2 + \Phi^4 ds^2_{(3)}, \quad N = \frac{2-\phi}{\phi} \tag{C.37}$$

turns out to be a double cover of the exterior Schwarzschild solution (C.1).

The result (C.37) can alternatively be derived from (C.36) and (C.1) by the conditions

$$\frac{dr}{\alpha} = \phi^2 d\rho, \quad r = \phi^2 \rho. \tag{C.38}$$

It defines an explicit expression for ρ upon integrating

$$\begin{aligned}\frac{d\rho}{\rho} = \frac{dr}{r\alpha} = \frac{dr}{\sqrt{r}\sqrt{r-2M}} &= \frac{d(r-M)}{\sqrt{(r-M)+M}\sqrt{(r+M)+M}}\\ = \frac{d(r-M)}{\sqrt{(r-M)^2-M^2}} &= \frac{dy}{\sqrt{y^2-1}},\end{aligned} \tag{C.39}$$

where we put $y = (r-M)/M$. Alternatively, we proceed by $\left(\sqrt{r}\sqrt{r-2M}\right)^{-1} dr = 2(\sqrt{r-2M})^{-1} d\sqrt{r} = 2dx/\sqrt{x^2-1}$, $x = r/M$. Recall $y = \cosh x$:

$$\frac{dy}{dx} = \sqrt{y^2-1}, \quad \frac{dy}{\sqrt{y^2-1}} = d\mathrm{arccosh} y = d\ln\left(y + \sqrt{y^2-1}\right), \tag{C.40}$$

so that

$$\frac{\rho}{\rho_0} = y + \sqrt{y^2-1} = \frac{r-M+\sqrt{r^2-2Mr}}{M} = \frac{\left(\sqrt{r}+\sqrt{r-2M}\right)^2}{2M}. \tag{C.41}$$

Consequently,

$$\phi^2 = \frac{r}{\rho} = \frac{M}{2\rho_0}\left(1 + \frac{\rho_0}{\rho}\right)^2. \tag{C.42}$$

The asymptotic condition $\phi \sim 1$ at infinity fixes $\rho_0 = M/2$, so that

$$\phi = 1 + \frac{M}{2\rho}, \tag{C.43}$$

as in (C.35).

The fact that (C.37) gives a double cover can be made more explicit by putting it in symmetric form with $M/(2\rho) = e^{-\lambda}$ [8]:

$$ds^2 = -\tanh^2\left(\frac{\lambda}{2}\right) + 4M^2\cosh^4\left(\frac{\lambda}{2}\right) ds^2_D \;\; (-\infty < \lambda < \infty), \tag{C.44}$$

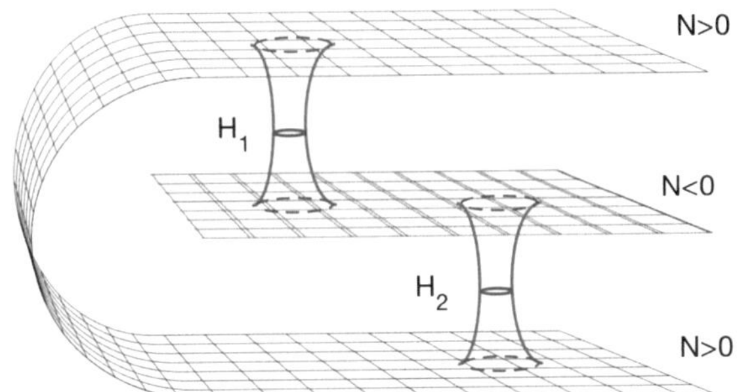

Fig. C.3 A wormhole in a universe ($N > 0$, flat bent sheet) established by a double throat over an intermediate world-sheet ($N < 0$). Stability is inherited from the analytic two-sheet solution (C.44). Negative lapse is restricted to the intermediate sheet between two bifurcation horizons H_1 and H_2. World-lines passing through the double throat have $N > 0$ at both ends, consistent with a positive time-positive energy relation in the universe

where $ds_D^2 = d\lambda^2 + d\theta^2 + \sin^2\theta d\varphi^2$ is known as the donut line-element. *The fact that the throat defined by the bifurcation horizon $\rho = M/2$ is not a wormhole follows from reversal of the direction of time according to a chance in sign in the lapse function $N = \tanh(\lambda/2)$.* Instead, exotic solutions can be envisioned with two throats connected over an intermediate sheet (Fig. C.3). A trajectory that enters a black hole can now return safely return (preserving direction of time) through a white whole elsewhere (in the same sheet), following a double change in sign of the lapse function.

References

1. van Putten, M.H.P.M., AM, 2014, 5, 2547; arXiv:1104.3617.
2. Spallicci, D.A.M., & van Putten, M.H.P.M., 2016, IJMMP, 13, 1630014
3. Chandrasekhar, S., 1992, *The Mathematical Theory of Black Holes* (Oxford University Press), Ch. 19
4. Henry, R.C., 2000, ApJ, 535, 350.
5. Wald, R.M., 1984, *General Relativity* (Chicago, IL: University Chicago Press)
6. van Putten, M.H.P.M., 2012, Phys. Rev. D, 85, 064046
7. Brill, D.R., & Lindquist, R.W., 1963, Phys. Rev. D, 131, 471; Lindquist, R.W., 1963, Phys. Rev., 4, 938.
8. van Putten, M.H.P.M., 2010, Class. Quant. Grav., 27, 075011.

Index

M.H.P.M. van Putten, *Introduction to Methods of Approximation in Physics and Astronomy*, Undergraduate Lecture Notes in Physics,
DOI 10.1007/978-981-10-2932-5

Printed in the United States
By Bookmasters